ECOLOGY
AND
MANAGEMENT
OF
CENTRAL
HARDWOOD
FORESTS

ECOLOGY AND MANAGEMENT OF CENTRAL HARDWOOD FORESTS

RAY R. HICKS, Jr.

West Virginia University

John Wiley & Sons, Inc.

New York · Chichester · Weinheim · Brisbane · Singapore · Toronto

This publication is designed to provide accurate and authoritative
information in regard to the subject matter covered. It is sold with
the understanding that the publisher is not engaged in rendering
legal, accounting, or other professional services. If legal advice or
other expert assistance is required, the services of a competent
professional person should be sought.

Library of Congress Cataloging-in-Publication Data:

Hicks, Ray R.
    Ecology and management of central hardwood forests / Ray R. Hicks,
Jr.
      p. cm.
    Includes bibliographical references (p.   ) and index.
    ISBN 0-471-13758-8 (alk. paper)
    1. Forest management--United States. 2. Forest ecology--United
States. 3. Hardwoods--United States. I. Title.
SD143.H43 1998
634.9'2'0973--dc21   98-6286

Printed in the United States of America

10 9 8 7 6 5 4 3 2

To my wife, Darron,
and my sons, Jonathan and Timothy,
for their love, support, and encouragement

Contents

Foreword

In the early 1950s, when I first started teaching silviculture in the central hardwood region, little was known of ecological relationships and effective forest management practices in hardwood stands. Certain hardwood forestry literature was available. Perhaps the most pertinent and best known were bulletins of the Yale University, Harvard University and Duke University forestry schools and those of the Black Rock Forest. Questions of the best cutting practices and silvicultural systems for central hardwood forests were perplexing and led to lengthy field discussions and disagreements.

Many foresters who came to teach silviculture or forest management in forestry schools in the central hardwood region were trained primarily in pine silviculture and management. Much was known about proper management of pine stands. Accumulating accurate ecological information on the major southern pines and on white and red pines had been relatively simple, since these species usually grew in pure stands. Unfortunately, the familiarity of foresters with pine ecology and silviculture was initially misleading when it came to making realistic decisions for complex hardwood mixtures.

Only through trial and error did it become obvious that practices used in managing the southern pines had no direct application to hardwood stands. Foresters found early that crown thinnings were not necessarily the best first thinning for hardwoods. Pine basal-area guides and rules were of little use, since most pole-stage hardwood stands could not be thinned back to 100 square feet/acre basal area — they had, at best, only 80–90 square feet/acre at the time of the first thinning or improvement cutting. Furthermore, hardwood mixtures were rife with unpromising, unmerchantable stems, due to their sprout origin, frequent wildfires, and repeated high-grading, which had removed the better trees and left the worst to dominate the site.

Another presumed source for good hardwood management practices was the wealth of silvicultural research in European journals. The Germans, Danes, and French had been managing hardwood forests for centuries. Should not their experiences supply answers to our problems? American foresters, however, quickly learned that European forests had little in common with the complex natural stands in the eastern United States. European forest literature was interesting to read, their forests fascinating to visit, but their management methods and experiences were rarely directly applicable to our forest conditions. It became apparent that American foresters would have to do their own ecological research and develop their own silvicultural systems to meet the needs and goals of American forest landowners.

In spite of these obstacles, during the fifties many significant research programs began to emerge. Their publications laid the foundation for our present knowledge of hardwood ecology, silviculture, and management. Such species as yellow-poplar and black walnut received the initial attention. The ubiquitous oaks were among the last to be tackled, yet even in this area U.S. Forest Service experiment station scientists and forestry school faculty have made enormous strides.

Developments in related fields have also facilitated in amassing needed information on hardwood forest ecology. A better understanding of forest soils and site quality led to insights concerning the site factors correlated with the tree growth and quality in the southern pine region. The realization that site productivity in the Appalachian mountains and plateaus was more closely correlated with such topographic features as aspect, slope position, and steepness of slope led to a new understanding of hardwood stands and their variations in composition and capabilities.

During this period, great strides were also made in the statistical field, with new methods of gleaning and identifying obscure relationships from scientific data. Foresters were quick to recognize the value of statistical design for their research projects and took full advantage of these tools to enhance their research programs. Forestry schools, in turn, produced many leading statisticians who have worked on the leading edge of statistical theory and computer programming.

Forest geneticists, originally occupied primarily with improving the southern pines, have continued to make strides in improving the form, growth rate, and quality of hardwood species, and have even developed individuals resistant to certain forest insects and diseases. Furthermore, foresters cooperating with the chemical industry have developed new herbicides and safe, efficient, and inexpensive application methods for controlling undesirable and unwanted vegetation. This has allowed them to establish and mold the composition of hardwood stands so that they better meet the needs of the forest land owner.

Thus, in the last half-century, literature related to central hardwood ecology and silviculture has appeared at an ever increasing rate — often too fast for field foresters to absorb and incorporate into their practices. On the other hand, new insights, understandings, and opportunities brought about by this research are now available for application when the knowledge is needed.

Hardwood forest literature is widely scattered through a variety of publications. Although much has appeared in major refereed journals, much else has been published as U.S. Department of Agriculture bulletins and technical papers. Not to be overlooked are land-grant university experiment station papers and circulars, and forestry school theses and dissertations. Many pertinent forest-related articles also appear in botanical and ecological journals, and journals of closely allied fields. It is a great challenge to locate and bring all of this information together.

Scientific publications that organize, evaluate, and assimilate this voluminous research information are a valuable contribution to the field. In this book, Ray Hicks has identified the most pertinent papers, interpreted their findings, and explained inconsistencies in their results. By synthesizing these findings, he has made this knowledge more readily available to the student, practicing forester, and land manager, and by interpreting these findings he has put them in forms that the forest scientist and forest land manager can use to effectively reach their research and management goals.

Kenneth L. Carvell
Professor Emeritus, Silviculture
West Virginia University

Preface

This book is an attempt to place central hardwoods silviculture and management in the context of the historical and environmental development of the resource while incorporating new technologies and ecological perspectives. It is specific to a large and significant forest region in North America and takes a regional approach in dealing with this unique and very important resource.

The significance of the central hardwoods can be illustrated in pure quantitative terms. The extent of the central hardwoods, as defined here, is 235,000 square miles, within which over 75 percent of the forest is deciduous hardwoods, more extensive than anywhere else in the world. About 50 percent of the area within the central hardwood region is forestland and more than 80 percent of the land is privately owned, generally in small, individually owned tracts. The forests of the central hardwood region have often suffered from abuse, neglect, and exploitation, the logging boom of the early 1900s being one of the most significant episodes of such exploitation. In spite of this, the forests of the central hardwood region have continued to grow at a rate that exceeds harvest since the 1920s, and many stands are now approaching 75 years of age and older. By conservative estimate, the maturing central hardwood forest currently contains in excess of 300 billion board feet of merchantable timber. At current market prices for stumpage, the value of standing timber in the central hardwood region would easily exceed 50 billion dollars.

The economic climate for management of this resource is further modified by the general timber picture in North America. First, Canadian timber, due to supply and demand, has been continually increasing in price. The U.S. federal lands in the Rocky Mountains and Pacific Northwest are currently being harvested at lower rates than in the recent past, and the southern pine region, for the first time since the Great Depression, is showing a harvest volume that exceeds growth. The foregoing facts pertain to areas that are traditional producers of conifer timber, which, until recently, did not compete in the same markets with hardwoods. But with the advent of composite materials and new lamination processes, hardwoods are now competing in the same markets with conifers for construction materials. Pulp and paper technology has also adapted well to the incorporation of greater proportions of hardwood pulp. Additionally, with the consistently high demand for quality hardwood sawtimber and veneer, an economic picture emerges that indicates a strong future role for central hardwoods in the American forest economy. An added facet is the location of the central hardwood region with the large urban population centers of the East, South, Midwest, and Great Lakes forming a virtual ring around its exterior. This has the effect of placing demands on the resource for tourism, aesthetics, and wildlife that have a direct impact on forest resource management.

The central hardwoods also represent a very complex mixture of species and sites. Although the traditional silvicultural treatments generally apply to hardwoods, the mixed stands of the central hardwood region pose some interesting and unique challenges and opportunities for foresters. There already exists a large body of knowledge relating to the culture of central hardwood forests, but this information is dispersed in numerous papers, reports, and books. Through this book I hope to bring this information together in a cohesive unit.

In summary, the central hardwood forest is a unique and significant resource that is approaching several thresholds simultaneously. The maturing central hardwood forest has tremendous value and diversity. Such value and versatility can lead to exploitation, but at the same time they can provide economic incentives to achieve the goal of sound forest management. My hope is that this book will be instrumental in achieving the latter.

Morgantown, West Virginia   Ray R. Hicks, Jr.
April 1998

Acknowledgments

I wish to gratefully acknowledge and thank all those who have helped, supported, and encouraged me in the completion of this project.

First, I wish to acknowledge the talent, patience, and perseverance of Darlene Mudrick, who typed and edited the entire manuscript and provided the excellent graphics throughout the text. Her background and skill in forestry, English, and computing is a rare combination, which is perfectly suited to this undertaking.

Second, I wish to thank the West Virginia University, College of Agriculture, Forestry and Consumer Science for its support. I was granted a Faculty Development Leave for the Spring semester of 1996 and awarded a travel grant from the Anderson Enrichment Fund, which enabled me to get off to a good start on the project. I also wish to thank my colleagues in the West Virginia University Division of Forestry for their confidence and support during this project. I especially thank Bill Grafton and Ken Carvell for photographs they allowed me to use.

I owe a great deal to the experts in central hardwood ecology, silviculture, and management who generously shared their knowledge and expertise. The following is a list of some of these: Edward Buckner, University of Tennessee (retired); Kenneth Carvell, West Virginia University (retired); Robert Cecich, USDA, Forest Service; Wayne Clatterbuck, University of Tennessee; Deborah Hill, University of Kentucky; Steve Horsley, USDA, Forest Service; Paul Johnson, USDA, Forest Service; David Loftis, USDA, Forest Service; Ralph Nyland, SUNY, ESF; Jim Patric, USDA, Forest Service (retired); Glen Smalley, USDA, Forest Service (retired); Susan Stout, USDA, Forest Service; Jeff Stringer, University of Kentucky; Wayne Swank, USDA, Forest Service.

I am greatly indebted to a number of colleagues who reviewed sections of the manuscript. These include Prof. Joy Berkeley, West Virginia University,

English Department (retired); Dr. Kenneth Carvell, West Virginia University, Division of Forestry (retired); Dr. Jack Coster, West Virginia University, Division of Forestry; Dr. Martin Christ, West Virginia University, Biology Department; Dr. Andrew Egan, University of Maine, Forestry Department; Mr. Tony Jenkins, USDA, NRCS; Mr. H. Clay Smith, USDA, Forest Service (retired); and Dr. Harry Wiant, West Virginia University, Division of Forestry (retired). I especially want to thank people who reviewed the entire manuscript: Mr. Wade Dorsey, Maryland DNR and Dr. George Rink, USDA, Forest Service. All the above reviews resulted in changes that greatly improved the final document.

Finally, I wish to acknowledge all my students — past, present, and future. It is their exuberance and curiosity that have been a constant source of inspiration to me over the past 28 years.

R. R. H.

Introduction

This book provides an in-depth examination of the central hardwoods through a regional perspective. It is intended as a text and reference book but is not meant to replace general silviculture texts or references, such as the USDA Agricultural Handbook series. Focusing on the central hardwood region permits a depth of coverage that would be impossible for books that cover more than one region. It also allows for better development of the background necessary for an understanding of how the hardwood ecosystem functions. For example, the sections on physiography, geology, soils, and climate provide a background for the phytogeographic descriptions. Forest cover types can readily be correlated with the environment in which they grow, which leads to an appreciation for how the forest arose.

The contemporary forests of the central hardwood region are not only a result of their geologic and climatic environment; they have also been radically affected by humans. The historical discussion in Chapter 2 describes the preconditions for the current forest and how it developed as a result of both geoclimatic and human influences.

An understanding of how the central hardwood resource developed enables the reader to place the ecological relationships in context. General ecological concepts are illustrated with instances specific to central hardwood forests, and the silvical characteristics of the major species are discussed. This sets the stage for the silviculture portion. The unique qualities of the mixed species stands typical of the region are explored and appropriate silvicultural treatments are discussed, especially those appropriate to small nonindustrial private landowners. But this is not intended to be a cookbook for hardwood silviculture. Rather, the concepts are emphasized with the expectations that students and working foresters can develop the understanding needed to apply and adapt silvicultural treatments under a wide array of circumstances.

Chapter 6 of the book attempts to place silviculture in the broader context of management, with emphasis on methods appropriate to the small nonindustrial private landowner. That is, what can the landowner afford to do? How can landowners manage responsibly and ensure a perpetual yield of products (both commercial and noncommercial) and do it within limits that will not cause irreparable damage to the ecosystem?

In Chapter 7, a summary and synthesis of the previous chapters is developed, which leads to recommendations for managing central hardwoods. In this chapter an attempt is made to better define the role of the professional resource manager in implementing biologically sound management of central hardwood forests.

Hopefully, this book will help foresters in the region to be better resource managers through better understanding of the resource and will promote sound silvicultural-based management — management that is both profitable and protective of long-term forest health and productivity.

1

❦

The Central Hardwood Region

GEOGRAPHIC EXTENT
AND EARLY OBSERVATIONS

For the purpose of this book, the **central hardwood region** is defined as the region south of the beech-maple forest, east of the Great Plains, and north and west of the southern pine forests of the Coastal Plain and Piedmont (Fig. 1). The central hardwood forest covers an area of approximately 235,000 square miles and is centered along the axes of the Appalachian Mountains east of the Mississippi River and the Ouachita/Ozark Mountains west of the Mississippi. Central hardwoods range from New York to Georgia and from Maryland to Missouri. Hardwoods predominate in this region, although conifers, such as shortleaf (*Pinus echinata* Mill.), Virginia (*P. virginiana* Mill.), eastern white (*P. strobus* L.), and pitch (*P. rigida* Mill.) pines, as well as eastern hemlock [*Tsuga canadensis* (L.) Carr.] and spruces (*Picea* spp.), may be important components locally within the region. Brooks (1965) states: "My native state of West Virginia, when the white man first saw it, was forested throughout, a fifteen-million-acre stand of trees. By the best calculations we can make, only about 11 percent of this area was in needle-foliaged species—pines, spruce, hemlock, cedar, and the like. All the rest, 89 percent of the entire state, was covered by deciduous hardwoods." No single tree species defines the central hardwood region although oak species occur throughout the region, and, as a group, they constitute a greater proportion of the growing stock than any other species or genus (Beltz et al. 1992).

The central hardwood region of the United States is the largest contiguous assemblage of deciduous tree species found anywhere in the world (Fig. 2). They share many common genera with remnant deciduous forests of Europe and Asia. The American species probably share a common ancestral lineage to their counterparts in Europe and Asia, dating to a time before the continents began to drift apart.

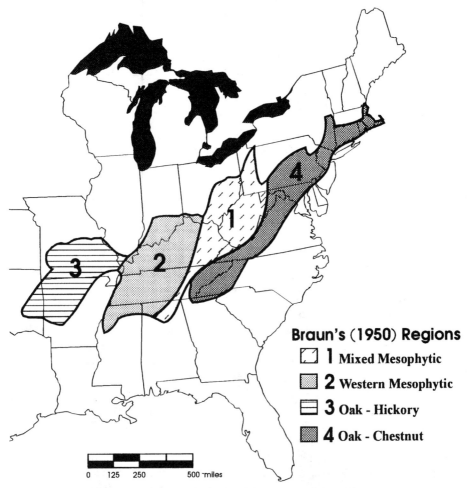

FIGURE 1   The central hardwood region of the United States.

In the continental United States, 43 percent of the forest growing stock is hardwood (57 percent softwood), but 90 percent of the hardwood is located in the central hardwood region (Powell et al. 1993). Due, in part, to the high proportion of private land ownership, 40 percent of the unreserved (available for harvest) forestland in the United States occurs within the eastern hardwood region and 74 percent of all eastern forests are classed as hardwoods (Powell et al. 1993). Although the central hardwoods are very diverse, certain species tend to predominate regionally. For example, oaks (*Quercus* spp.) and oak-hickory forests are found throughout the region but are most abundant in the middle latitudes. Maples [sugar (*Acer saccharum* Marsh.) and red (*A. rubrum* L.)] are most abundant in the northern portion of the region. Yellow-poplar (*Lirioden-*

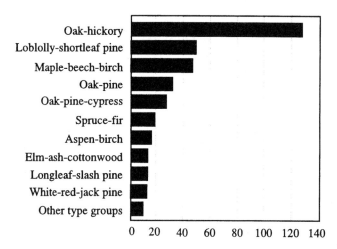

FIGURE 2   Forest type groups on unreserved forestland in the eastern United States (from Powell et al. 1993).

dron tulipifera L.) is most abundant in the central Appalachians and is the key species in the broad cover type referred to by Braun (1950) as the "mixed mesophytic hardwoods."

According to Powell et al. (1993), the oak-hickory forest type covers 127 million acres in the East and the maple-beech-birch (northern hardwoods) occupy 46 million acres. But the reality is less clear than this would indicate. In areas, such as the Blue Ridge Mountains and the dissected Appalachian Plateau, the effects of elevation, aspect, and slope position produce a complex mosaic across the landscape with northern hardwoods at higher elevations, oaks on south-facing aspects and lower ridges, and mesophytic hardwoods predominating on north and east aspects and coves.

Early surveyors and explorers described the forests of the central hardwood region that existed at the time of, and soon after, European colonization. From these reports, it can be seen that a lush forest existed throughout much of the central hardwood region, but many areas had already been substantially modified by human activities. Brooks (1910) cites an earlier report of the West Virginia Conservation Commission as stating: "When white men first came into the state, it was all forest except for a few cliffs and rocky peaks and two or three old fields where Indians had probably cultivated corn."

Surveyors' field books are a good source of information about vegetation, although not scientifically documented. Thomas Lewis, one of the surveyors of the Fairfax Line that delineated a land grant from King George II to Lord Fairfax, kept a journal as he crossed the Blue Ridge and Allegheny Mountains. Lewis recorded 82 witness trees for the 1746 survey. Stephenson (1993) listed the following species among those indicated in Lewis's records: "spruce, pine" (red spruce), "sugartrees" (sugar maple), "beach" (American beech), "birch," "cherry" (black cherry), and "pine." Notably missing from this list is oak, although Lewis's route took him through higher elevations where northern hardwoods would predominate.

Another surveyor's account is provided by Robert Love (1795), regarding his observations in present-day western North Carolina. He states:

The land consists of Vallies [sic] and Mountains and not more than one fifth of which can be called poor land and is all the best watered Country I ever saw. The General Growth of the Timber is walnuts black and white, Locusts, Sugar Trees, Buck Eyes, Lymes, poplars and Oaks of every description; the black and white Hickory grows very plentiful—a Great Number of Wild Cherry and Cucumber Trees are to be found throughout our lands. I have observations in my field Books of Buck Eyes measuring thirteen feet round the Body and black walnuts near the Same Size. These were found in plenty in running the line without searching out on either side as well on mountains as in the Vallies.

Early botanists and explorers, such as John Bartram and his son William, provide insight into the early forests of the central hardwood region. In a published journal of his 1751 trip through Pennsylvania, John Bartram describes the "production" of the Ridge and Valley section as "some wild grass, abundance of oak and chesnut [sic] trees." In a nearby valley, he described "good low land with large trees 5-leaved white pine, poplar and white oak." He described an area in New York as "a great white pine-spruce swamp full of roots and abundance of old trees lying on the ground, or leaning against live ones, they stood so thick that we concluded it almost impossible to shoot a man at 100 yards distant."

Using the number of times a species is mentioned in John Bartram's (1751) journal as indicative of the species' abundance, oaks were most abundant (20 percent of references), followed in order by white pine (10 percent), spruce, chestnut [*Castanea dentata* (Marsh.) Borkh.], and hickory (*Carya* spp.) (7 percent each), sugar maple, basswood (*Tilia* spp.), and pitch pine (6 percent each), and birch (*Betula* spp.), beech (*Fagus grandifolia* Ehrh.), elm (*Ulmus* spp.), white walnut, and poplar (4 percent each).

William Bartram (son of John) provided a well-documented account of his travels through the southeastern portion of North America, including the so-called Cherokee Mountains. He noted the similarity of vegetation in the area (present-day northern Georgia, eastern Tennessee, and western North Carolina) with that of more northern latitudes. He states:

I passed again steep rocky ascents and then rich levels, where grew many trees and plants common in Pennsylvania, New-York and even Canada, as *Pinus strobus, Pin. sylvestris, Pin. abies, Acer saccharum, Acer striatum, s. Pennsylvanicum, Populus tremula, Betula nigra, Juglans alba*; but what seems remarkable, the yellow Jessamine (*Bignonia semper-virens*), which is killed by a very slight frost in open air in Pennsylvania, here, on the summits of Cherokee mountains associates with Canadian vegetables and appears moving with them in perfect bloom and gaiety.

The evidence provided through these reports indicates that the forests encountered by the first European settlers were similar in species composition and in many other ways to our present-day forests in the central hardwood region. Oaks are apparently more abundant and widespread today than they were, and according to Spalding and Fernow (1899), eastern white pine was more abundant in the Allegheny Mountain region than it is today. Certainly the original forest was more extensive than it is today, being almost unbroken, but perhaps the most pronounced difference would be the size of trees. Brooks (1910) refers to a yellow-poplar cut in McDowell County, West Virginia, that contained 12,500 bd. ft. of lumber. By the 1940s nearly all the old growth timber in the central hardwood region had been cut over, except for small areas at the highest elevations of the southern Appalachians (Fig. 3) (Buxton and Crutch-field 1985).

FIGURE 3 "Old growth" forest in the southern Appalachians of North Carolina.

PHYSIOGRAPHY

The central hardwood region is situated in an area that, in the main part, is associated with the Appalachian system and, except for the northwestern portion, is unglaciated (Fig. 4). The primary **physiographic regions** of the central hardwoods are the Blue Ridge, Ridge and Valley, Appalachian Plateaus, Central Lowland, Ozark Plateau, and Ouachita provinces (Fenneman 1938).

The **Blue Ridge** is the easternmost province in the central hardwood region and begins in the south in northern Georgia and extends into southeastern Pennsylvania. Through Virginia, the Blue Ridge narrows to form a prominent ridge, hence the name "Blue Ridge." The highest elevations in the eastern United States occur in the Blue Ridge section of western North Carolina. Mount Mitchell in the Black Mountains just northeast of Asheville rises to a height of 6711 ft. above sea level and, like many of the tallest peaks in the Appalachians, is vegetated with a relic spruce-fir forest.

Immediately north and west of the Blue Ridge is the **Ridge and Valley** Province. It extends from central Pennsylvania in the north to northern Georgia and Alabama. The Ouachita Mountains of Arkansas, Oklahoma, and Missouri are similar topographically to the Ridge and Valley Province, and several authors believe that they are part of a common feature, part of which is buried in northern Alabama and Mississippi (King 1950; Petersen et al. 1980; Hubler 1995). The Ridge and Valley section is very old and geologically complex, having been uplifted, faulted, folded, and eroded. The ridges are long and narrow with narrow valleys through Pennsylvania, western Maryland, and West Virginia (Fig. 5). They broaden southward to form the valley of the Shenandoah in Virginia and the Tennessee farther south. Few rivers have breached the Ridge and Valley, but the New River and Susquehanna and Tennessee Rivers have done so. These are all very old rivers that have developed spectacular water gaps in cutting through the mountains as they were uplifted (Fig. 6). Most of the rivers that flow through the Ridge and Valley form a distinctive trellis drainage pattern with parallel streams running down the main valleys and tributaries intersecting at right angles flowing off the ridges (Fig. 7). The Pine Mountain overthrust, located in northeastern Tennessee and extending into eastern Kentucky, is an interesting feature in the Ridge and Valley. In this region, a 17-mile-wide piece of the Ridge and Valley has ridden over onto the Appalachian Plateau. This is the rugged headwaters of the Cumberland River and the location of an important corridor for plant, animal, and human migration, the Cumberland Gap. The Ouachita Mountains that extend west of the Mississippi River from Arkansas into eastern Oklahoma are similar in topography to the Ridge and Valley and share many floristic similarities.

Lying north and west of the Ridge and Valley is the **Appalachian Plateau**. A prominent escarpment (the Appalachian Front), extending from Pennsylvania through Maryland, West Virginia, Virginia, and Tennessee, divides the Plateau from the Ridge and Valley. The northern portion of the Plateau is called the

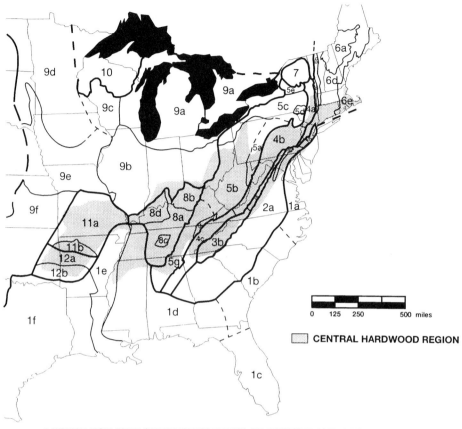

LEGEND FOR MAP OF PHYSIOGRAPHIC PROVINCES AND SECTIONS

1. Coastal Plain
   a Embayed section
   b Sea Island section
   c Floridian section
   d East Gulf Coastal Plain
   e Mississippi Alluvial Plain
   f West Gulf Coastal Plain

2. Piedmont Province
   a Piedmont Upland
   b Piedmont Lowland
3. Blue Ridge Province
   a Northern section
   b Southern section
4. Ridge and Valley Province
   a Hudson-Champlain section
   b Middle section
   c Southern section

Appalachian Highlands

5. Appalachian Plateaus Province
   a Allegheny Mountain Section
   b Unglaciated Allegheny Plateau
   c Glaciated Allegheny Plateau
   d Catskill Mountains
   e Mohawk section
   f Cumberland Mountains
   g Cumberland Plateau
6. New England Province
   a White Mountain section
   b Green Mountain section
   c Taconic section
   d New England Upland
     (including Reading Prong)
   e Seaboard Lowland
7. Adirondack Province

8. Interior Low Plateau Province
   a Highland Rim Section

Appalachian Highlands

   b Bluegrass section
   c Nashville Basin
   d Shawnee section
9. Central Lowland
   a Great Lake section
   b Till Plains section
   c Driftless section
   d Western Young Drift section
   e Dissected Till Plains
   f Osage section
10. Superior Upland--a part
    of the Laurentian Upland
11. Ozark Plateau
    a Ozark Plateau (including
      Salem and Springfield)
    b Boston Mountains
12. Ouachita Province
    a Arkansas Valley
    b Ouachita Mountains

Interior Highlands

FIGURE 4 Map showing physiographic provinces and sections of the eastern United States (from Fenneman 1938).

FIGURE 5   The Ridge and Valley physiographic province. The linear ridges resulted from faulting and folding. This province extends from Georgia to Pennsylvania and is a significant feature in the central hardwood landscape.

FIGURE 6   The New River Gorge of West Virginia is one of the oldest physiographic features in the Appalachian landscape. The New River still maintains its entrenched meanders originating from the time the river flowed over a level plain.

FIGURE 7   Aerial view of the folded Ridge and Valley physiographic province in the central Appalachians.

Allegheny Plateau and is characterized by gentle folding and moderate elevations (Fig. 8). In the southern section, through Tennessee and into Kentucky, it is called the Cumberland Plateau. The eastern section of the Plateau, especially prominent in southern West Virginia and eastern Kentucky, is very rugged and steep, a factor that has had a major influence on the economic development of these areas (Fig. 9). The Appalachian Plateau flattens into the Cumberland Plateau, typical of the area around Crossville in middle Tennessee. A prominent escarpment occurs to the west of Crossville, extending south into Alabama forming the so-called Highland Rim. This feature is brought about by the presence of a hard Pennsylvanian capstone underlain by softer shale and the location of this area near the headwaters of the Cumberland River. With a drainage pattern typical of the Appalachian Plateau (dendritic or tree-like), the headward cutting of such eroding streams has left part of the level plateau as a vestige of an earlier peneplain.

Moving northwestward along the Plateau, the rugged eastern section gives way to the lower hills in the north, typical of much of northcentral West Virginia, eastern Ohio, and southwestern Pennsylvania (Fig. 10). Farther north and west, the Plateau was fractured and slightly upraised in a northeast-southwest transect called the Cincinnati Arch. The fracturing was sufficient to cause more rapid gradation over the whole region and especially in two prominent areas (the Nashville Basin in Tennessee, and the Bluegrass Region of

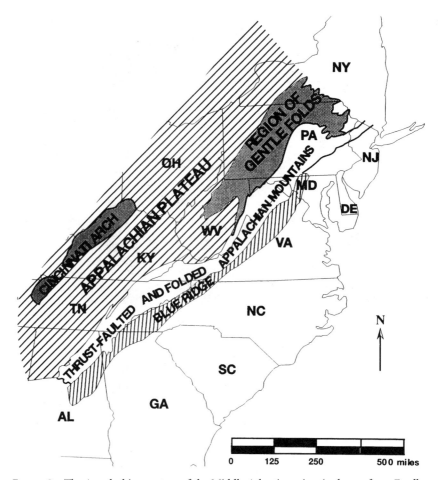

FIGURE 8   The Appalachian system of the Middle Atlantic region (redrawn from Eardley 1951).

Kentucky). Both of these areas are characterized by level to rolling topography and contain many karst areas with sink holes, caves, and underground streams. Immediately surrounding these karst zones are the rolling hills of the Pennyroyal sections of middle Tennessee and Kentucky. Along the southwestern side of the Cincinnati Arch in western Tennessee and Kentucky is the gently rolling topography of the Interior or Low Plateau sloping off toward the Mississippi River (Fig. 11).

To the north lay the **Central Lowlands** that form hilly sections in southern Indiana and southern Illinois. The topography levels northward toward the Great Lakes, and, in this zone of Indiana, Illinois, and Ohio, the effects of glaciation are evident in the topography, with level to rounded features.

FIGURE 9   Aerial view from 20,000 ft. of steep terrain in the dissected Appalachian Plateau Province.

FIGURE 10   A section of the Appalachian Plateau known as the Cumberland Plateau lays to the right in this view looking out into the Nashville Basin of Tennessee.

FIGURE 11   The gently rolling landscape of central Kentucky is characterized by agricultural fields and farm woodlots.

In northern Arkansas and southern Missouri, the Boston Mountains and **Ozark Plateau** form a topography that corresponds to that of the Appalachian Plateau. Like the Appalachian Plateau, the Ozark Plateau is a dissected peneplain with a dendritic (tree-like) stream drainage pattern where dissection has progressed headward along major streams, leaving the least eroded zone toward the interior.

GEOLOGY

The physiographic features described previously are a function of several factors (underlying rocks, tectonic movement, climate, and time). All these factors have been at work in the central hardwood region and have strongly influenced the present topography, soils, and vegetation.

Although the earth's crust (lithosphere) appears solid, it is underlain by semifluid and liquid materials around the center of the earth, and, through imperceptibly slow movement, the crust of the earth is constantly changing. These changes are described by the plate tectonic theory of crustal movement.

Geologic time has been divided using "unconformity" periods, or breaks in the stratigraphic record, to define the beginning and ending of the major periods (Table 1). Although the earth is believed to be about 4.6 billion years old, most of the rocks that occur under the central hardwood region are derived from sediments that were deposited over the area between 300 and 500 million

TABLE 1 Geologic Time Chart with Dates Given in Millions of Years

ERAS	PERIODS	EPOCHS	MILLIONS OF YEARS
Cenozoic 65–Present	Quaternary 2–present	Holocene Pleistocene	(Recent) 0.01–Present (Ice Age) 2–0.01
	Neogene or Late Tertiary 24.6–2	Pliocene Miocene Oligocene Eocene	5–2 24.6–5 38–24.6 55–38
	Paleogene or Early Tertiary 65–24.6	Paleocene	65–55
Mesozoic 248–65	Cretaceous		144–65
	Jurassic		213–144
	Triassic		248–213
Paleozoic 590–248	Permian		286–248
	Carboniferous		360–286
	Devonian		408–360
	Silurian		440–408
	Ordovician		505–440
	Cambrian		590–505

Source: Adapted from Hubler (1995).

years ago. These sediments resulted from erosion of high volcanic mountains and were deposited in deep basins that occurred where our present-day mountains exist. Sediments included calcareous minerals dissolved in acidic rainwater and deposited into the basins as limestone. Sand particles that were eroded and recemented formed sandstone after settling into the basins and mud or clay became shale and slate (Hubler 1995). These depositional basins were very deep. For example, Silurian rocks apparently covered most of present-day New York, Pennsylvania, and West Virginia to a depth of 1500 meters, and Mississippian rocks covered the present-day Ozark region to a similar depth (Petersen et al. 1980).

Sometime during the Permian period, the continents all drifted together to form the supercontinent Pangaea (Fig. 12). It was during this time that much of the force that created the Appalachian and Ouachita systems was brought to bear (Table 2). The continents of Africa and South America were pressing on North America from the east and south. This forced the land to raise and uplift, exposing the sediment basins as land masses, and through intense and constant pressure caused the faulting and folding that developed into the mountain systems of the eastern United States. The faulting and folding of the Appalachians are correlated strongly with similar events that formed the Ozark/Ouachita systems at about the same time (King 1950). However, the former probably resulted from pressure of the African continent and the latter from South America. The original uplifted mountains were many times higher than

FIGURE 12   Approximate relationship of continental land masses within the supercontinent Pangaea during the Permian Period (redrawn from M. S. Petersen, K. J. Rigby, L. F. Hintze, *Historical Geology of North America,* Wm. C. Brown Co., 1980; reproduced with permission of McGraw-Hill Companies).

TABLE 2 Geologic Time Scale

ERA	PERIOD[a]	ANIMAL LIFE	MOUNTAINS	PLANT LIFE
Cenozoic 60 million years	Quaternary Present−1.6 - - - - - - - - - - - - - Tertiary 1.6−66.4	Humans dominant Animals and birds		Flowering plants: grass, deciduous trees
	(Rocky Mountains Uplifted)			
Mesozoic 135 million years	Cretaceous 66.4−144 - - - - - - - - - - - - - Jurassic 144−208 - - - - - - - - - - - - - Triassic 208−245	Dinosaurs, other reptiles, insects	Appalachians high; Rockies a lowland under water	Gingko, cycads (conifers such as pine, spruce and fir)
	(Appalachian Mountains Uplifted)			
Paleozoic 350 million years	Permian 245−286 - - - - - - - - - - - - - Pennsylvanian 286−320 - - - - - - - - - - - - - Mississippian 320−360 - - - - - - - - - - - - - Devonian 360−408 - - - - - - - - - - - - - Silurian 408−438 - - - - - - - - - - - - - Ordovician 438−505 - - - - - - - - - - - - - Cambrian 505−570	Amphibians Age of fish Invertebrate animals in water	Present-day mountains and inland areas often under water	Coal forests: tree ferns, horsetail, rushes, club mosses Land plants appear Simple plants: algae, bacteria, seaweeds
Precambrian 4 billion years	No conspicuous fossils, but carbon deposits and imprints of simple forms			

[a]Period figures in millions of years ago.
Source: Adapted from Hubler (1995).

our present-day mountains, which after 150 million years of erosion have become the landscape that exists today (Fig. 13).

There are many interesting and unique features of the Appalachian system that, when viewed in the context of the geologic history of the region, begin to seem more logical. For example, the occurrence of hot springs from New York through the Carolinas and out to Arkansas gives testimony to the now extinct volcanos that helped build the early precursors of the Appalachians. There are a few remnant monadnocks that, due to their hardness, have resisted erosion. Features, such as Stone Mountain in Georgia and Mount Katahdin in Maine (Fig. 14), are granite mountains of extruded Precambrian material that was associated with the region's volcanic past. Mount Mitchell, the highest point in the eastern United States, is a monadnock of quartzite, a metamorphic rock formed from sandstone buried deep underground and subjected to intense heat and pressure. Metamorphic rocks like quartzite, feldspar, and marble are more common in the Blue Ridge regions, as well as in the Ouachita Mountains, providing evidence of their metamorphic origins. Near Mena, Arkansas, in the Ouachitas, is the site of one of a handful of areas where diamond can be found in the United States. Diamonds are formed when carbon (coal) is subjected to intense heat and pressure, giving further evidence of the forces that created these mountains.

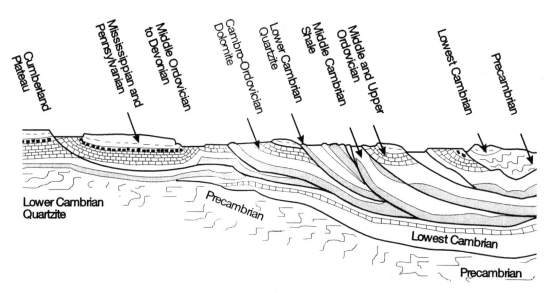

FIGURE 13   Structural cross section of the Appalachian region showing the thrust faults of the southern part of the orogenic belt. The Ridge and Valley Province is composed of overlapping slices of Cambrian and younger Paleozoic formations. Late Precambrian and Cambrian rocks are exposed in the Smoky Mountains to the east (redrawn from M. S. Petersen, K. J. Rigby, L. F. Hintze, *Historical Geology of North America*, Wm. C. Brown Co., 1980; reproduced with permission of McGraw-Hill Companies).

FIGURE 14   View across the granite surface of a prominent monadnock (Stone Mountain, Georgia) situated at the southern tip of the Appalachians.

Outcrops of Silurian and Ordovician age limestones occur extensively in the Great Appalachian Valley due to the action of folding and thrust-faulting. The same layers are exposed in the Cumberland/Appalachian Plateau in the Nashville Basin of Tennessee and Bluegrass Region of Kentucky, but here they are exposed due to the erosion of the harder Pennsylvanian-period materials weakened by uplifting of the Cincinnati Arch. Interestingly, these areas share several common features. One such feature is the occurrence of rounded sandstone hills called "knobs" around their perimeter (Fig. 15). Another similarity is the presence of karst topography with sinking creeks, sink holes, and caves. The vegetation is also similar with eastern redcedar (*Juniperus virginiana* L.) as a prominent old-field invader (Fig. 16).

Glaciation was directly involved over only a small part of the central hardwoods, that being the central lowlands of southern Illinois and Indiana (Fig. 17). In these areas, a thin layer of till covers much of the soil and here it is still the unglaciated hilly areas that remain forested, while the lower lands are cultivated cropland. Although glaciation had a limited direct effect on the central hardwood region, it had sweeping indirect effects. For example, many of the rivers that flow into the Ohio system were clogged by ice dams, and huge lake systems developed along much of the western and northern edge of what is now West Virginia and southern Ohio. Soils in these areas often are formed from significant lakebed sediments, making them finer in texture. Rivers have

FIGURE 15   The "knobs" of eastern Tennessee persist as features on the landscape because of the slower weathering rate of their sandstone rocks in comparison to the surrounding limestones and shales.

FIGURE 16   Eastern redcedar often invades old fields on soils with high base cation content like these in the Bluegrass Region of Kentucky.

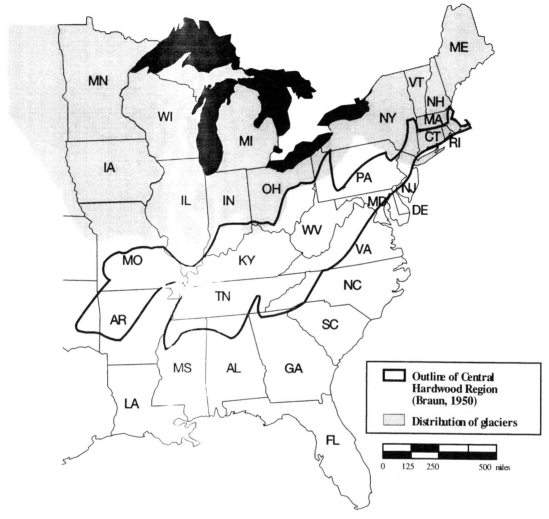

FIGURE 17   Map showing extent of glaciation in the eastern United States (from M. B. Davis, "Holocene Vegetational History of the Eastern United States," in *Late-Quaternary Environment of the United States*, Vol. II, H. C. Wright et al. eds. University of Minnesota Press, 1983).

left indelible marks on the landscape throughout the central hardwood region (Fig. 18). Perhaps one of the most interesting is the New River that heads in North Carolina and cuts its way across the Ridge and Valley and Allegheny Front. It empties into the Kanawha River (upriver of Charleston, West Virginia), which ultimately flows into the Ohio near Huntington. The fact that the New River cuts across these mountainous features indicates that it predated the

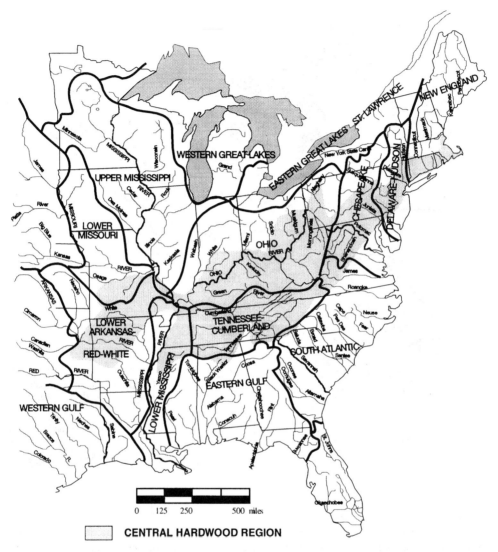

FIGURE 18   Principal rivers and drainage basins in the eastern United States (from J. J. Geraghty, D. W. Miller, F. van der Leeden, and F. Troise 1973).

uplifting of the mountains. Indeed, its meandering course, more reminiscent of a slow-flowing river in level terrain, is an indication that the mountains rose around the river leaving the so-called entrenched meanders as evidence of its past history. Some geologists speculate that the New River once had its headwaters in what are now the Atlas Mountains of northern Africa, dating back to the time when Pangaea was a single supercontinent (Hubler 1995). Other rivers, such as the Susquehanna and Delaware Rivers, like the New River, have also cut impressive gaps through the Appalachians. Such gaps that are presently occupied by a stream are called "water gaps." Other gaps, giving witness to the age of the Appalachians, occur where no river currently flows. These "wind gaps" were the sites where a river used to flow, but its passage was diverted when another stream pirated its waters. For example, the Blue Ridge has numerous wind gaps, such as Fancy Gap in Virginia, where rivers flowed at one time. But as the softer limestone in the valley was eroded away faster than the river channel could cut through the metamorphic strata of the Blue Ridge, the river began flowing down the valley. The Potomac/Shenandoah drainage system is a good example of this. The present tributaries all flow down valleys of the Ridge and Valley Province, but wind gaps in the Blue Ridge are left as evidence that many of them once flowed east.

Another interesting feature of the Appalachians, including the Ouachitas, is the presence of geosynclines and anticlines in the folded regions. In the strongly folded Ridge and Valley, the synclines (downward folds) often form the current mountain tops (Fig. 19). This is because compression at the bottom of the fold made the rocks more difficult to erode whereas fracturing at the top of the fold

FIGURE 19   The roadcut on Interstate 68 at Sideling Hill in western Maryland exposes the geosyncline that, due to compression of the layers, was more resistant to erosion and ultimately became the top of the ridge.

FIGURE 20   The Cheat River in northern West Virginia has cut this impressive gorge through the resistant Pennsylvanian-era sandstones in the Chestnut Ridge Anticline.

makes them more easily erodible. In the region of gentle folds to the northwest of the Ridge and Valley Province, anticlines have persisted as mountains where hard Pennsylvanian sandstones formed a resistant cap. The Chestnut Ridge Anticline that forms Chestnut Ridge in northcentral West Virginia and the Laurel Highlands of southwestern Pennsylvania is an example of this (Fig. 20).

Another geologic feature that has had an impact on the central hardwood region is the occurrence of fossil fuel deposits. Coal, gas, and oil are all abundant under much of the Appalachian Plateau. These deposits occurred during the lush vegetative growth of the Mississippian and Pennsylvanian periods (Carboniferous). Coal seams outcrop along hillsides in the dissected plateau where streams have eroded away the intervening valleys. Contour surface mining has been a common practice used to extract coal in the Appalachian Plateau. More recently, with the advent of larger equipment, surface mining by mountain-top removal/valley fill is being employed in steep terrain (Fig. 21).

SOILS AND SITE

Soil is a complex of mineral and organic fragments that supplies minerals, water, and support for plant growth (Wilde 1946). The characteristics of a given soil are a result of the interaction of the parent material (geologic substrate),

FIGURE 21 Photograph of surface mining for coal by mountain-top removal in the central Appalachians. This type of mining is practiced in steep terrain in the Appalachian Plateau from Tennessee through Virginia and West Virginia and often results in the conversion of forests to other post-mining land uses.

topographic position, climate, vegetation, and time. As discussed previously, the geology and topography of the central hardwood region are very complex. This complexity is reflected in the soils. Soils from limestone parent materials are generally finer textured, deeper, and higher in basic cations (Ca, K, Mg) and phosphorus than other soil types. For this reason, and the fact that they usually occur on more manageable slopes, these soils are predominantly occupied with agricultural activities (Fig. 22). This is true of the Nashville Basin, the Shenandoah and Tennessee Valleys, and the Bluegrass sections of Kentucky and Tennessee, as well as many of the more narrow valleys of the folded and faulted Ridge and Valley Province. These areas are often surrounded by sandstone-capped knobs, an example of which is shown in Figure 23. The occurrence of eastern redcedar is usually a good indicator of limestone deposits (Fig. 24).

Soils derived from shale (primarily siltstone and claystone) parent materials are generally the least productive in the central hardwood region due to lower fertility and water-holding capacity, depending somewhat on the climate where they occur. For example, the soils derived from Devonian shales on side slopes

FIGURE 22   A view of the Tennessee Valley in Monroe County showing a mix of agriculture and forests that occurs in this gently rolling landscape.

FIGURE 23   Knobs in the "Hiwassee" range near Madisonville, Tennessee illustrate the resistance of these Pennsylvanian-era sandstones.

FIGURE 24   Eastern redcedars vigorously colonize limestone outcrops along a road cut in central Kentucky.

of the Ridge and Valley Province are among the poorest in the central hardwood region. These soils, usually shallow and poorly developed, support scrubby oak forests (Fig. 25). By contrast, shale-derived soils on the broader and more moist landscapes of the Allegheny Plateau often support good growth of important forest species. Also, shales of the Mississippian and some younger strata are higher in weatherable nutrients, and thus more productive than those formed from other more acidic Pennsylvanian and Devonian rock. Sandstones of various periods are common capstones of ridges and plateaus throughout the central hardwood region. They occur interbedded with shales but produce soils of higher sand content and sometimes greater acidity. As with shale-derived soils, slope position, aspect, and climate will strongly influence productivity of sandstone-derived soils. These soils are not as sought after for agriculture but are capable of supporting very productive forests.

Because shales and sandstones occur interbedded in layers of varying thickness throughout the plateaus and folded portions of the central hardwood region, their influence on soil often mixes. On ridges and certain slope segments single parent material types form soils in place, called **residuum**. However, vast areas of sideslopes and footslopes in these physiographic regions commonly have soils formed in colluvial materials from both sandstone and siltstone.

FIGURE 25   This Ridge and Valley oak stand is typical of many in the central hardwood region where poor growing sites favor oaks over other species.

Where sufficient moisture exists, these soils form the basis for an important timber growing area within the central hardwood region (Fig. 26).

As discussed previously, the effect of climate is as important to soil productivity as geology (Eyre 1963). For most of the central hardwood region, precipitation exceeds evapotranspiration for a significant portion of the year. Its effect on soil development is seen in the moderate to high degree of weathering that has taken place in most forest soils in the region. Many temperate millennia of wet-dry and freeze-thaw cycles in the presence of carbonic and organic acids have resulted in considerable weathering of soil materials. Structural development and leaching of cations have occurred in most soils of the region, and soils on more stable landscape positions have well-developed argillic horizons that are enriched in illuvial clay particles translocated from overlying eluvial horizons. Organic horizons, though often ignored, are important in forest soils because they commonly contain much of the nutrient reserves available to forests in a management time frame. They are also critical to water, gas, and microbial dynamics and are most susceptible to impact from human activities.

In a few high-elevation areas on sandy parent rock, podzolization processes are important. Precipitation approaches or exceeds evapotranspiration for most of the year on these sites where spruce-fir forests often dominate. The very acid

FIGURE 26   A mixed mesophytic hardwood stand in the Appalachian Plateau Province. These stands are typically found on above-average sites such as concave slopes and north- and east-facing aspects. They contain a mixture of species including yellow-poplar, maples, and oaks.

leachates from this detritus chelate, translocate, and reprecipitate iron and aluminum, forming spodic horizons. These soils are called spodosols.

Temperature also affects forest soils of the central hardwood region. The soil temperature regime ranges from thermic (mean annual soil temperature 15–22 degrees Celsius), with a few areas of frigid soils above 3000–4000 ft. elevation (mean annual soil temperature < 8 degrees Celsius). Generally, as temperature increases, so does weathering, other things being equal. Therefore, lower elevations and latitudes are more likely to have soils with strong argillic horizons (ultisols and alfisols) and more reddish subsoils due to oxidation of iron. As can be seen in Figure 27 and Table 3, these soil orders occur extensively in the central hardwood region.

Slope aspect affects temperature and moisture relations in the sloping terrain of the central hardwood region. The relatively cooler and more moist northeast-facing coves usually have higher amounts of organic matter incorporated in the soil and thus more available plant nutrients. These coves have a higher site index and are less subject to burning than other aspects. This interplay between topography, microclimate, soils, and vegetation is an important feature of the hilly terrain of the central hardwood region.

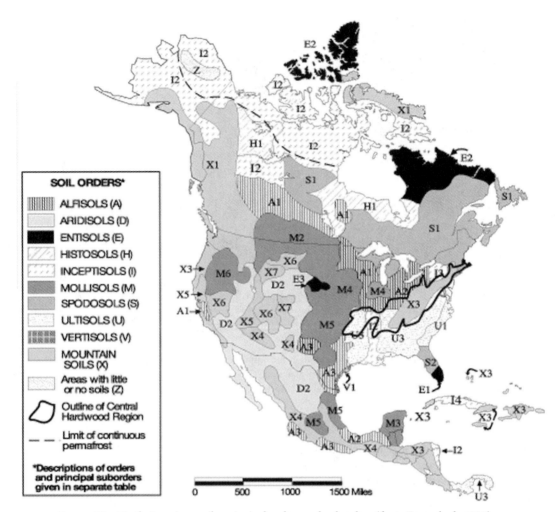

FIGURE 27   North American soils: principal orders and suborders (from Espenshade 1995).

The impact of vegetation on soils is generally related to the type of litter produced and deposited at the surface. The chemical and physical properties of litter affect the decomposition rates (Melillo et al. 1982) and this, in turn, affects the amount of organic matter and leaf litter fractions of the soil. It also affects the rate of mineral cycling (Hicks and Frank 1984). For example, Mudrick et al. (1994) found that when comparing the rates of decomposition (mass loss) of leaves of yellow-poplar, red maple, and chestnut oak (*Quercus prinus* L.), the yellow-poplar and maple decomposed almost completely during a single season in northcentral West Virginia, whereas the oak leaves were only about 40 percent decomposed. This, in part, explains the thicker litter layer that is common under oak stands as compared to other mesophytic hardwoods. Within the central hardwood region, there are a few areas that are predominantly forested by conifers. Examples of this are high-elevation Appalachian spruce or spruce/fir stands (Fig. 28), pine stands on south-facing ridges (Fig. 29), cedar brakes in the Nashville Basin and Bluegrass sections, and hemlock stands in the central Appalachians. These conifer stands produce litter that is quite different from hardwood leaves in physical properties and chemical content. Conifer leaves are generally more lignified than hardwood leaves and often produce a more acidic litter. Conifer litter, due to the above, tends to accumulate at the surface in deep layers of relatively undecomposed material (Fig. 30).

The entire region of the central hardwoods is geologically very old (in excess of 250 million years). Thus, the soils present reflect a long history of weathering and erosion. The well-developed profile of most soils within the region gives testimony to the long history of their development. The rounded features of the landscape indicate the continuing effect of erosion. This effect has produced zones of sideslope, footslope, colluvial, and alluvial soil throughout the region.

FIGURE 28   Red spruce stand near Grandfather Mountain, North Carolina. These high-elevation forests are remnants of a boreal forest that was more widely distributed throughout the present-day central hardwood region when the climate was much cooler.

TABLE 3 Description of Soils: Principal Orders and Suborders

This is a new classification known as the 7th Approximation. The system emphasizes the properties of the soils themselves in differentiating the orders, suborders, and so on. The nomenclature is systematic and uses Greek and Latin elements, which describe some of the soil characteristics.

Names of soil orders end in *sol* (L. *solum*, soil) which with a connecting vowel (o or i) is preceded by a descriptive prefix. The latter contains a formative element, part of which is used as an ending for names of the suborders (see syllable in parentheses below).

Arid (id)	L. *aridus*, dry		**Spod (od)**	Gr. *spodus*, wood ash
Hist (ist)	Gr. *histos*, tissue		**Ult (ult)**	L. *ultimus*, last
Incep (ept)	L. *inceptum*, beginning		**Vert (ert)**	L. *verto*, turn
Moll (oll)	L. *mollis*, soft			

Names of suborders have two parts. The first suggests diagnostic properties of the soil (see below), and the second is the formative element from the order name, for example, Id (Arid).

Aqu	L. *aqua*, water; soils that are wet for long periods		**Rend**	from Rendzina; high carbon content
Arg	L. *argilla*, clay; soils with a horizon of clay accumulation		**Torr**	L. *torridus*, hot and dry; soils of very dry climate
Bor	Gr. *boreas*, northern; cool		**Ud**	L. *udus*, humid soils of humid climate
Cry	Gr. *kryes*, icy cold; cold		**Ust**	L. *ustus*, burnt; soils of dry climates with summer rains
Ochr	Gr. *orchras*, pale; soils with little organic matter		**Xer**	Gr. *xeros*, dry; soils of dry climates with winter rains
Psamm	Gr. *psammas*, sand; sandy soils			

Only dominant orders and suborders are shown and each area delineated may include other kinds of soils.

Alfisols Podzolic soils of middle latitudes: soils with gray to brown surface horizons; subsurface horizons of clay accumulation; medium to high base supply	**Inceptisols** Immature, weakly developed soils; pedogenic horizons show little alteration but little illuviation; usually moist	**Spodosols** S1 Undifferentiated (mostly high latitudes)
		Aquods S2 Seasonally saturated with water; sandy parent materials
Boralfs A1 Cool to cold, freely drained	**Aquepts** I2 Seasonally saturated with water (includes some Humic Gley, alluvial tundra soils[a])	**Ultisols** Soils with some subsurface clay accumulation; low base supply; usually moist and low inorganic matter; can be productive with fertilization
Udalfs Temperate to hot; usually moist (Gray-brown Podzolic[a])	**Ochrepts** I3 Thin, light-colored surface horizons; little organic matter	**Aquults** U2 Seasonally saturated with water; subsurface gray or mottled horizon
Ustalfs A3 Warm subhumid to semi-arid; dry >90 days (some Reddish Chestnut and Red and Yellow Podzolic soils[a])	**Tropepts** I4 Continuously warm to hot; brownish to reddish	**Udults** U3 Low in organic matter; moist, temperate to hot (Red-Yellow Podzolic; some Reddish-Brown Lateritic soils[a])

Ardisols [ ] Pedogenic horizons lower in organic matter and dry for >6 mo. of the year (Desert[a] and Reddish Desert[a]). Salts may accumulate on or near surface.

Argids D2 With horizon of clay accumulation

Entisols [ ] Soils without pedogenic horizons on recent alluvium, dune sands, etc.; varied in appearance

Aquents E1 Seasonally or perennially wet; bluish or gray and mottled.

Orthents E2 Shallow; or recent erosional surfaces (Lithosols[a]); a few on recent loams

Psamments E3 Sancy soils on shifting and stabilized sands

Histosols [ ] Organic soils; bogs, peats and mucks; wholly or partly saturated with water

[ ] **Areas with little or no soil; ice fields and rugged mountains**

Mollisols [ ] Soils of the steppe (incl. Chernozem and Chestnut soils[a]); thick black organic rich surface horizons and high base supply

Borolls M2 Cool or cold (incl. some Chernozem, Chestnut and Brown soils[a])

Rendolls M3 Formed on highly calcareous parent materials (Rendzina[a])

Udolls M4 Temperate to warm; usually moist (Prairie soils[a])

Ustolls M5 Temperate to hot; dry for >90 days (incl. some Chestnut and Brown soils[a])

Xerolls M6 Cool to warm; dry in summer; moist in winter

Spodosols [ ] Soils with a subsurface accumulation of amorphous materials overlain by a light-colored, leached sandy horizon

Vertisols [ ] Soils with high content of swelling clays; deep wide cracks in dry periods, dark colored

Uderts V1 Usually moist; cracks open <90 days

Mountain soils [ ] Soils with various moisture and temperature regimes; steep slopes and variable relief and elevation; soils vary greatly within short distance

X1 Cryic great groups of Entisols, Inceptisols and Spodosols

X3 Udic great groups of Alfisols, Entisols and Ultisols; Inceptisols

X4 Ustic great groups of Alfisols, Entisols, Inceptisols, Mollisols and Ultisols

X5 Xeric great groups of Alfisols, Entisols, Inceptisols, Mollisols and Ultisols

X6 Torric great groups of Entisols; Aridisols

X7 Ustic and cryic great groups of Alfisols, Entisols; Inceptisols and Mollisols; ustic great groups of Ultisols, cryic great groups of Spodosols

[a]Terms refer to Great Soil Group terminology.
Source: Espenshade (1995).

FIGURE 29   Mixed pine-hardwood stands in the southern Appalachians of eastern Tennessee, showing pines on south-facing slopes.

There are also several areas in the central hardwood region with locally unique soils. As mentioned earlier, at high elevations in the Appalachians (generally above 4000 ft.) frigid soils occur. These soils are typical of those appearing much farther north (e.g., eastern Canada) but they exist here due to the effect of elevation on climate. Another unique local soil occurs in the so-called shale barrens of the Ridge and Valley Province of Pennsylvania, Maryland, and West Virginia. These soils, formed from shale materials, are low in fertility, and the tilted and interbedded strata channel water to drain into the slope, creating a droughty environment as well. These barrens are vegetated with scattered drought-tolerant oaks and a ground cover of xerophytic plants, including prickly pear. A local modification of some significance is the inclusion of glaciated topography, overlain by till, in parts of southern New England, the lower Hudson Valley, and the Pocono region of Pennsylvania, as well as areas of southern Indiana and Illinois. Till soils typically contain a substantial content of stone and gravel that was transported by the glacier. Although not directly affected by glaciers, portions of southeastern Ohio, including some portions of West Virginia along the Ohio River, were covered by extensive glacial meltwater lakes, and the soils at the surface contain the benthic sediments from these lakebeds. Loess beds of wind-carried sand results in local alterations in soils of

FIGURE 30   Conifer litter, such as this under a hemlock forest, is slower than hardwood litter to decompose; therefore, it builds up to greater depth on the forest floor.

western Kentucky and western Tennessee. These loess soils are generally fertile but very prone to erosion and have been deeply gullied in some areas. Finally, a significant local soil modification throughout the Appalachian coalfields from Pennsylvania and Ohio through West Virginia and into Kentucky, Tennessee, and Virginia is the occurrence of minesoils. These are anthropogenic in origin and result from surface mining for coal (Fig. 31). Reclamation is mandated after mining, but minesoils are very different from their parent soils since the subsurface geology and drainage are altered. They are usually reclaimed to a hayland-pasture cover as opposed to forest (Torbert and Burger 1995).

Figure 45 and Table 5, included under "Forest Cover" in Chapter 1, show the distribution of soils throughout the central hardwood region. These soils generally belong to the soil orders of Ultisols, Alfisols, and Inceptisols (Bailey et al. 1994).

The preceding discussion and Table 5 provide an overview of forest soils throughout the central hardwood region. But at best, this discussion is only an introduction to the subject. Natural resource managers may require diverse information about soils. For example, a silviculturist may be concerned primarily with fertility, productivity, or rooting support whereas a recreationist may be interested in compaction, erodibility, or hydraulic properties, and a forest

FIGURE 31   Prior to the 1970s, surface mine reclamation only involved leveling the spoil material and seeding with grass or other vegetation, leaving stark highwalls. This example is in Harrison County, West Virginia.

engineer would be concerned with engineering properties (slope stability, erodibility, etc.).

An excellent source reference for the above is the published soil surveys from the USDA, Natural Resource Conservation Service. These surveys include data for one or groups of counties and are generally available for the entire central hardwood region. They can be obtained by contacting local USDA, NRCS offices or from the state offices.

Site represents the sum total of all environmental factors that affect the functioning of a forest community. Factors, such as soil, climate, and vegetation, all affect site as shown in the simple model in Figure 32. Foresters have devised numerous systems in an effort to assess site quality (usually equated to forest productivity). Jones (1969) reviewed these various methods and discussed the advantages and disadvantages of each. Most of these methods of site quality evaluation have been applied to central hardwood stands. Use of tree height growth as a measure of site quality is one of the most enduring systems, and oak **site index** (height of dominant or codominant oaks at 50 years of age) is still the standard for the central hardwood region. Schnur (1937) was among the first to develop tree growth models based on data collected from a

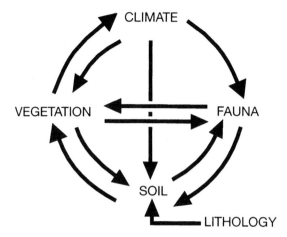

FIGURE 32   This diagram shows the interrelationships among the various components of the ecosystem (from Eyre 1963).

comprehensive sample of oaks growing on upland sites in the central hardwood region. Many others, including Carmean (1971) and Lamson (1980, 1987), have used similar methods to develop specific site index models for smaller geographic units where fewer climatic variations allow for more precise estimation. In addition to oaks, several other species have been used to develop similar site indices. Carmean et al. (1989) reviewed site index for upland oaks and 26 other hardwood species. Perhaps the most compelling reason to use oaks is the ubiquitous distribution of upland oaks throughout most of the central hardwood region.

Several components of the environment are intuitively associated with site quality. For example, soil factors, topographic factors, and climatic factors are all variables associated with site quality. Budyko (1974) suggested that climatic factors could be used to directly assess site quality rather than using site index based on tree growth. Tajchman and Lacey (1986) compared bioproductivity of a small watershed in the Appalachian Plateau of West Virginia with that of a site in Australia. Although reversed, the affect of aspect as measured by the radiative index of dryness (ratio of net radiation to the latent heat equivalent of precipitation) appeared to have the same association with bioproductivity in both the northern and southern hemispheres. This "**biophysical**" approach to site quality evaluation was also suggested by Lee and Sypolt (1974). The objective of this direct approach to site quality evaluation is to attempt to directly measure the environment that is associated with site productivity.

Considerable research has been conducted in the central hardwood region using **soil and topographic factors** to predict 50-year height growth of trees (Munn and Vimmerstedt 1980). Topographic factors include aspect (Trimble and Weitzman 1956; Yawney 1964; Yawney and Trimble 1968; Auchmoody and Smith 1979). Slope position and shape (concave, convex, etc.) have also been used as topographic variables to predict site quality (McNab 1987). Topographic variables capture some of the microclimatic factors (radiation, temperature, etc.) as well as soil moisture, since upper slopes are presumably drier than lower slopes.

In addition to topographic variables, soil factors, such as depth and drainage (Phillips 1966), have been included in site index prediction models. Hicks et al. (1982) used 38 soil variables, including minerals in the topsoil and subsoil as well as depth, pH, stoniness, and aspect to predict biomass productivity of sites in a 600-acre watershed in West Virginia. Among all these variables, aspect had the greatest association with bioproductivity.

Another approach to evaluating sites is to use **vegetation frequency** or occurrence to predict site quality. For example, Fountain (1980a, b) used understory vegetation (presence/absence) and frequency of tree species (stand composition) to predict site quality. Like site index, these methods involve using vegetative measurements as an indication of site quality. The common thread that links all the methods of site evaluation is the fact that they all attempt, either directly or indirectly, to account for the effect of environment on the forest.

Intuitively, site is a function of macroclimate, microclimate, and soil characteristics. Direct measurement of these environmental factors to predict site quality would be desirable. But due to the complexity of environmental factors, it is very difficult to identify variables that accurately depict the environment and are readily measurable. This has prompted foresters to turn to using vegetation as an index to site quality. Use of presence or abundance of indicator plants is confounded by the fact that previous land use history (agriculture, timber harvesting, fire, etc.) can have a major impact on the plant community present and, therefore, may lead to erroneous determinations of site quality.

The site index method, using tree height as an indicator of site, although still the standard, comes with its own set of problems. For example, appropriate trees (species, age, crown class) must be present before site index can be measured. When using oak site index, it must be kept in mind that, although different oak species share many similar attributes, species such as chestnut and white (*Quercus alba* L.) oaks grow at inherently different rates than scarlet oak (*Q. coccinea* Muenchh.) or northern red oak (*Q. rubra* L.) (Carmean et al. 1989). Thus, site quality determinations remain problematic for foresters and continuing efforts are under way to develop better methods of forecasting site quality in the central hardwood region (Wiant and Fountain 1980).

As a footnote, the effect of environment can readily be seen when comparing the growth curves of the same species in different environments. Carmean et al. (1989) reported the site index curves of Beck (1962) for yellow-poplar growing in the North Carolina mountains and North Carolina Piedmont while Schlaegel et al. (1969) reported yellow-poplar site index for the Appalachian region of West Virginia (Fig. 33). Growth curves for the southern

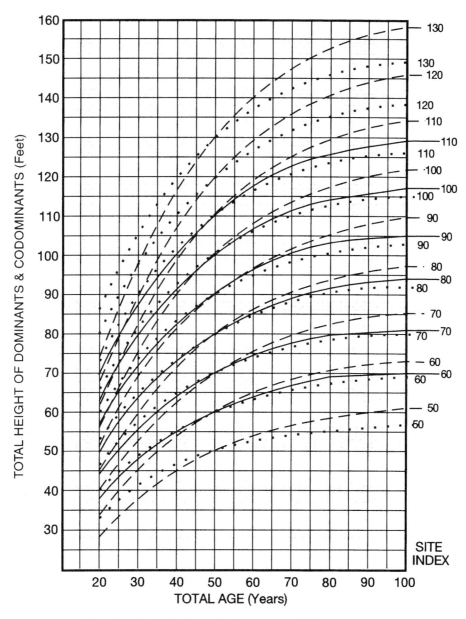

FIGURE 33 Site index graphs for three populations of yellow-poplar (from Carmean et al. 1989).

TABLE 4 Yellow-Poplar Site Index at Three Locations and Associated Climatic Variables

LOCATION	AVERAGE ANNUAL PRECIPITATION (in.)	MEAN GROWING SEASON (DAYS)	PREDOMINANT SOIL FORMATION
North Carolina Piedmont	44–48	210	Ultisols
North Carolina Mountains	56–62	160	Ultisols, Inceptisols
West Virginia, Appalachian Mountains	50–56	140	Inceptisols, Ultisols

Source: U.S. Department of Commerce (1983).

Appalachians of North Carolina were steeper than those for West Virginia or the North Carolina Piedmont. Climatically, the southern Appalachians of North Carolina have a longer growing season than West Virginia, and the soils of all three regions are somewhat different (Table 4). Barring inherent genetic variations in these populations of yellow-poplar, the differences in growth can be attributed to differences in climate and/or soils, but since several things are varying simultaneously (temperature, growing season, precipitation, soil), it is impossible to sort out these effects without replicated experiments. Another example of the effect of climate, soil, and geology on site productivity comes from examining the site index data from the USDA, Forest Service Forest Inventory and Analysis plots for West Virginia (DiGiovanni 1990). In Figure 34, the site index data were used to construct isopleths of equal site quality. This graph shows that the highest average oak site index in West Virginia occurs in the southwestern portion of the state where the growing season is longest and it receives about 44 inches of precipitation annually. The area of lowest site index occurs in a belt corresponding with the Ridge and Valley Province. This region is characterized by shale soils and an annual precipitation of about 32 inches per year.

CLIMATE

The climate of an area is the single most important factor controlling the type of vegetation that occurs in that area. Indeed, the distributions of certain global properties (including vegetation and soils) are described as being "zonal" relative to the effect of climate on them. Climate is generally described in terms of temperature, moisture relations, and movement of air masses, which, at the **macro scale**, are governed by latitude, elevation, and continental position (Bailey 1996). The effect of **latitude**, at a global scale, is evident in both temperature and precipitation. Because tropical latitudes receive more energy

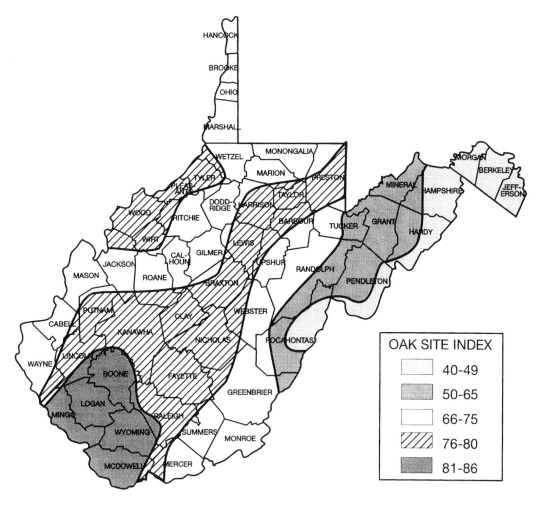

FIGURE 34   Isopleths of oak site index in West Virginia. This illustrates that the best sites occur in the Appalachian Plateau in the warmer southern part of the state, while the poorest sites are in the Ridge and Valley Province.

from solar radiation over the year, the tropics are generally the warmest places on earth, and, conversely, the polar regions are coldest. Temperate zones oscillate between seasonal extremes based on the declination of the earth relative to the sun. Precipitation is also affected by latitude at the macro scale in that tropical air near the equator tends to rise, which initiates precipitation. Near the Tropics of Capricorn and Cancer (23.5° north and south latitude), air sinks back to earth, forming more-or-less permanent high-pressure zones along this latitude and producing dry climates. As can be seen on Figure 35, zones of the

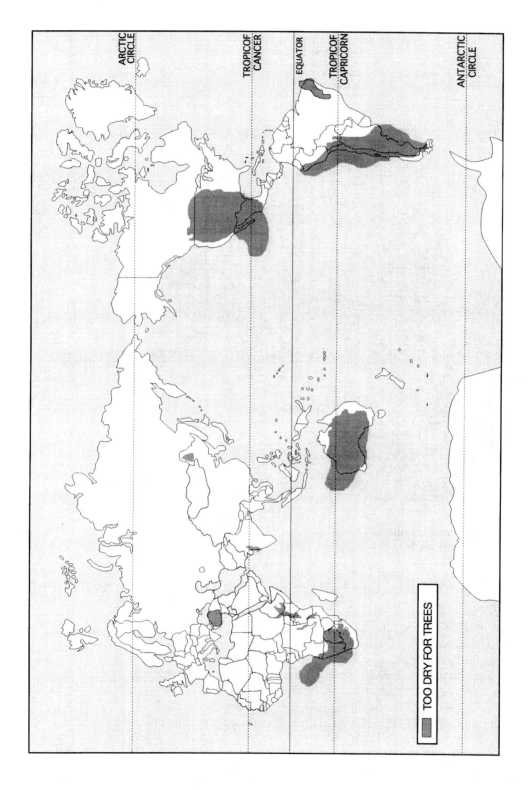

FIGURE 35   Zones of the earth too dry for tree growth (from Bailey 1996).

earth where it is too dry for tree growth generally coincide with these latitudes. The central hardwood region is in a so-called temperate zone, which has a moderate climate in both temperature and precipitation.

Continental position (or perhaps more aptly described as position relative to the ocean) has the effect of amplifying or dampening the extremes. For example, summer temperatures in the central plains of the United States (e.g., Chicago) can be very hot, while winter lows are also extreme in the opposite direction. Places near the ocean, such as Seattle and San Francisco, are generally moderate throughout the year. But the effect of oceans is generally more pronounced on the western side of the continents, due in part to global air movement patterns. Thus, the central hardwood region, including the portion near the Atlantic Ocean, generally has a continental, rather than a maritime, climate.

At the **mesoscale**, factors such as topography become more important in controlling climate. For example, elevation can induce altitudinal zonation. In the central hardwood region this effect is most obvious in the higher Appalachians where oaks and mesophytic hardwood species are replaced by northern hardwoods (beech-maple). In West Virginia this occurs at elevations above 3000 ft., and at the highest elevations (generally above 4000 ft.), a boreal spruce or spruce-fir forest occurs (Fig. 36). Topography also exerts a strong influence on precipitation. The orographic effect of mountains is apparent in the central hardwood region where higher levels of precipitation occur down the spine of the Appalachians, the highest being 80 inches per year in western North Carolina (Fig. 37). In eastern West Virginia and western Virginia is an area that receives less than 35 inches of annual precipitation. This condition results from a "rain shadow" associated with the mountains immediately to the west, where some locations receive in excess of 50 inches per year.

At the **microscale**, climate is controlled by factors such as **slope inclination, slope position, aspect** (Budyko 1974), and **soil color** (if exposed) or **vegetative cover** (Geiger 1965; Lilly et al. 1980). The south–facing slopes are generally warmer than north-facing slopes in the northern hemisphere while dark surfaces, such as organic soils and dark minesoils, heat much more rapidly than lighter-colored soils when exposed to solar radiation.

Climate classification has been attempted by various persons. The system that is now the accepted standard for scientific use is Köppen's (1931) system as modified by Trewartha (1968). Using this system, the climate of the central hardwood region is generally classed as "Subtropical Humid" (Fig. 38). In a refinement of this system, Bailey (1994) has developed a classification hierarchy for "ecoregions" of the United States. Using Bailey's system, the central hardwood region lies within the Humid Temperate Domain and the Hot Continental Division, except for the Ouachita Mountains, which fall within the Moist Subtropical Division (Fig. 39). These broadest subdivisions of Bailey's hierarchy are based on climatic variables whereas lower subdivisions are based on topography and vegetation. A more thorough discussion of Bailey's classifi-

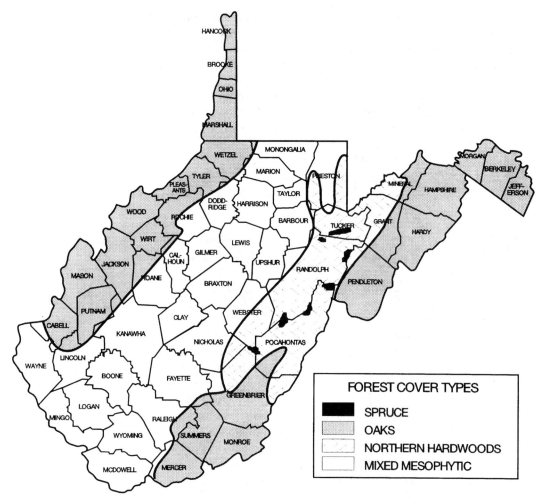

FIGURE 36 Forest cover types in West Virginia. The effects of the Appalachian highlands are apparent in producing a wide range of diversity in a relatively small area.

cation hierarchy is given in Chapter 1 of this book, under the subtitle "Forest Cover."

A qualitative description of the climate that prevails over most of the central hardwood region shows that it is characterized by a seasonally fluctuating climate with hot humid summers followed by relatively cold open winters (National Oceanic and Atmospheric Administration 1980). Tropical air from the Gulf of Mexico predominates during the summer, and polar air masses invade the area during winter, being alternately displaced by warmer air. Precipitation is evenly distributed throughout the year and generally averages

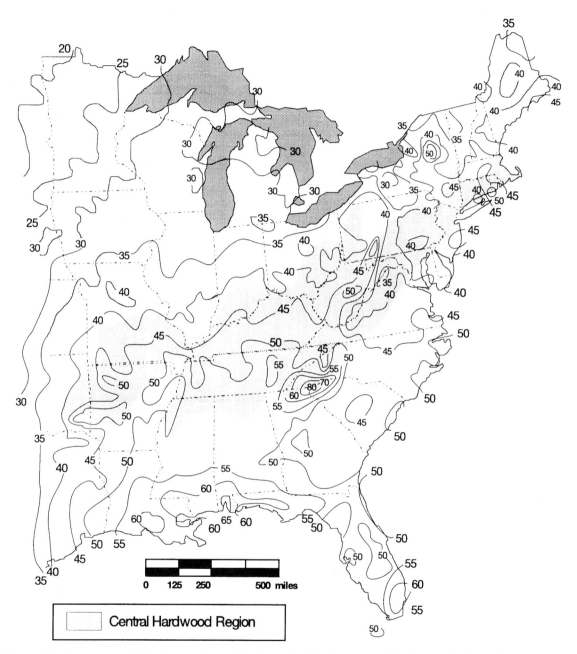

FIGURE 37   Average annual distribution of precipitation (inches) in the eastern United States (redrawn from J. J. Geraghty, D. W. Miller, F. van der Leeden, and F. Troise 1973).

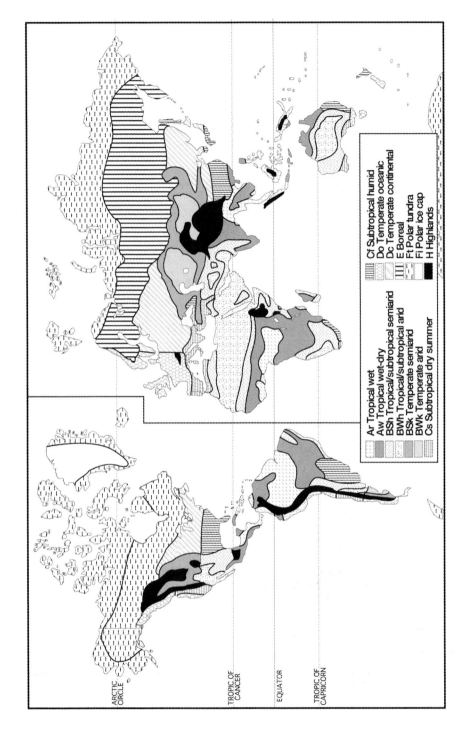

FIGURE 38 Ecoclimate zones of the earth (from G. T. Trewartha, *An Introduction to Climate*, McGraw-Hill, 1968, reproduced with permission of McGraw-Hill Companies).

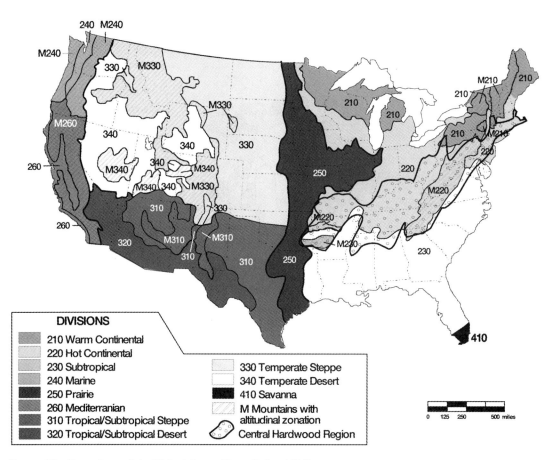

FIGURE 39   Ecoregions of the United States (from Bailey 1994).

from 40 to 50 inches, with locally higher or lower amounts (Fig. 37). Most of the precipitation in the southern and western portion of the regions falls as rain, with annual snow amounts exceeding 20 inches along the northern tier of the region and down the Appalachians into western North Carolina (Fig. 40). The frost-free growing season in the central hardwood region ranges from about 120 to 210 days, generally increasing from northeast to southwest, again with a prominent effect of elevation where higher elevation areas have a shorter growing season (Fig. 41).

Precipitation is a factor that has received much recent attention, and the presence of acids of sulfate and nitrate make up what is termed "acid deposition," which has been associated with the so-called forest decline phenomenon in the eastern United States (Eagar et al. 1992; Stephenson and Adams 1993). Based on results from the National Atmospheric Deposition Program

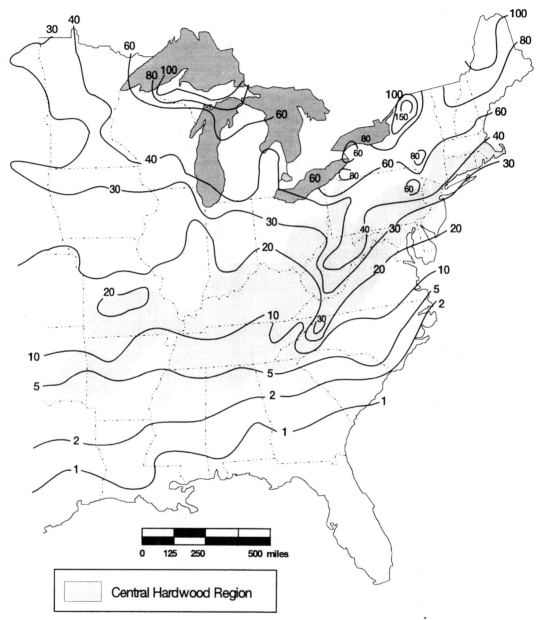

FIGURE 40   Average annual distribution of snowfall (inches) in the eastern United States (redrawn from J. J. Geraghty, D. W. Miller, F. van der Leeden, and F. Troise 1973).

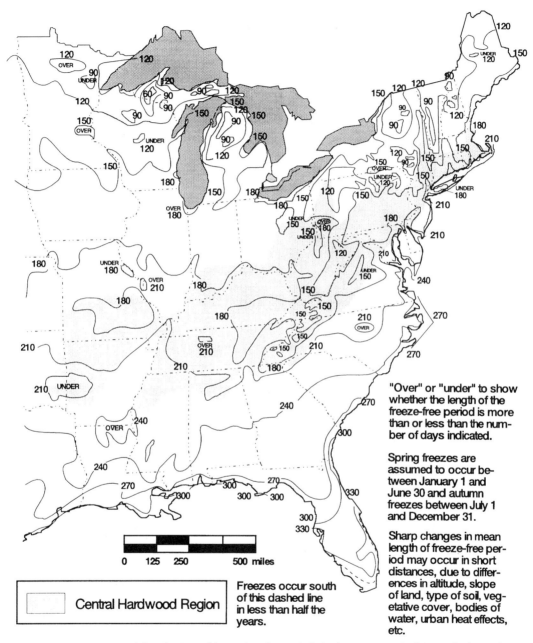

"Over" or "under" to show whether the length of the freeze-free period is more than or less than the number of days indicated.

Spring freezes are assumed to occur between January 1 and June 30 and autumn freezes between July 1 and December 31.

Sharp changes in mean length of freeze-free period may occur in short distances, due to differences in altitude, slope of land, type of soil, vegetative cover, bodies of water, urban heat effects, etc.

Freezes occur south of this dashed line in less than half the years.

0   125   250   500 miles

☐ Central Hardwood Region

FIGURE 41   Average annual distribution of freeze-free day periods in the eastern United States (redrawn from J. J. Geraghty, D. W. Miller, F. van der Leeden, and F. Troise 1973).

and other sources (Hicks 1992), the highest acid loading in precipitation in North America occurs in the region of northern West Virginia/central Pennsylvania, where rain pH averages approximately 4.2. The recent national trend in acid loading of precipitation has generally been toward a reduction in precipitation acidity (Lynch et al. 1995).

In summary, the climate of the central hardwood region is generally a favorable one for plant growth with moderate temperatures, a strong seasonal effect, adequate growing season precipitation, and a prominent effect of topography. It is these characteristics that have led to the dominance of the area by deciduous angiosperm species. Such species can exploit the warm, moist summers and use their broad horizontally-oriented leaves to capture a maximum amount of solar radiation for photosynthesis. Their ability to store photosynthates in root tissue and to remain dormant during the winter months, which are unfavorable to growth, enables deciduous angiosperms to have a competitive advantage over other life forms in the particular climate regime that exists in the central hardwood region.

FOREST COVER

As previously stated, macroclimate is the overriding characteristic that dictates the type of vegetation occupying an area. Bailey (1996) recounts: "Climate would appear to be, then, the initial criterion in defining ecosystem boundaries." Braun (1950), referring to the deciduous forest formation of North America, put it this way: "the formation is the long-time expression of climatic control. Through the vast area occupied by this formation, the seasonal distribution of rainfall and the length of the growing season are sufficiently alike to favor the dominance of deciduous forest." Vankat (1979) indicates that "the dominant characteristic of climate in the deciduous forest region is the division of the year into distinct seasons, including a winter with continuous below-freezing temperatures in the north and at least occasional frosts in the south." Distribution of plant communities is further affected by local differences in climate (usually associated with topography or bodies of water), differences in soils (Eyre 1963), and land use history. The central hardwoods, as they exist today, are a product of all these factors.

For those of us who live and work in the "great forest" (Buxton and Crutchfield 1985), it is easy to accept the superlatives often used to describe it. But in spite of Lucy Braun's (1950) reference to the "vast area occupied by the formation," on a global scale, temperate deciduous forests occupy only a small fraction of the land area. Bailey (1996), using primarily climate and vegetation, classified the "Ecoregions of the Continents." Using his classification system, the central hardwoods, as well as the entire eastern deciduous forest formation, occur within two divisions (the Hot Continental and Subtropical) of the Humid Temperate Domain. These divisions are mostly found within plus or minus 15°

of the 40th parallel in latitude and mostly in the northern hemisphere. An interesting fact in reference to the extent of these types of ecoregions is that the Humid Temperate Domain occupies about 22.5 percent of the global land-surface area, and the Hot Continental and Subtropical Divisions (including mountainous sections) account for a combined total of about 5 percent of the global land surface. These are the regions on the earth where deciduous forests are best adapted to grow. But due to agricultural land clearing in these ecoregions in places like China, Korea, Japan, western Europe, and New Zealand, the only deciduous hardwood ecoregion of the world that remains extensively forested is in the eastern United States.

To further place the extent of the central hardwoods in a global context, tropical rainforest ecoregions occupy almost 10 percent of the earth's surface. Tundra and Subarctic ecoregions combine for about 16 percent of the global land area, and lands falling within the Dry Domain (steppes and deserts) account for almost 47 percent of the total land area.

Forests of the central hardwood region, as implied by the name, are predominantly **deciduous angiosperms**. Clements (1916) referred to the eastern deciduous forest as the *Quercus-Fagus* formation, and he recognized that oaks were found throughout the region and that beech is frequently a climax species, especially in the northern portion. Braun (1950) in one of the most comprehensive phytogeographical studies to focus exclusively on the deciduous forest of North America states: "The Deciduous Forest Formation is made up of a number of climax associations differing from one another in floristic composition, physiognomy and in genesis or historical origin." For purposes of this book, the central hardwood region includes all of Braun's Mixed Mesophytic Forest Region, the Interior Highlands (within the southern Division of the Oak-Hickory Forest Region and the Southern Appalachians), Northern Blue Ridge, and Ridge and Valley, all within the Oak-Chestnut Forest Region (Fig. 42). This delineation of the central hardwood region matches up very well with Bailey's (1994) Provinces of the United States, where the central hardwood region includes the Eastern Broadleaf Forest (oceanic), the Central Appalachian Broadleaf Forest–Coniferous Forest–Meadow, the Ozark Broadleaf Forest–Meadow and the Ouachita Mixed Forest–Meadow. The central hardwood region, as defined, divides Bailey's Eastern Broadleaf Forest (continental) to include the southern part (generally the unglaciated portion) (Figs. 43–45). Focusing down to the Section level of Bailey's hierarchy, an even more exact fit between Bailey's and Braun's classification systems is possible (Fig. 46). Table 5 references the soil orders present in the central hardwood region with Bailey's Ecoregions of the United States (1994). Tables 5a–5j provide a more detailed description of Bailey's Provinces in the eastern part of the United States (1994). Thus, the central hardwood region, as defined, seems to fit well within accepted systems of vegetative and ecoregion classification.

To delineate the vegetation of the central hardwood region in more detail, the descriptions provided by Braun (1950) for the appropriate divisions and

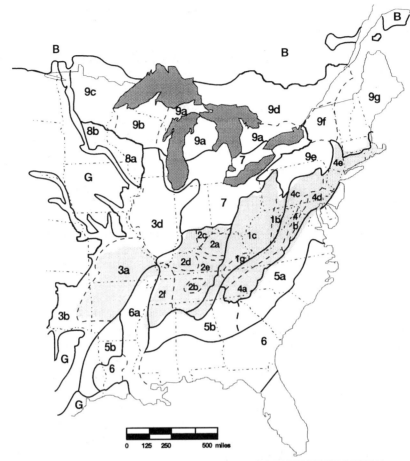

FIGURE 42   Braun's (1950) vegetational regions that comprise the central hardwood forest.

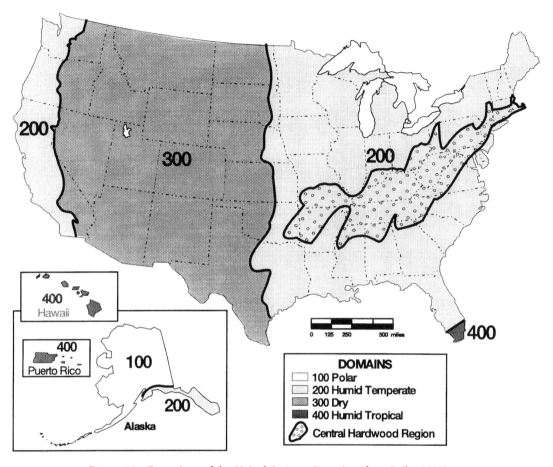

FIGURE 43   Ecoregions of the United States — Domains (from Bailey 1994).

sections are useful. For the Mixed Mesophytic Forest, she says it is centrally located within the unglaciated Appalachian Plateaus. She indicates that the climax association is shared by a number of species [beech, yellow-poplar, basswoods, sugar maple, buckeyes (*Aesculus* spp.), oaks (chestnut, red, and white), and hemlock]. The Western Mesophytic Forest occurs on the so-called Interior Low Plateau and climax associations include representatives of the Mixed Mesophytic association, with components of oak-hickory, oak-yellow-poplar, and beech. The Oak-Hickory Forest as described by Braun is centered in the Ouachita/Ozark Highlands, and climax communities are predominantly oaks and hickories. The description of the Oak-Chestnut Forest obviously preceded the almost complete elimination of American chestnut from the forest by chestnut blight. But this forest region includes the Ridge and Valley and Blue

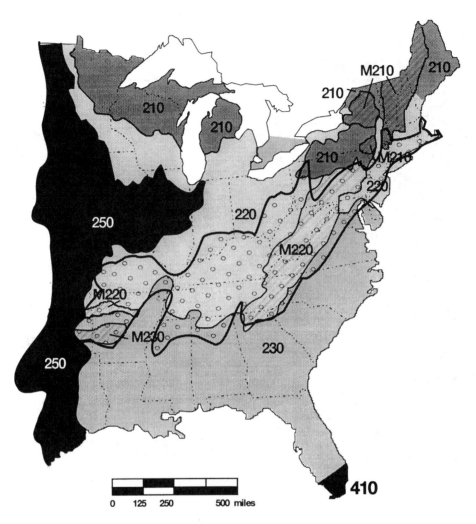

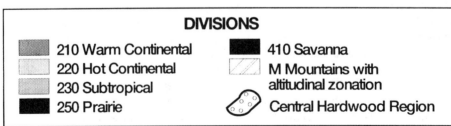

DIVISIONS

210 Warm Continental
220 Hot Continental
230 Subtropical
250 Prairie

410 Savanna
M Mountains with altitudinal zonation
Central Hardwood Region

FIGURE 44   Ecoregions of the United States — Divisions (from Bailey 1994).

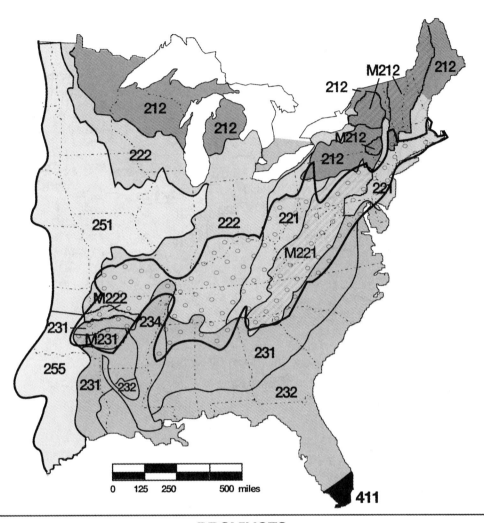

FIGURE 45   Ecoregions of the United States — Provinces (from Bailey 1994).

PROVINCES

212 Laurentian Mixed Forest

M212 Adirondack-New England
 Mixed Forest-Coniferous Forest-
 Alpine Meadow

221 Eastern Broadleaf Forest (Oceanic)
222 Eastern Broadleaf Forest (Continental)

M221 Central Appalachian Broadleaf
 Forest-Coniferous Forest-Meadow
M222 Ozark Broadleaf Forest-Meadow

411 Everglades

231 Southeastern Mixed Forest
232 Outer Coastal Plain Mixed Forest
234 Lower Mississippi Riverine Forest

M231 Ouachita Mixed Forest-Meadow

251 Prairie Parkland (Temperate)
255 Prairie Parkland (Subtropical)

M Mountains with altitudinal zonation

Central Hardwood Region

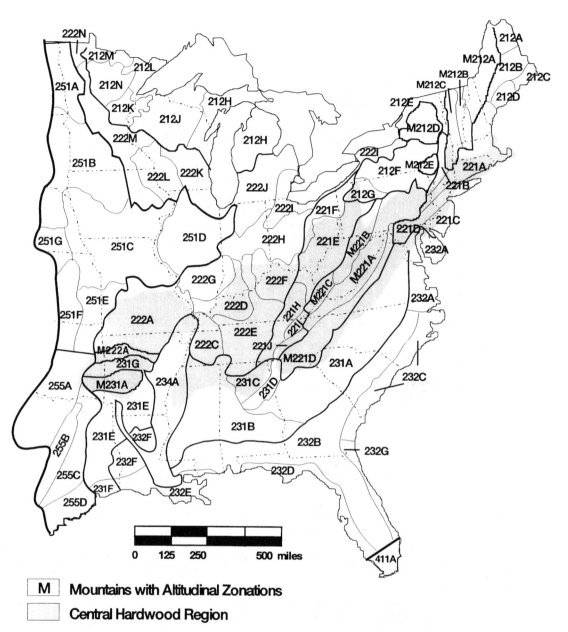

FIGURE 46   Ecoregions of the United States—Sections (from Bailey 1994).

TABLE 5 Soil Orders Present in the Central Hardwood Region, Referenced to Figures 45 and 46

PROVINCE[a]	SECTION[a]	SOIL ORDERS[b]	MODIFIERS[b]
221 (Eastern Broadleaf Forest, Oceanic Province)	221B	Inceptisols	
	221D	Ultisols, Inceptisols	Mesic, Udic
	221E	Ultisols, Inceptisols	Mesic, Udic
	221H	Inceptisols, Ultisols	Mesic, Udic
	221I	Ultisols, Inceptisols	Mesic, Udic
	221J	Ultisols, Inceptisols	Mesic, Thermic, Udic
222 (Eastern Broadleaf Forest, Continental Province)	222A	Ultisols	Mesic, Udic
	222C	Alfisols, Ultisols	Thermic, Udic, Aquic
	222D	Ultisols, Alfisols	Mesic, Udic
	222E	Ultisols, Alfisols	Mesic, Udic
	222F	Alfisols, Inceptisols	Mesic, Udic
	222G	Alfisols	Mesic, Aquic, Udic
M221 (Central Appalachian Broadleaf Forest Province)	M221A	Inceptisols, Ultisols	Mesic, Udic, Aquic
	M221B	Inceptisols, Ultisols	Mesic, Udic
	M221C	Inceptisols, Ultisols	Mesic, Udic-Aquic
	M221D	Ultisols, Inceptisols	Mesic, Frigid, Udic
M222 (Ozark Broadleaf Forest Province)	M222A	Ultisols	Thermic, Udic
231 (Southeastern Mixed Forest Province)	231A	Ultisols	Thermic, Udic
	231G	Ultisols, Mollisols	Thermic, Udic
M231 (Ouachita Mixed Forest Province)	M231A	Ultisols	Thermic, Udic

[a]See maps (Figs. 45 and 46) for location of provinces and sections.
[b]Brief description of soil orders and modifiers (from Kingsley 1985)
 Orders:
  Ultisols—Soils that are low in bases and have subsurface horizons of clay accumulations; usually moist, but during the warm season of the year, some are dry part of the time.
  Inceptisols—Soils that have weakly differentiated horizons; materials in the soil have been altered or removed but have not accumulated. These soils are usually moist, but during the warm season of the year, some are dry part of the time.
  Alfisols—Soils that are medium to high in bases (base saturation at pH 8.2) and have gray to brown surface horizons and subsurface horizons of clay accumulation; usually moist, but during the warm season, some are dry.
  Mollisols—Soils that have nearly black, friable organic-rich surface horizons, high in bases; formed mostly in subhumid and semiarid warm to cold climates.
 Descriptions of soil modifiers:
  Mesic—Pertaining to moist climates with adequate precipitation to support plant growth
  Udic—Relating to humid, or humid climates
  Thermic—Soil formation is modified by warm to hot climatic conditions
  Aquic—Soils formed in seasonally wet sites, either due to surface water or due to poor subsurface drainage
  Frigid—Referring to soils formed in cool to cold climates
Source: Bailey (1994).

TABLE 5a Eastern Broadleaf Forest (Oceanic) Province

Section	Geomorphology	Stratigraphy/ Lithology	Soil Taxa (Temperature; Moisture Regimes)	Potential Natural Vegetation	Elevation (ft.)	Precipitation (in.)	Temperature (°F)	Growing Season (days)	Surface Water Characteristics	Disturbances (Land Use)
221A (R9)	Glaciated peneplain, monadnocks	Cenozoic sediments; Paleozoic-Mesozoic sediments; Proterozoic granite	Inceptisols (Mesic; Udic)	N. hardwoods–hemlock–white pine forests	0 1500	35 50	45 50	120 180	Medium-gradient streams, lakes, coastal bays	Fire, wind (urban, forestry)
221B (R9)	Glaciated peneplain	Cenozoic sediments; Paleozoic shales, limestone	Inceptisols (Mesic; Udic)	Central hardwoods, N. hardwoods forest	0 1000	38 42	45 50	160 180	Medium-gradient streams	(Agriculture, urban)
221C (R9)	Low dissected plains	Cenozoic-Mesozoic sand, silt, clay	Ultisols (Mesic; Udic, Aquic)	Mixed pine, hardwood forests	0 300	35 50	50 55	170 210	Low-gradient streams, coastal beaches	Fire (forestry, urban)
221D (R9)	Series of peneplains	Mesozoic-Paleozoic sedimentary, igneous	Ultisols, Inceptisols (Mesic; Udic)	Oak-hickory forest	100 1000	35 45	50 57	160 200	Medium-gradient streams	Land clearing (agriculture, urban)
221E (R9)	Maturely dissected plateau	Paleozoic sandstone, siltstone, shale, coal	Ultisols, Inceptisols (Mesic; Udic)	Appalachian hardwood forest	660 1350	35 45	39 55	120 170	Medium-gradient streams	Fire, land clearing (agriculture, urban)
221F (R9)	Maturely dissected glaciated plateaus	Cenozoic till; Paleozoic sediments	Alfisols (Mesic; Udic, Aquic)	Maple-ash-oak Beech-maple forests	650 1000	35 40	48 50	150 160	Medium- to low-gradient streams, few bogs	Land clearing (agriculture, urban)
221H (R8)	Maturely dissected plateau	Paleozoic sandstone, shale, coal	Inceptisols, Ultisols (Mesic; Udic)	Mixed mesophytic, Appalachian oak forests	650 980	46	55	175	Many perennial streams	Fire (forestry)
221I (R8)	Faulted and folded monoclinal mountains	Paleozoic sandstone, shale, limestone	Ultisols, Inceptisols (Mesic; Udic)	Appalachian oak, mixed mesophytic forests	800 1000	46	55	175	Many perennial streams	Fire (forestry)
221J (R8)	Faulted and folded parallel ridges	Paleozoic carbonates, shale	Ultisols, Inceptisols (Mesic, thermic; Udic)	Appalachian oak forest	650 1950	36 55	55 61	170 210	Many perennial streams	Fire (agriculture)

TABLE 5b M221 Central Appalachian Broadleaf Forest–Meadow Province

SECTION	GEOMORPHOLOGY	STRATIGRAPHY/ LITHOLOGY	SOIL TAXA (TEMPERATURE; MOISTURE REGIMES)	POTENTIAL NATURAL VEGETATION	ELEVATION (ft.)	PRECIPITATION (in.)	TEMPERATURE (°F)	GROWING SEASON (days)	SURFACE WATER CHARACTERISTICS	DISTURBANCES (LAND USE)
M221A (R9)	Faulted and folded parallel ridges	Paleozoic limestone, sandstone, shale	Inceptisols, Ultisols (Mesic; Udic-Aquic)	Appalachian oak, oak-hickory-pine forests	300 4800	35 50	46 60	120 170	Parallel drainage network	Fire (agriculture, forestry, urban)
M221B (R9)	Severely dissected plateau	Paleozoic sandstone, shale	Inceptisols, Ultisols (Mesic; Udic)	Mixed hardwoods, spruce-fir forests	1000 4600	40 60	39 54	110 160	Deeply dissected plateau	Fire (forestry)
M221C (R8)	Faulted, folded monoclinal mountains	Paleozoic sandstone, shale	Inceptisols, Ultisols (Mesic; Udic-Aquic)	Mixed mesophytic, Appalachian oak forests	2000 2600	40 47	45 50	140 160	Incised dendritic drainage network	Fire (agriculture, forestry)
M221D (R8)	Tectonic uplifted mountain ranges	Proterozoic-Paleozoic igneous, metamorphics	Ultisols, Inceptisols (Mesic; Frigid; Udic)	Appalachian oak, spruce-fir, N. hardwoods forests	1000 6000	40 50	50 62	150 220	Many high-gradient, deeply incised streams	Ice, wind (forestry)

TABLE 5c Laurentian Mixed Forest Province

SECTION	GEOMORPHOLOGY	STRATIGRAPHY/ LITHOLOGY	SOIL TAXA (TEMPERATURE; MOISTURE REGIMES)	POTENTIAL NATURAL VEGETATION	ELEVATION (ft.)	PRECIPITATION (in.)	TEMPERATURE (°F)	GROWING SEASON (days)	SURFACE WATER CHARACTERISTICS	DISTURBANCES (LAND USE)
212F (R9)	Glaciated, dissected plateau	Cenozoic till; Paleozoic shale, sandstone	Inceptisols (Frigid; Mesic; Udic)	Appalachian oak-hickory, oak-pine forests	650 2000	30 50	46 50	100 160	Medium stream density, few lakes	Fire, land clearing (agriculture)
212G (R9)	Mature, dissected plateau	Paleozoic sandstone, s.ltstone, shale	Ultisols, Inceptisols (Mesic, Frigid; Udic, Aquic)	Hemlock–N. hardwoods, Appalachian oak-pine forests	1000 2400	40 50	46 48	120 150	Very few lakes, many streams	Fire (forestry, agriculture, oil production)

TABLE 5d M212 Adirondack–New England Mixed Forest–Coniferous Forest–Alpine Meadow Province

SECTION	GEOMORPHOLOGY	STRATIGRAPHY/ LITHOLOGY	SOIL TAXA (TEMPERATURE; MOISTURE REGIMES)	POTENTIAL NATURAL VEGETATION	ELEVATION (ft.)	PRECIPITATION (in.)	TEMPERATURE (°F)	GROWING SEASON (days)	SURFACE WATER CHARACTERISTICS	DISTURBANCES (LAND USE)
M212C (R9)	Subdued glaciated mountains, plateaus	Proterozoic-Paleozoic igneous, metamorphics	Spodosols (Frigid; Udic)	Montane spruce–fir, N. hardwoods–conifer forests	1300 4000	34 52	37 45	80 130	Headwaters, medium-gradient streams, lakes	Wind (forestry, recreation)

TABLE 5e Eastern Broadleaf Forest (Continental) Province

SECTION	GEOMORPHOLOGY	STRATIGRAPHY/ LITHOLOGY	SOIL TAXA (TEMPERATURE; MOISTURE REGIMES)	POTENTIAL NATURAL VEGETATION	ELEVATION (ft.)	PRECIPITATION (in.)	TEMPERATURE (°F)	GROWING SEASON (days)	SURFACE WATER CHARACTERISTICS	DISTURBANCES (LAND USE)
222A (R8)	Maturely dissected plateaus, low mountains	Paleozoic carbonates, sandstone; Proterozoic igneous	Ultisols (Mesic; Udic)	Oak-hickory, oak-pine woodland, Bluestem-cedar glades	300 1800	40 48	55 60	180 200	Streams, springs, losing streams	Fire (forestry, pasture, mining)
222C (R8)	Submaturely dissected alluvial plains	Cenozoic sand, silt, clay	Alfisols, Ultisols (Thermic; Udic-Aquic)	Oak-hickory, S. floodplain forests	80 330	45 60	61 68	200 280	Low- to medium-gradient streams	Fire (agriculture)
222D (R9)	Bluffs and low open hills	Paleozoic sandstone, siltstone, shale, coal	Ultisols, Alfisols (Mesic; Udic)	Oak-hickory forest, bluestem prairie	330 660	43 46	55 57	185 200	High stream density	Fire, drought, wind (agriculture, forestry)
222E (R8)	Maturely dissected plateau	Paleozoic sandstone, siltstone, shale, limestone	Ultisols, Alfisols (Mesic; Udic)	Oak-hickory forest, bluestem prairie	650 990	44 54	56 61	180 205	Medium stream density	Fire, drought (agriculture)
222F (R8)	Maturely dissected plateau	Paleozoic limestone, shale, sandstone	Alfisols, Inceptisols (Mesic; Udic)	Oak-hickory forest	650 1000	43 45	54 56	180	Medium stream density	Fire (agriculture)
222G (R9)	Glaciated plains	Cenozoic till; Paleozoic sandstone, shale	Alfisols (Mesic; Aquic, Udic)	Oak-hickory forest	400 825	35 45	54 56	180 200	Moderate stream density	Fire (agriculture)

TABLE 5f M222 Ozark Broadleaf Forest–Meadow Province

Section	Geomorphology	Stratigraphy/ Lithology	Soil Taxa (Temperature; Moisture Regimes)	Potential Natural Vegetation	Elevation (ft.)	Precipitation (in.)	Temperature (°F)	Growing Season (days)	Surface Water Characteristics	Disturbances (Land Use)
M222A (R8)	Maturely dissected plateau	Paleozoic sandstone, shale, limestone	Ultisols (Thermic; Udic)	Oak-hickory, oak-hickory-pine forests	650 2600	45 52	58 64	180 205	Many perennial streams	Fire (forestry)

TABLE 5g 234 Lower Mississippi Riverine Forest Province

Section	Geomorphology	Stratigraphy/ Lithology	Soil Taxa (Temperature; Moisture Regimes)	Potential Natural Vegetation	Elevation (ft.)	Precipitation (in.)	Temperature (°F)	Growing Season (days)	Surface Water Characteristics	Disturbances (Land Use)
234A (R8)	Dissected alluvial plain	Cenozoic stratified alluvium	Inceptisols, Alfisols (Thermic; Udic-Aquic)	S. floodplain oak-hickory forests	0 300	48 65	56 70	200 340	Low-gradient streams, artificial linear channels	Flooding (agriculture)

TABLE 5h 231 Southeastern Mixed Forest Province

SECTION	GEOMORPHOLOGY	STRATIGRAPHY/ LITHOLOGY	SOIL TAXA (TEMPERATURE; MOISTURE REGIMES)	POTENTIAL NATURAL VEGETATION	ELEVATION (ft.)	PRECIPITATION (in.)	TEMPERATURE (°F)	GROWING SEASON (days)	SURFACE WATER CHARACTERISTICS	DISTURBANCES (LAND USE)
231A (R8)	Maturely dissected plains	Precambrian granite; Paleozoic volcanics	Ultisols (Thermic; Udic)	Oak–hickory–pine, S. mixed forests	330 1300	45 55	58 64	205 235	Medium gradient streams	Fire (agriculture, forestry)
231B (R8)	Immaturely dissected alluvial plain	Cenozoic stratified marine deposits	Ultisols, Alfisols (Thermic; Udic)	Oak–hickory–pine, blackbelt forests	80 650	40 60	60 68	200 208	Many perennial streams	Fire (agriculture)
231C (R8)	Faulted and folded monoclinal mountains	Paleozoic stratified marine deposits	Ultisols, Inceptisols (Thermic; Udic)	Oak–hickory–pine, S. mixed forests	330 1300	50 55	60 62	200 210	Many perennial streams	Fire, drought (forestry)
231D (R8)	Faulted and folded parallel ridges	Paleozoic stratified marine deposits	Ultisols, Inceptisols (Mesic, Thermic; Udic)	Oak–hickory–pine, S. mixed forests	650 2000	36 55	55 61	170 210	Many perennial streams	Fire (agriculture)
231E (R8)	Little dissected alluvial plain	Cenozoic stratified marine deposits	Ultisols, Inceptisols (Thermic; Udic)	Oak–hickory–pine, S. floodplain forests	80 650	40 54	61 68	200 270	Many perennial streams	Fire (agriculture)
231G (R8)	Synclinal valley alluvial plain	Paleozoic stratified marine deposits	Ultisols, Mollisols (Thermic; Udic)	Oak–hickory, oak–hickory–pine forests	330 3000	44 50	61 63	200 240	Many perennial streams	Fire (agriculture)

TABLE 5i M231 Ouachita Mixed Forest–Meadow Province

SECTION	GEOMORPHOLOGY	STRATIGRAPHY/ LITHOLOGY	SOIL TAXA (TEMPERATURE; MOISTURE REGIMES)	POTENTIAL NATURAL VEGETATION	ELEVATION (ft.)	PRECIPITATION (in.)	TEMPERATURE (°F)	GROWING SEASON (days)	SURFACE WATER CHARACTERISTICS	DISTURBANCES (LAND USE)
M231A (R8)	Tectonic uplifted mountain ranges	Paleozoic stratified marine deposits	Ultisols (Thermic; Udic)	Oak-hickory-pine forests	330 2600	48 56	61 63	200 220	Many perennial streams	Fire (forestry)

TABLE 5j 232 Outer Coastal Plain Mixed Forest Province

SECTION	GEOMORPHOLOGY	STRATIGRAPHY/ LITHOLOGY	SOIL TAXA (TEMPERATURE; MOISTURE REGIMES)	POTENTIAL NATURAL VEGETATION	ELEVATION (ft.)	PRECIPITATION (in.)	TEMPERATURE (°F)	GROWING SEASON (days)	SURFACE WATER CHARACTERISTICS	DISTURBANCES (LAND USE)
232A (R8)	Little dissected alluvial plain	Cenozoic stratified marine deposits	Ultisols, Histosols (Mesic; Udic)	Oak-hickory-pine, S. floodplain forests	0 80	46 54	55 57	185 220	Estuaries, beaches, low-gradient streams	Fire (agriculture) urban)

Ridge Provinces. Climax vegetation here, according to Braun, is oak and American chestnut. The oak species that predominate are chestnut, northern red, and white. Occasionally yellow-poplar and other mesophytic species predominate, especially in coves.

Oaks are the species group that are consistently identified with the central hardwood region. White oak is the most widely distributed species that is an important component of forest stands throughout the whole region. Black oak (*Quercus velutina* Lam.) is also widely distributed but is seldom as important as white oak. Northern red oak is often the predominant oak species in the northern portions of the central hardwood region (Beltz et al. 1992). Southern red oak (*Quercus falcata* Michx.) becomes much more important in central hardwood stands along the southern tier of the region, from Arkansas to eastern Tennessee. Chestnut and scarlet oaks often predominate locally, especially on poorer sites, such as ridges and southwestern aspects. These species reach their greatest importance in the Ridge and Valley Section of the central Appalachian region (Fig. 47).

Oak species, such as burr (*Quercus macrocarpa* Michx.), blackjack (*Q. marilandica* Muenchh.), and post oaks (*Q. stellata* Wangenh.), are more important in the southwestern (Ozark/Ouachita) portion of the central hardwood region, and these species may be locally important components of stands, particularly on dry upland sites. Other oak species that occur as scattered components of upland forests in the central hardwood region are shingle oak (*Quercus imbricaria* Michx.), blackjack oak, and chinquapin oak (*Q. muehlenbergii* Englem.). Several oaks are locally important on moister sites throughout the region. Willow (*Q. phellos* L.) and Shumard (*Q. shumardii* Buckl.) oaks are more abundant in the southern portion of the region, and pin oak (*Q. palustris* Muenchh.) gains importance in the northern portion of the region, particularly associated with moist flats in glaciated topography.

In addition to oaks, a number of other hardwood species are typically associated with central hardwood stands. In fact, the typical situation for central hardwood stands is a mixture of species, often up to 20 species or more. Some of the common associates include **hickories** [pignut (*Carya glabra* (Mill.) Sweet), mockernut (*C. tomentosa* Nutt.), shagbark (*C. ovata* (Mill.) K. Koch), shellbark (*C. lacinosa* (Michx.f.)Loud.), etc.]. The affiliation of hickories with upland oaks has been noted by many authors, and an oak-hickory forest cover type has been identified by several authors. But others (Ware 1992) have questioned the ranking of hickories to such importance in making stands as to deserve being included in the oak-*hickory* title. In fact, Beltz et al. (1992) show the greatest distribution of hickory growing stock to occur in a band including most of Tennessee and Kentucky and extending into southern West Virginia, but even here, hickories often rank behind yellow-poplar and red maple in importance.

Another species with a broad distribution throughout the central hardwood region is **red maple.** This species, although distributed from Florida to Maine

FIGURE 47   A stand of chestnut oak/scarlet oak growing on a poor site in western Maryland.

and from Virginia to Oklahoma, is generally most abundant in the northeastern portion of the central hardwood region, extending down the higher Appalachians into western North Carolina (Beltz et al. 1992; Alerich 1993).

Sugar maple is somewhat similarly distributed in the central hardwood region as red maple but generally has a narrower overall distribution than red maple and seems to lack the amplitude of red maple to tolerate extremes in site conditions, sugar maple being more typically associated with the "northern hardwoods" [beech, birch (*Betula* spp.), maple]. This cover type is generally found to the north of the central hardwoods, but stands may occur within the region due to the climatic effect of higher elevation.

Another species that is frequently associated with central hardwood stands is **yellow-poplar.** This species' role in identifying the Mixed Mesophytic Hardwoods of Braun is analogous to the role of oaks in identifying the central hardwoods. Yellow-poplar is a major component of stands from the Allegheny Plateau of central Pennsylvania through West Virginia and into Kentucky and Tennessee. Yellow-poplar is usually associated with the so-called cove hardwoods type and is often found in pure stands on north-facing slopes, extending down into the coves (Fig. 48). Yellow-poplar's importance diminishes greatly in the Ridge and Valley Province, and although present in Braun's Western Mesophytic Forest, yellow-poplar is less important there than in the Mixed Mesophytic type.

FIGURE 48   A "cove hardwood" stand of yellow-poplar growing on a high-quality site in the central Appalachians.

It is not a component of the central hardwood stands in the Ozark/Ouachita region. Yellow-poplar makes its best development and is most important as a component of forest stands (greater than 20 percent) in southern West Virginia, ranging into eastern Kentucky.

Black cherry (*Prunus serotina* Ehrl.) is another species that is an important component of some central hardwood stands. Cherry is most important in the Allegheny Plateau of Pennsylvania and at higher elevations in the Appalachians (Alerich 1993), from West Virginia through Virginia and North Carolina, where trees of good quality grow. Cherry may be relatively abundant in other stands throughout the central hardwood region, but trees are frequently of poor quality and smaller size outside the prime growing areas.

American beech is a widely distributed species that is found throughout the central hardwood region, having its greatest importance in association with the northern hardwoods, where it often occurs with maples and birches. A number of other hardwood species are associated with central hardwood stands, often being locally important. These include pin cherry (*Prunus pennsylvanica* L.) (primarily in the northeast), blackgum (*Nyssa sylvatica* Marsh.), sweet (*Betula lenta* L.), river (*B. nigra* L.), and yellow (*B. alleghaniensis* Britton) birches, ashes (*Fraxinus* spp.) (especially from central Kentucky into eastern Ohio), black walnut (*Juglans nigra* L.) (often on alluvial soils), American sycamore (*Platanus occidentalis* L.) (particularly in riparian zones), basswoods (*Tilia* spp.), Carolina silverbell (*Halesia carolina* L.) (in the southern Appalachians), and cucumber magnolia (*Magnolia accuminata* L.).

Conifers are overall less than 10 percent of the growing stock within the central hardwood region. For example, in West Virginia, Zinn and Sutton (1976) indicate the conifer component, statewide, to be about 5 percent. But in spite of this, conifers are generally interspersed within the central hardwood region, being locally important. For example, eastern hemlock is often mixed in hardwood stands from the northeastern part of the region and down the Appalachians to northern Georgia (Fig. 49). Eastern white pine has a similar distribution to hemlock in the central hardwood region, and white pine is a major component of some stands, particularly in the mountainous regions of western Virginia and down the Blue Ridge through western North Carolina into northern Georgia (Fig. 50).

Several upland yellow pines are locally important components of central hardwood stands. Shortleaf pine is a significant component of stands in the Arkansas/Oklahoma/Missouri highlands. It is also locally important in the Tennessee Valley and Cumberland Plateau, where it spreads north in the Appalachian Plateau, eventually dropping out in the Ohio Valley about 75 miles north of Huntington, WV. Virginia and pitch pines are locally abundant in the Appalachian section of the central hardwood region, both species being found along dry ridges and south slopes in mountainous terrain (Fig. 51). The latter species is also important in the so-called pine barrens of the northeast from Cape Cod down through eastern Long Island. Table mountain pine occupies a

FIGURE 49   A hemlock-hardwood stand in central Pennsylvania.

niche similar to that of pitch pine but is generally found in the southern Blue Ridge Province.

There are several unique communities within the central hardwood region that are distinct from the prevailing vegetation. Most of these are brought about by unique or unusual soils or local climates resulting from topographic modifications. Several such areas are given the name "barrens" or "breaks." These include the **pine barrens** that are generally vegetated by pitch pine and scrub oaks and other so-called "pygmy forests." These communities usually occur on droughty soils and the species that occupy them are drought tolerant. But the community also owes its existence to the periodic occurrence of fire, which cleans out the competing vegetation and prepares the seedbed for

FIGURE 50   Eastern white pine-hardwoods growing in Raleigh County, West Virginia.

FIGURE 51   A view of central hardwoods showing pines growing on south-facing aspects.

germination of pine seeds. The oaks persist due to their ability to sprout from the root collar. **Shale barrens** are distributed in shale outcrop areas along the Ridge and Valley from Pennsylvania into West Virginia. These communities are dominated by widely spaced scrub oaks and other drought tolerant species, such as Virginia and pitch pines. The shale soils are generally thin, low in fertility, and droughty. An understory of xerophytic plants, such as prickly pear and Cates Mountain clover, indicates the droughty nature of the site. Eastern redcedar is frequently associated with so-called **cedar breaks**. The cedars of Lebanon State Park in the Nashville Basin area of middle Tennessee and similar sites in the Bluegrass Region of Kentucky, are good examples (Fig. 52). The condition that favors development of these communities is a dolomite limestone soil parent material that weathers rapidly and produces a thin soil with relatively high pH. Redcedars are common as an early successional stage on limestone soils, but are normally replaced by hardwoods in the absence of disturbance. Communities on these thin soils do not progress past the cedar stage, hence the cedar break community. Open and **grassy meadows** or prairies occur in some areas and are apparently part of the natural vegetation of the region. The Nashville Basin of Tennessee, Bluegrass Region of Kentucky, and the Ozark glades of Missouri represent such areas, many of which are currently agricultural. Like the cedar breaks, occurrence of these grasslands is related to soil and

FIGURE 52   Eastern redcedar invading old agricultural fields in eastern Tennessee.

parent material with the added influence of climate and periodic grass fires. Another type of grassland, called "**grassy balds**," occurs on scattered high mountain tops along the Appalachian highlands. Examples are Bald Knob (West Virginia), Stratton's Meadows (Tennessee), and Brasstown Bald (Georgia). The reason for the presence of these balds is still unexplained, since they do not occur above a true timberline. Several hypotheses have been put forward, including historic agricultural clearing and fire (Fig. 53). Finally, a unique community that is apparently a relic from the glaciation is the occurrence of **boreal spruce fir** communities at high elevations in the Appalachians. These red spruce (*Picea rubens* Sarg.) and spruce-fir forests, such as Mt. Mitchell in North Carolina, cover the highest peaks in the southern Appalachians (normally above 4500 ft) (Fig. 54). Red spruce stands occur at elevations above 3800 ft in West Virginia, and a thriving lumber industry was centered around this resource at the turn of the century (Fig. 55).

The understory of forests within the central hardwood region is just as varied, or perhaps more so than the overstory. There are a few species and genera that are ubiquitous associates of the deciduous hardwood forests of the central hardwood region, but most understory species, like their overstory counterparts, are regionalized (Harshberger 1911). Any list of understory plants

FIGURE 53   Bald Knob, a grassy bald or mountain meadow in West Virginia.

FIGURE 54   Red spruce growing above 4000 ft. elevation in western North Carolina.

is likely to be incomplete simply due to the large number of species involved. But a few species and genera are worthy of mention for their important role. Flowering dogwood (*Cornus florida* L.) is one of the most widely distributed understory species in the region and currently in jeopardy from dogwood anthracnose disease, especially in the northern and eastern portion of the region, where the disease was first introduced (Hicks and Mudrick 1994). Other prominent understory components that occur throughout the region are eastern redbud (*Cercis canadensis* L.) (especially on basic soils), hawthorns (*Crataegus* spp.), *Vacciniums*, *Viburnums* (black haws, etc.), spice bush (*Lindera benzoin* L. Blume), witch hazel (*Hamamelis virginiana* L.), and several species of ferns. Some understory plants are specific to certain regions, such as mountain laurel (*Kalmia latifolia* L.) and rhododendrons (*Rhododendron* spp.), which are found mostly in the Appalachian Mountains, and stripe maple (*Acer pennsylvanicum* L.), which is generally found in the northern portion of the region. Like the overstory, the understory species are affected primarily by climate and soils. But they have the additional influence of overstory (composition and density) with which to contend. Understory plants are also much more affected by disturbances, such as fire and past land use history, so the anthropogenic effect on the understory, if anything, has been greater than on the overstory.

FIGURE 55   Remnants of "ghost town" of Spruce, West Virginia. Such towns were in their heyday during the logging boom of the early 1900s.

Anthropogenic influence on the central hardwood forest is sweeping in its effect and takes several forms. First, much of the land was cleared and used for subsistence farming before regrowing to forest, only to be cut over and highgraded by loggers between 1900 and 1920. Forest burning was also an important anthropogenic effect. Thus, the forest we see today is a combination of climate, site, and disturbance effects.

But perhaps one of the most significant anthropogenic effects is human introduction (plants, diseases, and insects). Notable examples of these are plant introductions, such as Kudzu (Fig. 56), which is especially problematic in the southern and southeastern portion of the region. Japanese honeysuckle, multi-flora rose, and autumn olive are also examples of introduced pests in the region. One of the most significant introductions to the central hardwood region is that of chestnut blight, which has almost totally eliminated a species that was once important enough to have a cover type named for it [the oak-chestnut of Braun (1950)]. Another introduction that is having a major impact on an important understory species is dogwood anthracnose disease, and gypsy moth is an introduced insect that is threatening to alter stand compositions throughout the region, since they prefer oak hosts (Fosbroke and Hicks 1989).

In summation, the natural vegetation of the central hardwood region is dominated by deciduous hardwood species, especially oaks, and these species

FIGURE 56   Kudzu, an introduced plant from the Far East, has become a serious pest in the southern portion of the central hardwood region.

are acclimated to the region due to their adaptation to the strongly seasonal climate and other environmental conditions. Regional deviations in climate and soils bring about corresponding variations in vegetation. In fact, at the local or site scale, the forests of the central hardwood region present a pattern resembling a mosaic of the broader climatic cover types referred to by Braun (1950) as formations, although one of these types may predominate regionally. For example, in the mixed mesophytic region, aspect and slope position affect the climate at the site scale so as to create a mosaic where mesophytic hardwoods predominate on north- and east-facing slopes and in coves, and oak types predominate on ridges and southwest aspects (Hicks and Frank 1984). But the vegetation of the region is also strongly affected by human activities. In Chapter 2, the historical development of the central hardwoods will be explored, and this plus the current chapter should provide a context for future chapters on ecology, culture, and management of central hardwoods.

THE CENTRAL HARDWOOD REGION, SUMMARY

The central hardwoods occur in North America from Massachusetts to Missouri, south of the glaciated regions and north and west of the Piedmont Alluvial Plains. Most of the region occurs in the mountains or hilly terrain of the Appalachians, Ouachitas, and associated plateaus. Geologically, the central hardwood region is a product of ancient vulcanism, followed by an erosion/deposition cycle, uplift, faulting and folding, and finally an erosion cycle to produce the present-day physiographic features. Soils of the region are generally podzolic in origin and deep with clay deposition in the subsoil.

The climate of the central hardwood region is a humid temperate climate in the Hot Continental Domain with a strong seasonal cycle and precipitation that is evenly distributed throughout the year. The growing season varies from about 120 days to almost 200 days, with a pronounced elevational effect.

Because of the climate, soil, and topography, the vegetation of the central hardwood region is predominantly deciduous hardwoods. The central hardwoods generally coincide with Braun's (1950) forest formations: the oak-chestnut, mixed mesophytic, western mesophytic, and oak-hickory. Oaks are the most abundant trees in the region. Other important species include hickories, maples, and yellow-poplar with conifers, such as eastern white pine, shortleaf pine, eastern hemlock, and eastern redcedar, being locally important components of stands. The central hardwood region is the largest and most extensive area of deciduous hardwoods in the world and represents a rich mixture of valuable tree species that is approaching the threshold of economic maturity.

2

Historical Development

An appreciation of the historical developments that shaped the central hard-wood region is important to an understanding of the present forests, and knowledge of history enables us to better interpret past successes and failures. Thus, history provides not only background but also the context for future management. For purposes of this discussion, the history of the central hardwoods will mean the period after the last ice age (the period of human impact). Because the contemporary forest is the result mostly of events that have occurred during the life of the trees that presently comprise it, this period will be emphasized.

It was previously pointed out that the overriding factors that predispose a geographic region to a particular vegetation type are climate and soil (site), and this has been true throughout the hundreds of millions of years that vegetation has existed on earth. Furthermore, the prevailing vegetation creates habitat for animal forms that are adapted to it, and they, in turn, have a biofeedback relationship with the vegetation. This type of association is the basis for what is called an **ecosystem** (Odum 1971).

EVOLUTION AND THE DECIDUOUS FOREST

The ability of plants to capture solar energy by using readily available elements (carbon, hydrogen, oxygen, etc.) has been the key to all biological development on earth. Fossil evidence is our window to the past, and although incomplete, it appears that this process of **photosynthesis** first developed in primitive marine algae between 400 and 500 million years ago (Eyre 1963). From these evolved the primitive land plants that occupied the earth's surface prior to the

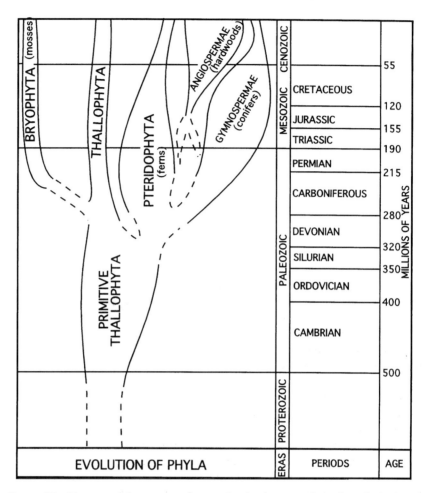

FIGURE 57   Diagram of the process of vegetative development through geologic time (from Eyre 1963).

Carboniferous period (more than 200 million years ago). During the Carboniferous period (215–280 million years ago), ferns and their relatives dominated
the earth (Fig. 57). Around 190 million years ago, the gymnosperms began to
take over as the principal vegetation, and they have given way to the dominance
of angiosperms (including deciduous hardwoods) during the last 50 or so
million years. The angiosperms, though relatively recent in their development,
have shown a remarkable ability to adapt to the changing conditions of the
earth. From this point of view, it might be said that the central hardwoods
represent the cutting edge of evolutionary change, as compared to the boreal
conifer forest.

Perhaps the most recent (and most significant) episode of climatic and vegetational change to affect the distribution of present-day vegetation was the so-called **ice age**. The continental glaciers extended down through North America into southern Illinois (Fig. 17), and a zone of tundra extended well down into much of the present central hardwood region (Davis 1983). As the Wisconsin ice sheet began to retreat northward about 14,000 years ago, a boreal forest type replaced the tundra, and vestiges of this forest still exist as the high-elevation spruce and spruce-fir communities in the Appalachians (Maxwell and Davis 1972). The deciduous forest community has gradually become the dominant vegetation of the region and now forms the most extensive forest of this type in the world (Buckner 1989).

IMPACT OF THE NATIVE PEOPLE ON THE PRE-COLUMBIAN FOREST

As important as the ice age and its retreat has been, the impact of humans has probably been just as significant in shaping the central hardwood forest. Indigenous people occupied the central hardwood region for at least 10,000 years before Europeans came (Lesser 1993). In the southern portion of the region, human occupancy probably dates back as long as 14,000 years ago owing to the more hospitable southern climate (Buxton and Crutchfield 1985). Archaeologists classify human occupancy of North America into Late Prehistoric, Mississippian, Woodland, Archaic, and Paleo periods (Lesser 1993). Table 6 shows the chronology of these periods and outlines their cultural life-style as well as broad climatic and vegetational environments (Buckner 1992). Prior to the Woodland period, indigenous peoples were nomadic hunter-gatherers. Their influence on the forest during this period was most likely through the setting of fires. Accidental human-caused fires no doubt occurred during this time, and perhaps intentional fires were set to drive out game or to retain habitat. MacCleery (1992) speculates that the early Europeans' observations of abundant game (bison, elk, etc.) in the central hardwood region indicate a greater proportion of grassland occurred than currently exists, which implies that burning probably took place. Such burning most likely would have been the work of the indigenous population.

Around 1000–500 B.C., the Native Americans adopted a more sedentary life-style, and at some point during that time, they began the practice of agriculture (Swanton 1979). They also constructed elaborate burial mounds, which can be found in places such as Moundsville, West Virginia, Marietta and Hillsboro, Ohio, East St. Louis, Illinois, and Cartersville, Georgia. During this agricultural phase, which extended into the Late Prehistoric period until European contact (about A.D. 1600), the Native Americans generally lived in permanent villages, often on the level land of river floodplains (Davis 1978). The village would often have a structural layout with houses of thatch or pole

TABLE 6 Chronology of Native People in North America

Cultural Event	Date	Years Ago	Climatic/Vegetational Stages
Historic Period			
Modern times	2000	0	Man-made forests widespread
Settlement times			Exploitation of forests and soil
High Indian mortality			
America discovered	1500	500	Indian impacts (cultivation and fire) mold forest character
Mississippian Period			
Indian cultures largely agrarian, large palisades	1000	1000	
Woodland Period			
Pottery	500	1500	
Corn cultivated; bow and arrow			
	A.D.		Northern pines had moved into
Burial mounds	0	2000	Canada while southern pines had moved into Tennessee—their
	B.C.		present distributions
Archaic Period	1000	3000	
Marked increase in Indian populations; exchange with other regions	2000	4000	Sea level rises to modern position
Beginnings of cultivation with fire as the only feasible tool for land clearing	3000	5000	"Southern pine rise" results in marked increase in dominance of southern pines in Southeast
Archaeological evidence that Archaic Indians used total landscape of southern Appalachians	4000	6000	
	5000	7000	Increased summer warmth and
	6000	8000	drought
	7000	9000	Central hardwood oak-hickory forests became established
	8000	10000	Periglacial climate extended as far south as an east-west line through Asheville, NC
Paleo-Indian Period			
Largely hunting/gathering tribes: fire was an available tool	9000	11000	Temperate, deciduous forest replace Jack pine/spruce/fir/larch in Tennessee and North Carolina
First evidence of humans in Southeast	10000	12000	
		15000	Jack pine, spruce, and fir are the primary forest types as far south as Tennessee
		18000	FULL GLACIAL MAXIMA

Source: Buckner (1992).

FIGURE 58   Diagrammatic representation of a Native American village, showing houses and gardens (from MacCleery 1992).

construction surrounded by gardens and fields (Fig. 58). During this time, the Native Americans used tools as well as fire to clear land. MacCleery (1992) states the case as follows:

The south was dominated by fire-created forests, such as long-leaf pine savannas on the Coastal Plain and Piedmont. The hardwood forests of the Appalachian Mountains were also burned frequently by native peoples. Virginia's Shenandoah Valley—the area between the Blue Ridge Mountains and the Alleghenies—was one vast grass prairie. Native peoples burned the area annually.

MacCleery further states: "On the western fringe of the eastern forest, fire-dominated forests, such as oak and oak-pine savannas, covered tens of millions of acres." Martin and Houf (1993) indicated that remnants of so-called "balds" still exist in the Ozarks of southwestern Missouri. These grassy glades usually occur on dolomitic limestone soils and require periodic fires for their maintenance. They tolerate intermittent grazing, as would have occurred when migratory herds of bison occupied the area, but under year-round grazing imposed by the European settlers, eastern redcedar began to overtake the glades. Beilmann and Brenner (1951) use several lines of evidence to indicate that extensive forest cover is a relatively recent development in the Ozark region. They refer to Houck's (1908) account of Ferdinand DeSoto's observations of the region in 1541, where he described fertile alluvial bottoms planted in maize, pecans, plums, and mulberry trees. The open, park-like countryside was in contrast to the hardwood and pine forests that exist today. Another description of park-like vegetation in the Ozark region was provided in 1541 by the explorer Coronado, where he noted the hunting of buffalo by the Osage Indians. Although substantial prairie-like areas probably occurred throughout the central hardwood region, Steyermark (1959) indicates that even within the Ozark region, where grasslands were common, substantial areas of hardwood forests existed prior to European settlement.

Because the indigenous people tended to settle on floodplains of major rivers, their influence was probably greatest in such areas. Therefore, in many of the regions that are dominated by uplands where rivers have narrow floodplains, such as interior West Virginia, western Kentucky, southeastern Virginia, and into the Cumberland Plateau of Tennessee, the evidence of Native American influence is more limited. However, their use of fire may have enabled them to exert an influence far removed from their settlements. The population density of Native Americans is still being debated, with estimates ranging from 1 million to more recent higher estimates, up to 18 million (Buckner 1992). At higher levels, their impact on the forest would have been greater, but in any event, they most certainly had a significant effect on the central hardwood region.

POST-EUROPEAN SETTLEMENT TO 1860

The first Europeans in North America, apart from settlers in colonial villages along the Atlantic seaboard, were essentially hunter/trapper/explorers. They had a limited direct impact on the forest, but their indirect impact was profound. They brought very little with them in the way of equipment, livestock, or supplies, but they did bring something that proved to be more significant than any of these—disease. As Williams (1989) points out, the epidemic of disease introduced by the Europeans affected the native peoples to such a degree that, between 1520 and 1700, there was a return to a more forested landscape in North America (Buckner 1992). It was this forest that most historians refer to as the "impenetrable ancient forests." Early explorers kept few records of their observations, but they opened the way for others including botanists and surveyors. Britain offered huge grants of land to individuals, such as Lord Fairfax and William Penn, to encourage settlement of the New World and to gain control of the land from the native population. By A.D. 1600 waves of European immigrants, looking for a better life and an opportunity to escape the feudal system, made their way to the New World.

Most early settlers were subsistence farmers, a technology the Europeans had adopted 1000 years earlier (Blethen and Wood 1985). The European model had proved very successful in their homeland; therefore, the immigrants set about clearing the North American forest to plant their crops and graze their livestock, in much the same way as they had "tamed" the European landscape about 1000 years earlier. They brought with them two things the native people lacked — metal tools (ploughs, axes) and draft animals. Their attitude was that the forest represented an obstacle to be conquered. The forest also harbored wolves, mountain lions, and bears, which posed a threat to domesticated livestock, such as sheep and hogs.

Although the forest was a challenge to be overcome, forests also provided many of the early settlers' needs. For example, nearly every family lived in a wooden house and kept their livestock in a wooden barn (usually log construction). Additionally, wood was the primary fuel for heating. It was not uncommon for a family to use 20–40 cords of wood annually to heat and cook (MacCleery 1992). Wood was used for fencing. MacCleery (1992) estimated that it required about 8000 fence rails to enclose a 40-acre square field, and by 1850 there were about 3.2 million miles of rail fence in the United States (mostly in the East). Potash was another important product from American forests in the early to mid-eighteenth century. The demand for potash in Britain was great during that time. It was used as a soil amendment and as an ingredient in industrial processing (Williams 1989). Hardwoods, especially oaks and maples, produced the highest amounts of potash per unit wood burned and were preferred by potash producers.

From the middle 1700s into the 1800s, many naturalists trekked across North America in pursuit of unknown, unnamed, or undescribed species of plants and animals. Some were students of Carolus Linnaeus, the Swedish botanist. Others, such as Andre Michaux (1805) and John Bartram (1751) and his son William (1791), published detailed accounts of their travels through the central hardwood region. From their descriptions and others, such as surveyors' field notes, we can get an idea of what the forest was like at that time.

By this time the Native American populations had been decimated by disease, and the deciduous forests had been virtually free from their effect for between 150 and 250 years. These observers wrote about vast forests of hardwoods. Stephenson (1993) quotes Diss Debar's description of the high-elevation forests of West Virginia on the "table-lands" of the Cheat and Greenbriar mountains: "Here, also neither Oak, Poplar nor Hickory are to be found, but in their room thrive noble specimens of Sugar Maple, Ash, Beech, Birch, Wild Cherry and Black Walnut." But observers who wrote about the

TABLE 7 Summary of Frequency of Mention by Species (or Species Groups) from Bartram's (1751) Trip Through Pennsylvania

SPECIES	NUMBER OF TIMES MENTIONED	RANKING
White oak and black oak	25	1
White pine	12	2
Chestnut	10	3
Spruce	10	3
Hickory	8	4
Sugar maple	8	4
Linden	7	5
Pitch pine	7	5
Elm	6	6
Beech	6	6
White walnut (butternut ?)	6	6
Birch (yellow ?)	5	7
Poplar (yellow-poplar ?)	4	8
Ash	4	8
Sugar birch (black ?)	3	9
Great magnolia (cucumber ?)	3	9
Locust (black ?)	2	10
Walnut (black ?)	1	11
Hophornbeam	1	11
Plane (sycamore ?)	1	11

Ridge and Valley and Appalachian Plateau forests of lower elevation noticed an abundance of oaks, hickories, maples, and yellow-poplar. Using the frequency of mention as an index to species abundance, from John Bartram's (1751) travels through Pensilvania [sic] to Onandaga, a ranking of species in order of importance was developed as shown in Table 7. Although the age and size of trees and extent of the forest was greater in 1750 than it is today, the species mix looks fairly typical of present-day stands in the northern part of the central hardwood region.

The early immigrants settled along the Atlantic coast and generally used rivers as a primary means of transportation. Westward progress of settlement was relatively slow from 1600 well into the 1700s, mostly due to the lack of a transportation system. The Fall Line of the Piedmont marked the limit of navigability on most eastern rivers; thus, many settlements were established at this position along the frontier. Innovative feats of engineering created water transportation systems, such as the C and O Canal, brainchild of George Washington, but it was only possible to proceed from the Chesapeake Bay to Cumberland, Maryland. The Allegheny Front proved too formidable for the technology of the day. It was the railroads that finally opened the land beyond

FIGURE 59   The Old Valley Furnace, Preston County, West Virginia, is a monument from the iron-making era in the central hardware region. Such furnaces were scattered across the region from Pennsylvania to middle Tennessee and over to Missouri.

the Alleghenies. Even though roads (such as the Drover's Road in North Carolina) were used during this time, they were not suitable for transporting heavy materials.

A fledgling iron industry was developing in the eastern United States, especially in the Appalachian Plateaus. Low-grade iron ore was processed in stone furnaces using local limestone for purification and charcoal for heating (Fig. 59). Charcoal was produced by clearing patches of forest 1–3 acres in size (Clatterbuck 1990). Luther (1977) estimated that the 11 furnaces operating in the mid-1800s on the Highland Rim of Middle Tennessee required 375 square miles (240,000 acres) of timber to support them. Similar operations in West Virginia, Pennsylvania, and elsewhere (Fig. 60) combined would have consumed several million acres of forestland in the central hardwood region.

THE INDUSTRIALIZATION PERIOD, 1861–1929

As is often the case in a postwar era, the Reconstruction period following the Civil War ushered in a period of sweeping change. The steam engine, which could be moved from place to place, was taking the place of water power for

FIGURE 60   Excavation pits from extraction of iron ore, Preston County, West Virginia.

milling. The factories of the North were being converted to peacetime production. And the development of a massive rail transportation system was well under way, spurred on initially to supply troops in the war (Fig. 61). With the end of slavery, many large plantation farms were unprofitable. Many subsistence farmers throughout the central hardwood region gave up farming for a more lucrative life-style, working in factories, mines, or logging camps. Extensive tracts of land were purchased and consolidated by large timber or mineral companies as a speculative enterprise (Eller 1985). The migration of farmers to logging, milling, or mining camps was taking place nationwide (Fig. 62) but was more pronounced in areas such as the central hardwood region, where small subsistence farming predominated.

In addition to the emergence of America as an industrial nation, the Industrial Revolution brought about significant changes in agriculture. Tractors powered by steam, and later gasoline, replaced draft stock, and the use of fertilizers and genetically improved plants and animals increased production per unit area. Up until about 1908, as the population of the United States grew, more land needed to be placed under cultivation, and, conversely, land was taken out of forest production (Fig. 63). The amount of forestland and cropland

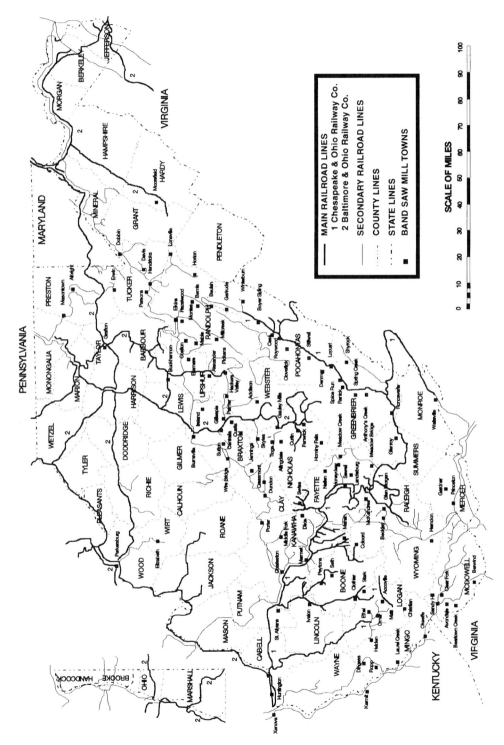

FIGURE 61 Clarkson's (1964) map of West Virginia railroads in 1917 and band saw mill towns, 1875–1920.

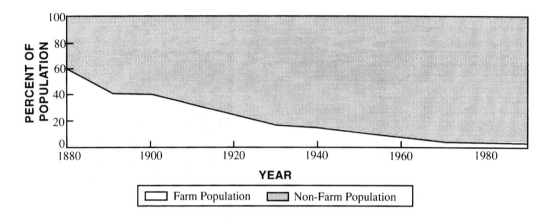

FIGURE 62   Farm and non-farm population, 1880–1988 (from MacCleery 1992).

stabilized with the small additional withdrawals of forestland coming from urbanization and rights-of-way for highways, pipelines, and power transmission (Fig. 64). Mechanization of agriculture also caused a shift in the type of land that could easily be cultivated. Much of the steeper land within the central hardwood region became submarginal and was abandoned. The capital investment required to buy large machinery put agriculture on more of a business level than that of a homesteading operation.

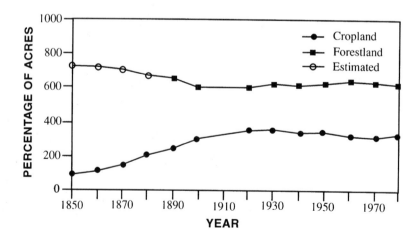

FIGURE 63   Crop versus forest area, 1850–1980 (from MacCleery 1992).

FIGURE 64   Electric transmission line right-of-way, Garrett County, Maryland.

Several secondary effects of the industrialization period on the central hardwoods were also felt. For example, the factories, machines, and vehicles needed fuel, which promoted rapid development of the fossil fuel industry. Coal, oil, and gas production was spurred on throughout the central hardwood region. Forests were withdrawn from production by surface mining, roads, well sites, pipelines, and so on. In addition, the booming economy required wood for construction of factories, new housing for factory workers (either immigrants or relocated farmers), mine timbers, and so on.

The booming economy ushered in by the industrial revolution and the vast resource base in North America swept over the timber industry as well. According to Frederick and Sedjo (1991), production of forest products tripled in the United States between 1860 and 1910 (Fig. 65), mostly as a result of increased production of lumber. Lumber production in West Virginia, where hardwoods predominated, showed a similar trend (Fig. 66). The economic boom in North America for consumer goods continued until the beginning of the Great Depression, but the timber boom had peaked in the central hardwood region by 1920, mostly as a result of overcutting. Ahern (1928) in his booklet "Deforested America" stated the situation as follows: "In 1919 the annual drain on our forest resources was estimated at four times the annual growth." Other

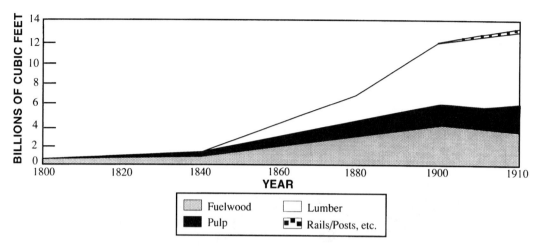

FIGURE 65   Domestic production of forest products, 1800–1910 (from MacCleery 1992).

estimates, although less dramatic, show the same trend where drain exceeded growth during this period (Fig. 67).

The history of the North American logging boom is the subject of several books (Brown 1923; Fries 1951; Blackhurst 1954; Clarkson 1964), and like similar events, such as the California Gold Rush, the timber boom is surrounded by a certain amount of folklore as well as fact. But, in any event, it had a more

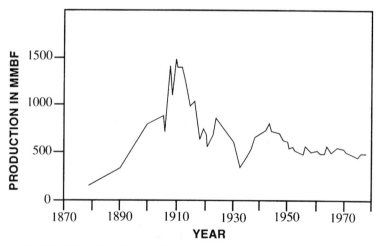

FIGURE 66   West Virginia timber production over time (from Zinn and Jones 1984).

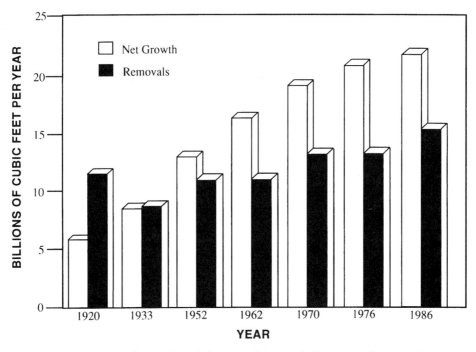

FIGURE 67   North American timber growth/removals (from MacCleery 1992).

dramatic effect on our present-day central hardwood forest than any single event. Almost all the forests that had any merchantable value were cut over during that period. Only small pockets remained, such as found in the Joyce Kilmer Memorial Forest in North Carolina (Lorimer 1976). The use of steam engines for logging and steam donkeys for skidding exacerbated the effect in causing numerous fires, which burned repeatedly through the logging slash and forest floor, exposing the mineral soil to erosion. Eller (1985) quotes Thomas Wolfe of Asheville, North Carolina, as follows: "The great mountain slopes and forest had been ruinously detimbered; the farm-soil on the hillsides had eroded and washed down; high up, upon the hills one saw raw scars of old mica pits, the dump heaps of deserted mines. . . . It was evident that a huge compulsive greed had been at work." Boom towns, such as Davis and Spruce, West Virginia, Tellico Plains and Gatlinburg, Tennessee, and Fontana, North Carolina, developed rapidly; some are completely gone today, and others turned to other industry or tourism to survive. But in spite of the apparent devastation, the forests of the central hardwood region started to regrow, resulting from natural regeneration (stored seed, seed distributed by wind or animals, seedling sprouts or stump sprouts). This period of recovery is the focus of the following section.

THE REGROWING FOREST, 1930–1996

Well before 1930, several persons in positions of leadership in American forestry had expressed grave concern about the state of the forest in North America. Gifford Pinchot, generally regarded as the father of American forestry, stated his opinion in Ahern's (1928) publication, as follows:

Forest fires are steadily growing worse in America, and fire prevention is absolutely indispensable. But the axe, carelessly used is the mother of forest fires. The axe and not fire is our greatest danger. Until the axe is controlled there can be no solution of the fire problem, or of the problem of forest devastation in America.

Pinchot, who was a consummate politician, had been trained in forestry in Europe. His first job had been as a forester on the huge Vanderbilt estate near Asheville, North Carolina, in 1892 (later to become the site of the Biltmore Forest School), but his greatest accomplishments were in the political arena (Miller 1994). He became a Washington political insider and the first Chief of the USDA, Forest Service. He was instrumental in establishment of the Yale School of Forestry and the Society of American Foresters. Pinchot's influence on the central hardwood forest came mostly through the establishment of the eastern National Forests. Earlier, Pinchot's influence on then President Theodore Roosevelt had resulted in substantial additions to the national forest reserve between 1905 and 1909. About 80 million acres of federal land in the western states had been set aside as forest reserves as Pinchot and others became increasingly concerned about overcutting in the East.

But the eastern lands had been purchased by private owners and were not so easily annexed by the federal government. The breakthrough came in 1911 with passage of the "Weeks Act," which authorized the federal government to purchase lands in eastern America. Initially, the purpose for acquiring the land was to protect headwaters of navigable streams. Lands that formed the nucleus of many of our current National Forests were purchased under the Weeks Act authority. Almost all the initial purchases were cut-over, and often burned-over, lands belonging to large corporate owners and were bought for prices of $2 to $4 per acre; hence, they were called "the lands that no one wanted."

In 1924, the Clarke-McNary Act added the production of timber to the mandate for National Forest lands. By 1930, the Great Depression was in full swing, and land prices, especially the cut-over forestlands, were extremely low, as corporate owners often viewed the land as a tax burden with no foreseeable harvest of timber in sight. For example, in 1933 a purchase of almost 330,000 acres for the Monongahela National Forest was made at an average cost of $3.43 per acre (USDA, Forest Service 1970). From these beginnings, the eastern National Forests developed (Fig. 68), many within the central hardwood region.

The Weeks and Clark-McNary Acts, in addition to mandating the acquisition of land, also provided funding to states for fire control. The Forest Service

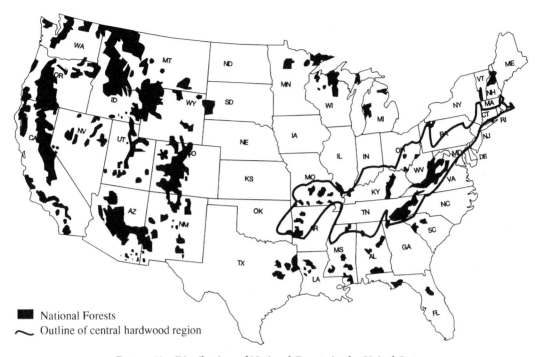

National Forests
Outline of central hardwood region

FIGURE 68   Distribution of National Forests in the United States.

also became the lead agency in promoting forest fire control at the federal level. The results came slowly at first, but by the 1960s, forest fires that consumed an average of 40–50 million acres in the 1930s were reduced to 2–5 million acres (USDA, Forest Service 1987) (Fig. 69). Although the central hardwood region is not as fire prone as conifer-dominated regions of the country, fire has always been a significant disturbance in the oak forests of the central hardwood region (Van Lear and Waldrop 1989). In spite of the national trend, there are fire-prone forests in the central hardwood region that, during extreme fire seasons, still burn with regularity. For example, in 1987 more than 50 percent of the land area in Mingo County, West Virginia, burned over (Atkins 1988; Gillespie et al. 1992; Hicks and Mudrick 1994) (Fig. 70).

In 1933, President Franklin D. Roosevelt signed the Civilian Conservation Corps bill into law. The CCC was part of Roosevelt's "New Deal," intended to solve two problems (unemployment and environmental destruction). It was estimated that nearly 250,000 young men, who were of employment age, were out of work during the 1930s and Major Stuart, Chief of the USDA, Forest Service, suggested that the army be used to build camps and generally administer the program. But unemployed civilians made up the work force.

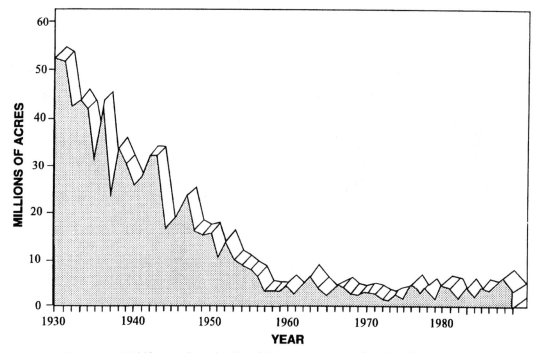

FIGURE 69   Wildfire trends in the United States, 1930–1989 (from MacCleery 1992).

Many CCC camps were established within National Forests and their activities included planting of trees, fire control, watershed improvements, road improvement, and construction of recreational facilities. Although their impact on today's central hardwood forests is largely undetectable, they planted millions of trees (mostly conifers) on denuded lands (Fig. 71), and many of the recreational structures they built are still usable today, almost 60 years later.

The Depression drove more people from marginal farms, and the trend in reversion of farmland to forest continued through the 1930s. In 1942, the Depression ended with America's entry into World War II. The war accomplished what governmental programs could not by employing millions of Americans, either as soldiers or in the factories that manufactured war materials. The war put technological development into high gear, and many of the devices that have significantly affected our society were developed during this period. The post-war economy was booming when soldiers returned home, many to take factory jobs, marry, and raise families. The impact of these developments on the central hardwood region was a shift in the life-style and objectives of landowners. Many people kept their residence in a rural setting while seeking

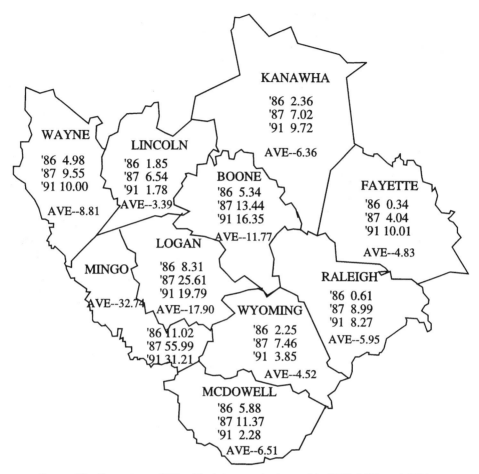

FIGURE 70 Percentage of West Virginia counties burned in 1986, 1987, and 1991.

employment in mills and factories elsewhere. The reason for owning land shifted away from utilitarian (production) objectives toward recreational/ aesthetic enjoyment/residential objectives (Birch and Kingsley 1978) (Fig. 72).

The changing land ownership characteristics continued into the most recent decades throughout the central hardwood region. This is illustrated by the changes that occurred from 1957 to 1975 in West Virginia, a state that experienced little or no total population growth over that period. In 1957, the Forest Industries Committee reported that there were 133,570 private owners of forestland in the state, who owned over 8.9 million acres. The average size of ownership was 66.6 acres in 1957. By 1975, the number of owners had increased to 207,500 and the forestland acres had also increased to 10.343 million acres.

FIGURE 71 A red pine plantation established in 1941 by the Civilian Conservation Corps.

But the average size of ownership had dropped to 49.8 acres (Birch and Kingsley 1978). Twenty percent of the landowners in 1975 were not from rural or farm backgrounds. Twenty-two percent of the landowners held their land primarily for aesthetic enjoyment or recreation, and another 23 percent owned the land primarily for their residence. Only 3 percent owned land primarily for timber production. Similar changes have taken place throughout the central hardwood region.

Alig et al. (1990) reported the trends in forested land by states. If the 11 states that constitute the bulk of the central hardwood region are summarized, the trend in total forestland is depicted in Figure 73. A decrease of about 2.5 percent in forestland is projected to occur by 2040, mostly due to urban expansion.

Agricultural policy continued to have an impact on forestry throughout the post-war period. The Conservation Reserve Program in the 1960s provided compensation to farmers to convert cropland to forest by planting trees. The Forest Incentives Program provided cost sharing to small private landowners for tree planting and other forest practices. The Stewardship Incentives Program of the 1990s provides federal cost sharing to small forest landowners to develop and carry out planned forest management. Some states have offered additional

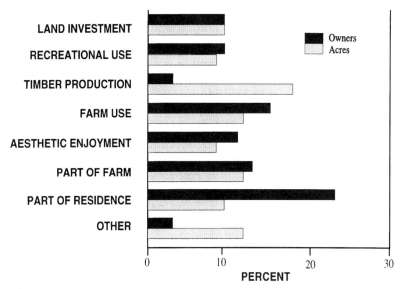

FIGURE 72 Reason for owning land in West Virginia (from Birch and Kingsley 1978).

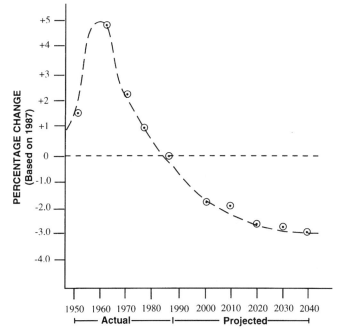

FIGURE 73 Change in amount of forestland over time.

property tax incentives to landowners enrolled in the Stewardship Incentives Program.

These programs are designed to promote good forest management among the many and varied small private landowners, a type of ownership that predominates the central hardwood region. Technical support for forestry is provided to private landowners through state forestry agencies, Agriculture Extension Agents, and the USDA Forest Service, State and Private Forestry. But in spite of this large effort, the level of participation among private forest landowners in the central hardwood region has generally lagged behind government expectations, partly due to the inherent independence of property owners, partly due to a lack of interest or economic incentives, and partly due to an inability to inform landowners of the availability of such programs.

The withdrawal of forestland to agriculture leveled off about 1933, but during the post-war decades, some decline in forest area has occurred. Most of this has resulted from development-related activities. One example is the interstate highway system, which was developed during the 1960s and 1970s for the most part. Rights-of-way for gas transmission and electric power transmission also consume large corridors through the forest, and surface mining and urban sprawl have taken up significant amounts of forestland (Figs. 74 and

FIGURE 74   Power transmission right-of-way in the hilly region of the Appalachian Plateau.

FIGURE 75 Urban development near Boone in western North Carolina. Much forestland is being converted to such developments in the central hardwood region.

75). In spite of this, topography and economic factors have resulted in the central hardwood region's being an area that is currently among the highest in the eastern United States for coverage with natural vegetation (Klopatek et al. 1979) (Fig. 76).

Some of the most significant changes to the central hardwood forest in recent years have been brought about as a result of human introduction of insects and disease organisms. These changes are often not dramatic, but their impact is unquestionably large. The list of important introduced pests includes dogwood anthracnose, Dutch elm disease, oak wilt, and beech bark disease. But the two with the most impact or potential impact on the central hardwood forest are **chestnut blight** and **gypsy moth**. The former is a disease that has effectively eliminated a species, which at one time was among the most widely distributed and economically important in the central hardwood region. The latter is an introduced insect that continues to spread southwestward through the central hardwood region and seems to have the potential to spread through the entire region over the next 50 to 75 years.

The **chestnut blight** fungus (*Cryphonectria parasitica*) was introduced to the New York Botanical Gardens in 1904 (Murrill 1904; Giddings 1912). By 1915

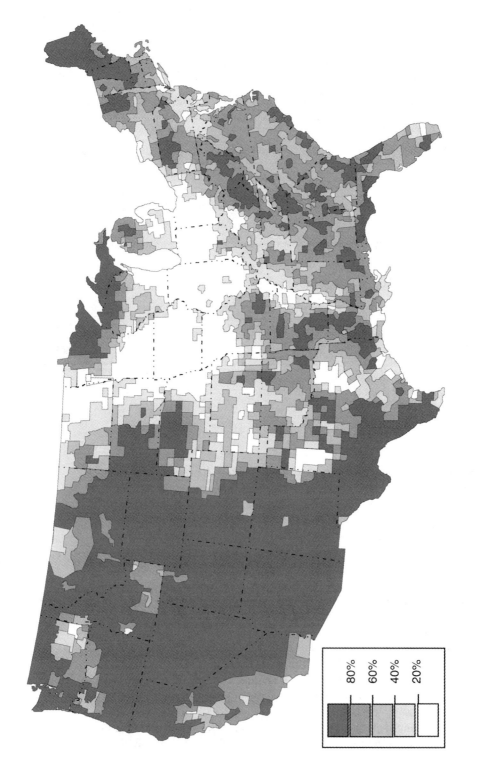

FIGURE 76 Percentage of area potentially covered by natural vegetation in the United States (from Klopatek et al. 1979).

80%
60%
40%
20%

most states east of the Mississippi River reported infected trees. By the late 1930s, most mature chestnut trees were dead throughout the eastern states. The following should provide a frame of reference for the significance of American chestnut. Chestnuts were a source of food and commercial enterprise for mountain farmers. During the settlement period, log cabins were often constructed of chestnut logs due to their durability and decay resistance. Most split-fence rails and posts were constructed of chestnut where it was available. And due to its durability, chestnut lumber was widely used in construction of barns, sheds, and even for the manufacture of furniture. The bark of chestnut was an important source of tannin for leather tanning factories.

Chestnut also occupied an important position in the hardwood ecosystem. Braun (1950) used the name **oak-chestnut** to identify one of the most significant forest formations in the eastern deciduous forest. American chestnut was an important source of mast for a variety of wildlife species (animals and birds), most of which served as prey for larger predatory mammals. The impact of the loss of American chestnut can never be fully assessed. But Brooks (1910) indicated the following facts regarding chestnut in West Virginia alone. The annual harvest of chestnut in West Virginia in 1910 was 118 million board feet. The entire volume of chestnut in West Virginia was estimated to be 5 billion board feet. At a stumpage value of $8 per thousand board feet, in 1910, the entire chestnut inventory was estimated to be worth $55 million. At current stumpage prices for red oak that value would be about $1.5 billion. The only evidence remaining of the once important chestnut are occasional root sprouts, which usually survive to the sapling stage before being killed by the blight, and the grey remains of stumps of salvaged trees (Fig. 77). Ironically, many of these stumps still have coatings of charcoal, evidence of the fires that burned over the area, probably after the chestnuts had already succumbed to the blight.

Gypsy moth (*Lymantria dispar*) is an introduced insect that potentially could have as great, or greater, impact on the central hardwoods than chestnut blight. Gypsy moth was inadvertently introduced into the Boston area in 1869 by Leopold Trouvelot (Leibhold 1989), who was interested in hybridizing the gypsy moth with native silkworms. The spread of gypsy moth has continued, and it is potentially capable of becoming established throughout the central hardwood region. Although gypsy moth can defoliate a variety of tree species, including conifers, deciduous species are their preferred hosts. Oaks are among the most preferred hosts of gypsy moth (Bess et al. 1947; Gansner and Herrick 1987). Black cherry, hickory, and maples are intermediate in preference while species, such as yellow-poplar and ashes, are virtually resistant. The significance of this to the central hardwood region is the fact that susceptible species make up the bulk of the forest composition throughout the region, the exceptions being yellow-poplar and ash, especially in the mixed mesophytic forest.

Currently, gypsy moth has become established in, and is endemic to, an area stretching from Michigan to North Carolina (Fig. 78). But isolated

FIGURE 77   Stumps of American chestnut in Webster County, West Virginia. Chestnut blight eliminated this species from its native range, which included a substantial proportion of the central hardwood region.

populations occur throughout the central hardwood region, having been spread by movement of vehicles, logs, or other objects containing egg masses.

As a defoliator, gypsy moth's impact on the tree is to reduce its photosynthetic surface. The major defoliation occurs during the spring (May and June), which is a period of critical importance to deciduous trees since the soil moisture is generally high and the temperature is moderate during this time. Usually one defoliation is not lethal to trees, but a single defoliation during a drought year or multiple years of defoliation can lead to extensive tree mortality (Fosbroke and Hicks 1989). The factors that precipitate extensive gypsy moth

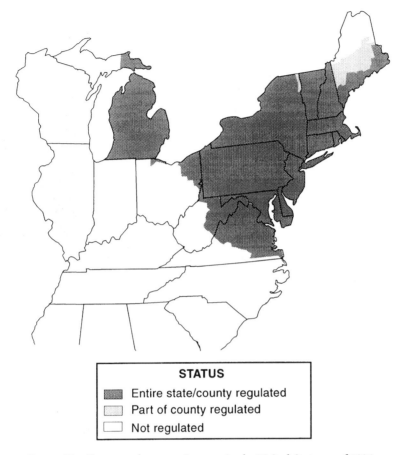

FIGURE 78   Gypsy moth quarantine area in the United States as of 1996.

outbreaks include presence of a suitable host plus climatic conditions that favor buildup. Since gypsy moth is situated in an oak-dominated region, the host is generally suitable, thus outbreaks can occur whenever weather triggers them. As can be seen in Figure 79, cycles of outbreaks seem to occur about every 10 years. The outbreaks in recent years have been affecting more acres than in previous outbreaks, mostly due to the spread of gypsy moth to occupy a larger endemic range.

The gypsy moth story is still unfolding, but its impact on the central hardwood forest has already been experienced throughout the portion of the region where it has spread. As reported by Gansner and Herrick (1987), oak experienced an average mortality rate of 24.2 percent between 1979 and 1984. Quimby (1987) reported a similar result where summer droughts plus gypsy

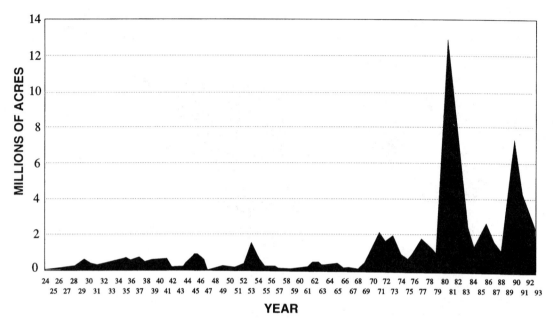

FIGURE 79   Outbreak history of gypsy moth in the northeastern United States.

moth defoliation in 1980–1983 resulted in the premature loss of 68 million trees in Pennsylvania, a mortality rate of 27.6 percent. Much of the mortality occurred in oak stands. Hicks and Fosbroke (1987b) reported similar rates of tree mortality in southwestern Pennsylvania and western Maryland. The forest invaded by gypsy moth as it entered northeastern Pennsylvania has been dubbed "the new frontier" by Herrick and Gansner (1988), and essentially this new frontier represents the oak-dominated forest of the central hardwood region. The reduction in oak stocking that has occurred in the part of the central hardwood forest where gypsy moth has become established will most likely represent what can happen throughout the entire region.

Another episode in recent history that has had a significant impact on the central hardwood forest is the decline and later reestablishment of **wildlife populations** across the region. In addition to white-tailed deer, elk and bison were found extensively throughout the region prior to European settlement. Those species were, in part, hunted into extinction but mostly were displaced due to habitat destruction by land clearing for farming. Farmers also eradicated predator species, such as eastern mountain lion, eastern timber wolf, and black bear, since these predators were a threat to their livestock. By 1910, bison, elk, mountain lion, and wolf had been eradicated from the central hardwood forest, and white-tailed deer and black bear were found only in small isolated populations in remote and mountainous regions.

Wildlife advocates and hunters pushed for, and obtained, stricter hunting regulations, such as seasonal hunting, bag limits, and bucks-only hunting. The results for white-tailed deer have been remarkable. In the absence of natural predators, and with the second-growth hardwood forest providing prime habitat, the deer herd has risen sharply throughout the central hardwood region. In fact, deer have generally become a serious problem to forest regeneration in many areas. Deer consume 1–1.5 kg of vegetation daily, consisting of twigs, acorns, and some grass and herbs (Smith 1993). Deer are selective and prefer seedlings of oaks and maples to those of black cherry (Marquis 1981a). Overbrowsing by deer can result in development of a fern-dominated understory where tree seedlings have a difficult time getting established due to competition and allelopathy (plant toxins). Thus, the long-term effect of deer on the forest may be to retard the process of regeneration, selectively change the species composition, and reduce the overall diversity of species (Fig. 80).

The modern chapter in forest history also includes a marked change in the **social context** of forest resources. This has been brought about by a combination of factors including the changing demographics in America, where people today are physically removed from the land. But other factors have come into

FIGURE 80   A deerproof exclosure near Warren, Pennsylvania. Abundance of seedlings inside the enclosure shows clearly how severe the impact of white-tailed deer is on forest regeneration.

play. For example, abusive and exploitive land uses, such as exploitive timbering, unregulated mineral extraction, air and water pollution, and uncontrolled development, are all sources of public concern. As a result, an environmental movement began to take shape as a cultural phenomenon during the late 1960s. Rachel Carson's book, *Silent Spring*, did for environmentalism what Harriett Beecher Stowe's *Uncle Tom's Cabin* did for the abolitionist movement. Against a backdrop of Kennedy altruism and polarizing issues, such as the cold war, racism, and a budding Viet Nam conflict, it seems, in retrospect, inevitable that the environmental movement, led by altruistic urban youth, should flourish. Most foresters, at that time, were from rural backgrounds and had come through their forestry training with a paternalistic approach toward resource management. Their view appeared to be that the forester is the trained professional and should be trusted to make the appropriate decisions regarding management of the resource. Thus, the elements of conflict were in place. The political arena became the venue for the struggle. Environmental activist groups gained strength. The **Wilderness Act** of 1964 represented a milestone in environmental policy legislation whereby areas in the National Forests could be set aside for "wilderness," and timber cutting in these areas would be prohibited (Dana and Fairfax 1980). Controversy over the use of clearcutting on public land resulted in the landmark **Monongahela Decision** of 1970 that recommended uneven-age management as a primary management policy for the Monongahela National Forest in West Virginia. This decision was more significant as a precedent than it was to the Monongahela National Forest. In effect, it imposed the will of the public over that of foresters, who had always been regarded as the experts in forest resource management.

The issue of the 1990s seems to revolve around regulation of private land. The "spotted owl" controversy in the Pacific Northwest set the tone. The perceived situation was one in which an endangered organism's habitat (old-growth forest) was not limited to a single landowner but extended to private land as well as public land.

The debate extending to small nonindustrial private forest (NIPF) owners progressed from the passage of regulations requiring the use of "best management practices" for logging to a more comprehensive and controversial debate over whether or not land use should be controlled at the **landscape level**. If such control is needed to facilitate what is called **ecosystem management**, how can it work in areas, such as the central hardwood region, where over 75 percent of the land is owned by NIPF owners? Proposed solutions include a combination of government educational and assistance programs plus incentive programs (Campbell and Kittredge 1996). The use of legislative regulation to achieve the goals of ecosystem management is a controversial approach (Argon 1996) and one that evokes strong emotions. The fear among private owners that they will lose prerogatives to use their land is not unfounded. Examples abound where the "general good" of the public has taken precedence over the desires of the individual.

TABLE 8 Summary of Historical Development of the Central Hardwood Forest

PERIOD	TIME FRAME	DESCRIPTION/EVENTS
I. Glacial	>14,000 years before present	As ice sheets moved across the northern tier of the North American continent, colder climates displaced deciduous forests to the south, except for isolated patches in low elevations. Tundra and boreal forests occupied the area that is the central hardwood region. Hardwoods reinvaded as ice sheets retreated.
II. Early Human	12,000–3000 years before present	Native people existed in North America as nomadic hunter/gatherers, with low population density and limited impact on the forest, except for occasional burning.
III. Sedentary Native	3000–500 years before present	Native people adopted agriculture and lived in organized villages, usually in level and fertile land along stream bottoms. With increased resources, their population increased rapidly. Fire was the primary tool for land clearing. The result was a dramatic impact on the forest with extensive prairies and savannas created by human-set fires.
IV. European	500 years–present	This period is divided into four subsections.
A. Explorer/ Hunter/Trapper	500–300 years before present	European explorers had little direct impact on the forest except to remove large numbers of furbearers. Their main impact was to introduce disease to the native population, causing a massive die-off, and to open the way for future settlers. Without the impact of native people, forests regrew and became the unbroken "primeval" forest discovered by later travelers.
B. Subsistence Farmers	300–130 years before present	Immigrants came with metal tools and draft animals and livestock. Forestland was cleared for agriculture (grazing and crops). Settlement proceeded inland from the east, first along rivers and later roads. The central hardwood forest was reduced from almost 100 percent coverage to less than 50 percent. Forests remained only in areas unsuited to agriculture either because of topography, fertility, or other limitation.
C. Industrial Development	130–70 years before present	Steam and internal combustion engines replaced draft animals for farming; thus, farming shifted to more level land. Railroads moved people and products, and population began to shift away from farming to a more urban population. Marginal farmland was abandoned to regrow into forest. Industrial development and prosperity fueled a construction boom, which precipitated exploitation of forests. Essentially, the entire central hardwood region was cut over, much of it burned over as well.
D. Modern	70 years before present–present	The Great Depression brought an end to the economic boom. A conservation movement was under way, and cut-over land was acquired in the central hardwood region to establish National Forests. Genetics and high-impact agriculture boosted per-acre production, and more marginal farmland reverted to forest. In the booming post–World War II economy, the shift from a rural to an urban population continued. Significant human-assisted pest introductions, such as chestnut blight and gypsy moth, began to have a serious impact on the central hardwood forest. The second-growth central hardwood forest we have today is a product of all the above events.

Source: Buckner (1992).

On the other side of the issue are those representing private landowners (Farm Bureau, National Woodland Owners Association, etc.) who contend that the loss of management options to a private landowner due to regulation constitutes a "taking" under the Fifth Amendment of the Constitution, for which the owner is entitled to be compensated. As these debates continue, the outcomes will most certainly affect the way forests can be, and will be, managed and exploited in the central hardwood region in the years to come.

HISTORICAL DEVELOPMENT, SUMMARY

In summary, the history of the central hardwood forests is one that is based on the actions and interactions of climatic and biological factors, and, in more recent times, it has been strongly affected by humans and their actions (Table 8). The retreat of the glaciers 12,000–14,000 years ago and the subsequent climatic warming created an ideal environment for the deciduous forests, which have flourished in the central hardwood region. Indigenous people used fire and cultivated land, and their activities peaked about the time European settlement began. The forest regrew for about 200 years, and then clearing for subsistence farming began in earnest. This proceeded until about the time of the Civil War and the onset of the Industrial Revolution. This period marked the beginning of a shift in agriculture to more level land, and abandoned hill lands in the central hardwood region began to regrow into forest. A period of exploitive logging around the turn of the twentieth century was followed by a regrowing forest. Forest landowners and the public changed in their attitudes concerning forest resources in recent times. Now with many hardwood stands in the central hardwood region approaching 100 years of age, the resource is maturing. Developers, speculators, industries, agencies, advocacy groups, and environmentalists are all becoming aware of this unique and diverse forest. Our actions today will be the history of the future, and foresters should be prepared to participate in, and hopefully guide, the decision-making process at this important crossroad in the central hardwood region's development.

3

❧

Ecological Relationships

GENERAL ECOLOGICAL CONCEPTS

This Chapter provides an introduction to the basic concepts of ecology, especially as they relate to forest ecosystems in the central hardwood region. This will set the stage for following Chapters dealing with silvical characteristics of important tree species and cultural practices for management of forest stands. In the central hardwood region, more than for most forest regions, silviculture is "applied ecology" because of the prevalent use of natural regeneration, the limitations of terrain, and forest ownership patterns.

An **ecosystem** is an assemblage of organisms that function in a particular environment, having interactions with their environment and with each other. Some of the concepts that will be discussed in this section include succession, disturbance, nutrient cycling, energy dynamics, competition, regeneration, and forest stand dynamics. Obviously, the space limitations of this treatment preclude a thorough discussion of these subjects. Therefore, any student of forest science will need to study these subjects in greater depth in order to gain a full understanding of them.

Ecologists have long recognized that certain types of vegetation tend to predominate in a given area. There could be several reasons for this (soils, historical management, topography, etc.), but the overriding factor appears to be the **macroclimate**. The influence of climate on vegetation includes the obvious direct effects, such as the supply of resources (water, heat, light) and the limitations imposed on plant functions (coldest temperature, most severe drought, seasonal cycles, wind, etc.). But the indirect impact of climate is also important in that soils and landforms are themselves a product of a given climate (Bailey 1996). If we try to recreate the "evolution" of a site, the sequence might start with a geologic event, for example, sediment deposition into a basin or volcanic activity. This geologic material is exposed to a particular climatic sequence (erosion cycles, freeze-thaw, etc.), which begins to weather the material, forming soil. Propagules of plant species adapted to the conditions (soil, climate) become established and their roots and organic matter further change the site in a feedback relationship. As the site changes, it becomes suitable for other plants and animals and the plants, animals, and landform may indeed also have an effect on the local climate. The sequence described above generally fits the description of a "**primary succession**" (Clements 1916). This view of succession implies a "relay floristics" pattern (Egler 1954), and although it does not totally capture the complexity that actually occurs in central hardwood systems, it serves as a starting point for describing some of the major elements. The changes that occur during site development are not discrete events, rather they are all taking place simultaneously, and they often occur over extremely long periods (thousands of years); thus, such changes may be imperceptible when viewed in reference to a human life span.

Lucy Braun (1950) referred to the "major vegetation unit" of an area as a **formation**, for example, deciduous forest. She used the term **association** to denote the "climax unit" of a formation. For example, the mixed mesophytic forest is an association. The term **climax** was used by Clements (1916) to refer to a vegetative community that perpetuates itself. The "**climatic climax**" is, therefore, an association that, due to the specific climate of a region, would ultimately predominate in the region in the absence of disturbance or other limitations. The climax is the end product of a process of **succession**, which is presumed to be an orderly change in vegetation over time. In attempting to conceptualize some of the unifying principles of ecology, Margalef (1963) described succession as follows: "any ecosystem not subjected to strong disturbances coming from outside, changes in a progressive and directional way." Succession may begin with a previously unoccupied site (sand dune, rock outcrop, lava flow, etc.), and such successions are termed "primary." Alternatively, succession may begin with the interruption of a **community**, resulting from some type of disturbance (forest harvesting, fire, wind or ice storms, insect or disease outbreaks). Such disturbance-mediated successions are termed "**secondary succession**." It is this type of process that is most typical in the forests of the central hardwood region, most of them being even-aged forests that regrew after logging, fires, or agricultural abandonment within the past 100 years.

Although the classical ecological concept of succession provides an important point of reference, in reality, the behavior of plant communities does not always fit the stereotypical pattern. For example, the early invaders of a site are called "**pioneer**" species, and they are typically shade intolerant, fast growing, short-lived, and capable of rapid and wide dissemination of propagules. Classically, pioneer plant species are herbaceous, but some tree species also function in this role, such as Virginia pine, pin cherry, and eastern redcedar. Species that are longer lived, shade tolerant, and slow growing would, therefore, fit the description of climax species. Definitive examples of this type of species would be American beech, sugar maple, or eastern hemlock. But many tree species do not appear to fit either of the above situations and play more opportunistic roles. Indeed, the model most appropriate to secondary forest succession appears to correspond to Egler's (1954) "initial floristics" model, where propagules (seed, seedlings, sprouts, etc.) originate from existing sources.

An example of a species that does not conform to a definitive role includes red maple, which fits many of the criteria for a climax species but has light windborne seeds, may invade old fields, and functions as an early successional species. The oaks seem to persist on poorer sites, relatively unchallenged by other competitors even though they may be less shade tolerant than some of their associates. Yellow-poplar can form a component of relatively old-growth forests in the fertile coves of the southern Appalachians, even though it often behaves as a pioneer in old-field situations. There are many examples where site limitations impose constraints on the ecosystem such that the climatic climax is never reached. Examples of this are so-called physiographic and edaphic climax communities, where either topography or soils prevent a true climax from being reached. Excessive and selective deer browsing can influence the species composition by selectively removing certain preferred species from the understory, therefore altering the progression of succession. Likewise, outbreaks of insects and diseases can selectively eliminate, or greatly reduce, certain species or groups of species, therefore altering the would-be successional pathway. The effect of gypsy moth on oak is an example of this sort of selective alteration in climatic climax.

Changes in climate can also alter the pathway of succession. Numerous models have been developed to predict vegetative changes that would result from a "global warming" scenario. An example of a community that has apparently experienced such a change is the relict spruce-fir forest along the highest elevations of the Appalachians. This forest has been drastically reduced in extent over the past 200 years, accelerated by the actions of human populations (Hicks and Mudrick 1994). The instability of this ecosystem is further demonstrated by the numerous insect, disease, and decline problems associated with it.

Ecology is a relatively young science, having developed after the turn of the twentieth century, and even its progenitors disagreed as to the mechanism of such fundamental processes as succession. For example, Clements (1916) visualized the ecosystem as a "complex organism," and for any given site (environment), there is a particular climax community that represents the "mature" form of that ecosystem. In a sense, this **superorganism** view of the ecosystem has at its core the view that the whole is greater than the sum of the parts. A contemporary of Clements, H. A. Gleason (1926), took a more reductionist view, that the community is simply a result of the interaction of the organisms in it, and each species or type has its own ecological role (**niche**) in the ecosystem. Using Gleason's logic, a community *is* the sum of its parts. Gleason also pointed to the need for seed and propagules of a species to find their way to the site, and he pointed out the role of chance in affecting the final outcome of succession. For example, the coincidence of a site disturbance, which makes the site available for a new community, occurring in synchrony with a bumper seed crop of some species present in nearby forests, would represent such a chance event. There seem to be elements of truth in both points of view, but most forest ecologists seem to lean more toward the reductionist perspective, which does appear consistent with the development of many current stands in the central hardwood region.

The role of **disturbance** as a component of the process of secondary succession is undeniable. And disturbance has undoubtedly played a significant part in the development of the central hardwood forest. Disturbances can take many forms and may come from within (**endogenous**), such as tree fall, native insect defoliation, or disease outbreak, or may come from outside agents (**exogenous**), such as fire, windstorm, and ice. Fire is an important type of disturbance that may not only kill vegetation but may alter the site in other ways by converting biomass into ash, for example. By contrast, insect defoliation, such as gypsy moth, can result in extensive tree mortality, but it may occur somewhat gradually, and the conversion of leaf biomass to insect biomass and frass has a unique effect on the mineral cycling process (Eagle 1993). Windstorms (hurricanes, tornados, etc.) can cause extensive blowdown of canopy trees, but they cause little disturbance to the forest floor, other than opening it up to additional light (Fig. 81). **Fire** as a disturbance is one of the most influential factors in the development of the current forest that occupies much of the central hardwood region. Van Lear and Watt (1993) indicate that fire has played a dominant role in sustaining oak forests. Oaks are resistant to the ground fires that are typical in the region, and fires help to create favorable conditions for oak regeneration. Lorimer (1993) pointed out that virtually no stands in the central hardwood region are without evidence of fire, and Overstreet (1993) discussed the frequent and extensive fires that occurred in Kentucky just after the exploitive logging era. Similar scenarios were generally occurring throughout the central hardwood region at the time. Thus, fire has played a significant part in the regeneration of the natural even-age stands that occur throughout the region. The logging era, with an abundance of slash, created ideal conditions for forest fires. The steam locomotives used for logging provided the necessary sparks to ignite fires, and fires were so commonplace in many parts of the central hardwood region that local residents took little notice of them. **Forest harvesting** can be a disturbance to the forest that may or may not occur in conjunction with fire. In subsequent sections of this book, the use of cutting as a silvicultural tool to emulate natural disturbances will be discussed in greater detail.

Agricultural disturbance was also a major occurrence in the central hardwood region, and the abandonment of subsistence farms has led to the development of many of today's stands (Kalisz 1993). In addition to cultivation of corn and other crops, woodlands were typically used for livestock grazing, which has further affected today's forests. Old fields typically have regrown to

FIGURE 81   Forest damage caused by Hurricane Fran in the mid-1990s in the Shenendoah National Park, Virginia.

pioneer species, such as hawthorns, sassafras, or redcedar. These shorter-lived species, on north- and east-facing slopes, often give way to species such as yellow-poplar, whereas on south- and west-facing slopes oaks usually replace the short-lived pioneers. This effect of **aspect** is related to the adaptation of these species to the particular environmental conditions of these sites.

The impact of **chestnut blight** as a disturbance, particularly in the Appalachians, has been to gradually remove chestnut from the forest with oaks playing an increasingly important role in the replacement of chestnut (Arends and McCormick 1992). Species such as oaks and hickories appear to be important in the replacement of American chestnut on poorer sites, while a mixture of mesophytic hardwood species have taken this role on the better sites. Thus, it appears that many of the elements that favored regeneration and survival of oaks were set into motion inadvertently around the turn of the century, and the result has been the oak-dominated forest that we find in place today in the central hardwood region.

Mineral nutrient cycling is a fundamental process in natural biological systems. Terrestrial forest systems typically conserve mineral elements needed for growth by returning them in biomass and releasing them slowly via the decomposition process (Bormann and Likens 1979). The released elements may be reabsorbed by the vegetation to continue the cycle. In deciduous hardwood ecosystems this process has a seasonal periodicity owing to the annual shedding of leaves. The forest floor is annually replenished with leaves to form a litter layer. The rate of decomposition depends on a variety of factors including temperature, moisture, and intrinsic characteristics of the litter itself. In general, warm, moist conditions promote more rapid decomposition, and in the central hardwood region, the conditions are generally favorable to decomposition, as compared to more arid regions or cooler ecosystems like the boreal forest. However, different species of hardwoods produce leaves that decompose at markedly different rates (Melillo et al. 1982; Mudrick et al. 1994). For example, oak leaves with higher lignin content and lower nitrogen content decompose more slowly than the leaves of maples or yellow-poplar. In the central hardwood region, oaks are often more abundant on specific sites, such as ridges and south- or southwest-facing slopes. On these sites, the drying effect of the more direct incidence of sunlight also contributes to lower rates of litter decomposition. Thus, oak litter tends to accumulate at the surface and release nutrients more slowly into the soil. The lower mineral output from a watershed draining a predominantly oak site as compared to one draining a mixed mesophytic hardwood site was documented by Hicks and Frank (1984).

ADAPTABILITY OF PLANTS

Site is defined as the sum total of all environmental factors affecting a given locale (soil, climatic and biotic factors). At the geographic scale of the central hardwood region, certain factors, mostly **macroclimatic**, are responsible for the dominance of deciduous hardwood species. A distinct summer/winter seasonal pattern, moderate winters, and adequate precipitation, well distributed throughout the year, are all characteristics of the climate of the central hardwood region. According to Bailey's (1996) map, "Ecoregions of the Continents," the central hardwood region falls predominantly in the **Hot Continental Division** of the **Humid Temperate Domain**. This type of macroscale climate occurs in portions of eastern Europe and the Far East, including Korea and Japan. Interestingly, these areas also support mixed deciduous forests, and plants exchanged between these areas are generally compatible with their introduced range, in some cases becoming pests. Deciduous species have the advantage of concentrating their growth and bioproductivity during a portion of the year when environmental conditions, such as soil moisture and temperature, are most favorable (Day and Monk 1977).

There are many examples of species adaptations that occur at smaller scales. These adaptations generally result from the so-called **mesoscale** phenomena (Bailey 1996) and are in response to general landform differences. For example, the geology, soils, climate, and vegetation all differ between the Ridge and Valley and Appalachian Plateau Provinces, and these differences translate into different resource management constraints.

At the **microscale**, adaptations of vegetation occur in response to soil and topographic differences. For example, in the Great Smoky Mountains, elevation can produce life zone changes that occur over a rather small horizontal distance scale but reflect climates that occur extensively many miles away. The southern spruce/fir forest that occurs on mountaintops of the Appalachians are similar to, but disjunctive from, boreal forests that exist hundreds of miles to the north (Fig. 82). The relationship between elevation and latitude has been recognized for many years and was described by Hopkins (1938). The mosaic of species related to aspect and slope position in mesophytic forest areas is a good example of microscale adaptation. Here, oaks predominate on tops of ridges and southwest slopes and mesic species, such as yellow-poplar, predominate in coves and northeast slopes. Soils may also influence microscale adaptations. For example, eastern redcedar generally occurs in conjunction with soils having a high calcium carbonate content. **Ecological classification** is an attempt to group sites with similar ecological characteristics. Johnson (1992) states: "Rigorously developed ecological classification systems can be important silvicultural tools for identifying potential problems and opportunities." Several workers have

FIGURE 82   A mixed red spruce–hardwood stand at high elevation in the southern Appalachians.

provided in-depth regional classifications of ecosystems within the central hardwood region (Smalley 1987; McNab 1991), and Bailey (1996) has synthesized this and other work to provide a national framework for classifying forest ecosystems.

Species adaptation can occur at scales even smaller than Bailey's (1996) "microscale." These adaptations usually relate to very precise adaptations of species that enable them to fill a specific **niche** in the ecosystem. For example, several woody species, due to their small mature height, are relegated to understory positions in the forest. These species (e.g., flowering dogwood, black haw, spice bush, witch hazel, rhododendron) are, of necessity, highly shade tolerant.

Berglund (1969) explains the ecological role and adaptability of organisms as being associated with three factors: (1) ecological **requirements**, (2) ecological **tolerance**, and (3) ecological **efficiency**. Viewing the ecosystem from the strict perspective of energy relationships, this is indeed true, but plants have also evolved specific survival strategies, such as seed dissemination mechanisms, seed dormancy, sprouting ability, and allelopathy, that cannot be accounted for using a perspective based strictly on energetics. That is not to minimize the importance of energy relationships in the forest ecosystem. On the contrary, green plants (**autotrophs**), with their ability to use solar energy in photosynthesis, form the energy base for all life. This association of organisms through energy flow is sometimes called the **food chain** or **food web** (Fig. 83). The autotrophic plants are the first step in collecting solar energy by converting oxidized carbon (CO_2) into a reduced form (carbohydrates), which can then be consumed and oxidized by **heterotrophs** to provide energy for their needs. Indeed, the total potential bioproductivity of an ecosystem is ultimately dependent on the amount of photosynthesis that is possible. Concepts, such as the **limiting factor concept**, are useful in understanding why certain systems are more productive than others. This concept states that the rate of a process (e.g., photosynthesis) is limited by the factor that is in shortest supply. For example, bioproductivity of a boreal forest may be limited by temperature whereas an oak forest growing on a shale soil may be limited more by mineral nutrients and/or by available water (Vose and Swank 1990).

Using ecological requirements, tolerances, and efficiencies to help explain the ecological role of species, the following questions can be asked: (1) Does the site supply, in adequate quantity, all the resources needed for the functioning of the species? (2) Is the species tolerant of extremes in site conditions (temperature, moisture, light, soil nutrients, pH, etc.)? (3) How efficient is the species at obtaining and using the resources of the site, and how well can a species compete with other species for these resources?

From the standpoint of requirements, all organisms have certain basic needs. Berglund (1969) refers to them as the "**direct factors**," and he lists six direct factors needed for the functioning of autotrophic plants. These are **heat,**

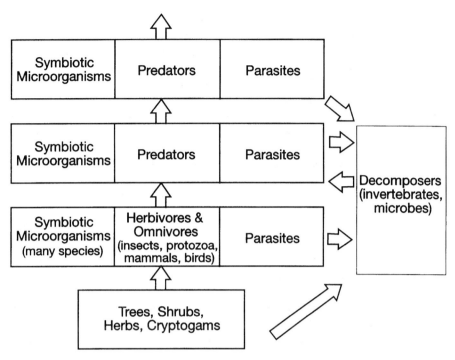

FIGURE 83   Diagram of a simplified food web (from D. A. Perry, *Forest Ecosystems*, 1996, p. 178, The Johns Hopkins University).

light, water, carbon dioxide, oxygen, and **mineral nutrients**. These factors are obtained from the environment (site) and, at this level, are referred to as the site **resources**. As stated previously, macroclimate is one of the most significant features of a given site. In addition to influencing the direct factors of heat, light, and water, climate has had a major effect on soil formation, thus is indirectly associated with mineral nutrient levels as well. The soil, as a medium for plant growth, serves as a reservoir for water, mineral nutrients, and oxygen (for root growth). The particular parent material of a soil strongly influences the kinds and amounts of mineral nutrients available and the ability of the soil to retain and supply minerals to plants. For example, coarse, sandy-textured soils have little water-holding capacity and little **cation exchange capacity** (the ability to adsorb and release positively charged ions). This partially explains the existence of so-called pine barrens in coastal plain areas of the northeast where deep sandy soils occur. Although there are differences in soils within the central hardwood region relating to parent material and microclimate, with some local exceptions, soils in the region are capable of meeting the requirements for tree growth.

Ecological tolerance has to do with a species' ability to cope with the extremes of a site. Foresters frequently use the word "tolerance" to refer to *shade* tolerance. But, in the broader context, plants must be tolerant of excesses and deficiencies of all the six direct factors. For example, the range of temperatures within which a species can survive is called the **cardinal limits**, and if this is exceeded (either high or low) the species cannot survive. Similar limits exist for other resources as well, and it is obvious that the limits of the geographic range of a species are set by the attainment of such extremes. For perennial woody plants that may live for many years, it is important that species have a wide amplitude of adaptability since they will be exposed to wide variations over time. Perhaps a major reason for the diversity of species that grow within the central hardwood region is the moderate climate and environment that exist within the region.

Within a given macroenvironment, light is one of the most variable factors at the stand level and the one over which managers can exert the greatest influence. For this reason, foresters focus much of their attention on the shade tolerance of plant species, since it is this characteristic that often determines how species will respond to openings in the canopy. Shade tolerance is related to a species' ability to utilize light at low intensity. This is described as **light compensation point**. Light compensation is defined as the light intensity at which the rates of photosynthesis and respiration are equal: that is, the point at which food production and consumption rates are equal. Other things being equal, species that are shade tolerant have a lower light compensation point than shade intolerant species. In addition to reduced light quantity, the **quality** of light reaching the forest floor is depleted in the most photosynthetically active wavelengths, those in the red and blue portions of the spectrum. Many shrub or small-tree species are obligate understory dwellers and, therefore, must be shade tolerant or possess some mechanism for coping with their reduced-light environment.

The issue of ecological efficiency becomes more significant when looking at the interaction of plants growing together and competing for the same resources. When resources are limited, such an association leads to **competition** for resources. Efficiency is important in a variety of situations but is most significant when one species is more efficient than others at using a limited resource on a particular site. For example, species, such as oaks and hickories with their better developed root systems, are more competitive on sites where moisture is limited. Conversely, hardwoods such as yellow-poplar, with their shallower and more fibrous root systems, become more competitive where moisture is not limited. The concept of **ecological optimum** is also useful when looking at competitive interactions among species. If the cardinal limits represent the range of conditions that a given species can tolerate, the optimum represents the point within that range at which the species functions best. When examining the adaptation of a species to a site, the optimum is usually more significant than the cardinal limit because, even though a particular species' cardinal limits may not be exceeded on a given site, the species usually lives in a competitive association with other species. When several species are in competition, the ones that are closest to their optimum will prevail. **Competitive advantage** regarding a limited resource can be gained by having either an advantage due to **position** or an advantage due to **vigor** and may occur above ground (crown competition) or below ground (root competition). Position advantages are strictly physical. For example, a species that arrives at the site first may achieve a position advantage relative to light by growing up more quickly and shading out potential competitors. An individual that is situated near a stream or in an above-average microsite may gain an advantage initially due to vigor and later, as it becomes proportionately larger than its competitors, due to position. Genotype may enable one plant to gain an advantage due to vigor over another, as well. In any event, in the forest, as in most ecosystems where competition occurs for limited resources, once one individual gains competitive advantage over its neighbors, the advantage usually compounds. For example, in stand structure and development, those individuals that slip into a subordinate canopy position will tend to continue their slide until ultimately they become suppressed and die. Foresters utilize a system of crown class designation based on a tree's canopy position (Fig. 84). Highly shade tolerant species may be able to persist in a subordinate position for many years, awaiting an opportunity when a canopy tree dies allowing them adequate sunlight to grow. Such **advance regeneration** is typical of highly shade tolerant species, such as maples, beech, hemlock, and basswood.

The work of Kolb and Steiner (1990) and Kolb et al. (1990) illustrates the relationships of ecological requirements, tolerances, and efficiency to the adaptability of certain species to specific sites. They studied seedlings of northern red oak and yellow-poplar grown under different environmental conditions. Both these species are important natural components of the central hardwood forest, but yellow-poplar tends to be more abundant on the mesic north-facing slopes and coves while northern red oak is more abundant on the south- and west-facing slopes. In one experiment, it was noted that when both species were grown in competition with grass, both grew slower, but the yellow-poplar was more affected than the oak (Kolb and Steiner 1990). When certain environmental resources (light, water, nutrients) were restricted in the seedling's environment, the yellow-poplar was more affected than the oak (Kolb et al. 1990). They further found that seedlings of both species compensated for environmental deficiencies, but in different ways, the yellow-poplar by producing less volume of functional xylem and the oak by restricting water flow (loss) in the existing xylem (Shumway et al. 1993). This mechanism may explain the results of Guthrie (1989) who found that tree species with ring-porous wood anatomy dominated on the more xeric sites while diffuse-porous species were more abundant on the more mesic sites in West Virginia.

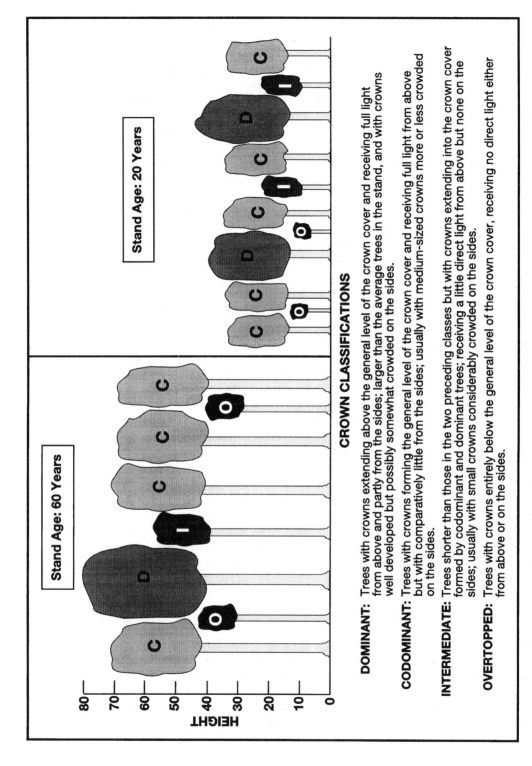

CROWN CLASSIFICATIONS

DOMINANT: Trees with crowns extending above the general level of the crown cover and receiving full light from above and partly from the sides; larger than the average trees in the stand, and with crowns well developed but possibly somewhat crowded on the sides.

CODOMINANT: Trees with crowns forming the general level of the crown cover and receiving full light from above but with comparatively little from the sides; usually with medium-sized crowns more or less crowded on the sides.

INTERMEDIATE: Trees shorter than those in the two preceding classes but with crowns extending into the crown cover formed by codominant and dominant trees; receiving a little direct light from above but none on the sides; usually with small crowns considerably crowded on the sides.

OVERTOPPED: Trees with crowns entirely below the general level of the crown cover, receiving no direct light either from above or on the sides.

FIGURE 84   The relative position of trees in the different crown classes (from D. M. Smith, *The Practice of Silviculture*, 1986; reprinted by permission of John Wiley & Sons, Inc.).

ECOLOGICAL INTERACTIONS

In addition to competition, discussed in the previous paragraph, there are numerous interrelationships that occur in the ecosystem between plant species as well as between plants and animals. These interrelationships may have as much to do with the success or failure of a given species on a given site as do the ecological requirements, tolerances, and efficiencies discussed previously. Some of these interrelationships are **allelopathy**, **phytophagy**, **parasitism**, **saprophytism**, and **symbiosis**. All these systematic interactions occur and are of varying importance in the central hardwood ecosystem. In addition to these systematic processes, there are random elements that cannot be predicted or modeled. Examples of random phenomena are weather events (droughts, floods, windstorms), bumper seed crops — which may themselves be related to weather events — fires, and insect and disease outbreaks. The regeneration and success of a species on a site may be affected dramatically by these random or non-systematic processes. Excellent examples of this can be seen in situations where

FIGURE 85   Hillside in Webster County, West Virginia showing various stages of old-field abandonment. The fencerow indicated by the larger trees extending up the slope on the left side divides an almost pure stand of yellow-poplar (left) from a mixed stand including a significant component of American sycamore. Note the larger sycamore seed tree in the right center of the picture.

old fields have been abandoned sequentially, and the invading species composition is quite different for the different patches of regrowth (Fig. 85). This phenomenon illustrates the effect of timing and all the random events that go with it, such as bumper seed crops and favorable seed beds.

Allelopathy results when one plant produces substances that escape into the environment which are toxic to other plants. These "allelochemicals" are released into the environment in at least four ways: leaching from leaves, root exudation, volatilization, and decomposition (Horsley 1988). The classic example is the production of the toxic chemical juglone by walnut trees (Bode 1958). Because the effect of allelopathy is often subtle and can be confused with competition, it may be overlooked and perhaps underestimated as a factor in controlling the rate of succession and the species composition of regeneration. It is particularly important in old-field succession (Rice 1974), which has been a significant factor in the development of central hardwood forests. Horsley (1977) documented allelopathic inhibition of black cherry seedlings by several species of herbaceous plants that are frequently part of the pioneer community of Allegheny hardwood forests following disturbance. These included grasses, ferns, goldenrod, and aster.

Phytophagy is the consumption of plants by animals. There are numerous obvious examples of phytophagy that result in extensive damage and/or destruction of plants. Notable in the central hardwood region is the browsing of white-tailed deer on tree seedlings. Michael (1988) summarized the situation in the Appalachian region by stating: "White-tailed deer populations in the Appalachians are increasing at an unprecedented rate and the impacts on hardwood forests are potentially disastrous." Deer seem to express preference for certain species (such as oaks, maples, and ashes), which has resulted in nearly pure black cherry regeneration in some areas of Pennsylvania (Tilghman 1987). Deer herds are increasing throughout the central hardwood region and their impact will most likely be greater in the future.

Another type of phytophagy is the consumption of plant tissue (or fluids) by insects. These run the gamut from bark beetles to sucking insects, seed insects, and defoliators. Perhaps the most dramatic of these in the central hardwood region are the defoliators. Native defoliators usually exist at endemic levels and coexist with trees, while doing minimal damage (Butler and Wood 1985). But occasionally native defoliators reach outbreak proportions, due to the availability of extensive areas of suitable habitat and triggering events, such as droughts. The outbreak of loopers that occurred in the early 1980s in West Virginia is a good example of such an outbreak where thousands of acres of oak forest suffered heavy defoliation and subsequent tree mortality (Etgen and Hicks 1987). Far more damaging, in general, are the introduced insects, such as gypsy moth. This species has already become established through about one-third of the central hardwood region and will most likely continue to spread throughout the region. With oaks being a favored host of gypsy moth, it is likely that the scenario that occurred in the central hardwood forests of Pennsylvania will be repeated elsewhere as the infested area spreads southwestward (Herrick and Gansner 1988).

Consumption of fruits and seeds is also a type of phytophagy that can be both beneficial and damaging to the forest. Several native and introduced insect species cause extensive loss of seeds. Acorns are particularly vulnerable due to their relatively large size and nutritive content (Gribko and Hix 1993).

Consumption of seeds by birds and mammals is also a common occurrence in the central hardwoods. In a Pennsylvania study, Steiner (1995) reported that vertebrates (mostly white-tailed deer) consumed up to 38.6 percent of the annual acorn crop. Healy (1988) summarized the impact of birds and mammals on regeneration of Appalachian hardwood forests. For the most part, consumption of seed was not normally regarded as a cause for regeneration failures. Several benefits accruing from bird and animal scatterhoarding of seeds were noted: removal of seed from direct competition with the parent tree, protection of seed from predation, placement of seeds where they are more likely to germinate, and facilitation of long-range dispersal. For some species that have drupaceous fruits, fruit consumption is necessary in order that seed may be scarified by stomach acids to improve germination. On the other hand, animals may also facilitate the dispersal of competing vegetation into regenerating forest areas. Examples of this are grape vines, hawthorns, spice bush, multiflora rose, brambles, and green brier. Birds and rodents are especially important vectors of long-range spread of seeds.

Parasitism is an important relationship that occurs in central hardwood forests. Parasites are organisms that reside in or on another organism and derive nourishment from their host. Many parasites are detrimental to their host. For example, diseases, such as chestnut blight, anthracnoses, *Nectria*, and Dutch elm disease, are all parasitic. There are also examples of parasitic plants, such as mistletoe, which sometimes become problematic to individual trees. Diseases, such as *Armillaria* root disease, can function as parasites when they attack living trees or as **saprophytes** (obtaining nutrition from dead tissue) after the tree has died. The leaf and wood decaying organisms are also saprophytic. The decay process of leaf litter is enhanced by processing of the tissue by **shredders** (Mudrick et al. 1994). Mites in the suborders Oribatei, Mesostigmata, and Prostigmata, as well as Collembola, are important shredders of organic residues in hardwood forests (Hoosein 1991).

Symbiosis is a type of interaction that occurs in forests where two organisms interact for their mutual benefit (Oosting 1956). Symbiotic relationships are important in the central hardwood region. Perhaps most important is the plant/fungal association that results in the formation of **mycorrhizae**. These fungal roots develop when plant roots are infected by certain fungi, and, as a result, the absorbing surface of the root is increased. The fungi benefit by deriving nutrition from the tree. Almost all species of woody plants exhibit mycorrhizal roots (Daubenmire 1959), and, in some cases, inoculation of plant roots with mycorrhizal fungi results in increased rates of growth (Marx and

Beattie 1977). Symbiotic nitrogen fixation is another example of symbiosis that is important in central hardwood forests, especially in early successional stages and in fire-dominated ecosystems (Boring and Swank 1984). Several woody genera are documented to fix atmospheric nitrogen through nodulated roots that are infected with nitrogen-fixing bacteria. These include *Ceanothus* and *Alnus*, as well as several species of legumes. Prominent among the latter is black locust. According to Boring and Swank (1984), black locust seedlings are often abundant in the early regeneration that occurs after clearcutting, and a 4-year-old stand of black locust regeneration at Coweeta Hydrologic Laboratory in Franklin, North Carolina, was estimated to fix 30 kg/ha/yr of N. The ability to fix atmospheric nitrogen can give such species a competitive advantage on sites where soil nitrogen levels are low.

REGENERATION AS AN ECOLOGICAL PROCESS

Nyland (1996) describes the ecological process of regeneration in terms of a chain of events that must remain more-or-less unbroken in order to be successful. Loftis (1993) cites Noble and Slater (1980) who suggest that tree species possess "vital attributes" that determine their ecological role. These include:

1. The method of arrival or persistence of a species at a site during or after disturbance.
2. The ability to grow to maturity in the developing stand.
3. The time needed for an individual of a species to reach critical life stages.

The first of these is related to the regeneration process and the remaining two are associated with the process of stand dynamics. Classical ecological theory generally advances a concept of regeneration best described as "relay floristics" (Clements 1916), where one community occupies the site, alters it, and is followed by another community. This model has limited applicability to forest regeneration where propagules or other potential sources of regeneration are often already present prior to removal of the overstory (Loftis 1993). In such systems, the concept of "initial floristics" advanced by Egler (1954) is more applicable.

As with many forest systems, the central hardwoods have a long history of disturbance, and the dominance of oaks is due in great measure to this history (Van Lear and Watt 1993). Oaks are uniquely able to persist on fire-disturbed or marginal sites (Abrams 1992). Oaks invest much energy in their large seeds and seedlings utilize much of their energy during the establishment phase to develop an extensive root system (Dickson 1994). Their strategy of regeneration involves the accumulation of propagules, or potential propagules over time — either seedlings, seedling sprouts, or stump sprouting potential (Johnson 1993).

The role of fire in the oak regeneration picture seems to be one of selectively removing species that would compete with the oaks and stimulating the development of vigorous seedling sprouts from the well-developed oak seedling root systems. These seedling sprouts are then able to gain a height advantage over would-be competitors. On xeric sites, oaks seem to possess an inherent advantage over other species and form relatively stable forests without a successional tendency toward more shade tolerant species (Johnson 1992).

On mesic sites, oak regeneration is often poor and when disturbances occur that promote large openings, aggressive shade intolerant species such as yellow-poplar and black cherry are common invaders, especially in the Appalachian and Western Mesophytic regions (Trimble 1973). These species often employ strategies whereby seed are widely disseminated by wind or animals or are accumulated in the forest floor as dormant seed. Such species, unlike the oaks, display rapid early height growth and are capable of exploiting the abundant resources of the post-disturbance site. When disturbances are less extensive, as would result from the death or removal of single large trees or small groups, the result on mesic sites is usually the development or release of advance regeneration of shade tolerant species such as maples, American beech, and hemlock. Swank and Vose (1988) in an effort to better explain the regeneration process diagrammed the microenvironmental changes that often accompany forest regeneration (Fig. 86).

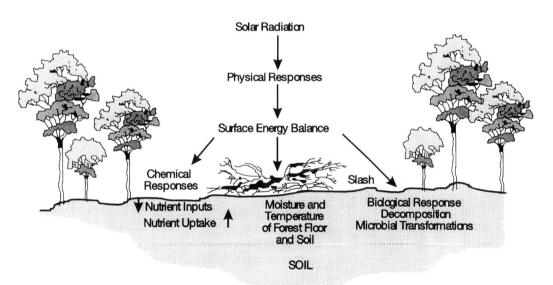

FIGURE 86   Microenvironmental components of the ecosystem that are modified by cutting (from Swank and Vose 1988).

FOREST STAND DEVELOPMENT

A **community** is defined as an assemblage of organisms living in an environment and interacting with each other and with the environment. As previously stated, plant succession is a change in community species composition over time, with each community comprised of species that are adapted to the specific conditions that exist at the time. The pioneer community of a secondary succession is the first community that follows a disturbance event. This community presumably gives way to what is termed the climax, and interim communities are called **subclimax**. Sometimes a community may, due to some site factor or other environmental condition, be prevented from reaching the "climatic climax" for the region and will be maintained in a subclimax state. Soil factors may impose what is called an **edaphic climax** (Braun 1950). The oak communities that occur in the shale barrens of the Ridge and Valley Province and in the Missouri Ozarks are a good example of this. The pine barrens that occur along the coastal plains from Massachusetts to New Jersey are maintained by periodic fires and are, therefore, a "fire climax." Indeed, many of the existing oak-dominated forests of the central hardwood region appear to be a subclimax and an artifact of the disturbance history of the region.

Community development, when viewed as succession, focuses on the species that occupy the site, but another way of viewing the chronological progress of an ecosystem is in terms of **biomass accumulation**. Bormann and Likens (1979) proposed four phases of biomass accumulation following clear-cutting of a northern hardwood ecosystem (Fig. 87). Using a biomass accumulation model (JABOWA, Botkin et al. 1972) to simulate a northern hardwood ecosystem, researchers found that the accumulation of biomass reaches its peak in about 170 years, which marks the end of the aggradation phase. Because of the favorable climate and prevalence of faster-growing species, this time frame is generally shorter in central hardwood forests (Shugart and West 1980).

Foresters, when dealing with timber production, often use analogous terms and concepts to the ecological terminology discussed previously. For example, the **stand** is "a spatially continuous group of trees and associated vegetation having similar structures and growing under similar soil and climatic conditions" (Oliver and Larson 1996) and is analogous to the community, but focuses more on the trees and vegetation. The term **stand dynamics** is, therefore, analogous to succession and focuses on the "changes in forest stand structure with time, including stand behavior during and after disturbances" (Oliver and Larson 1996). Forest **growth and yield** models would be the forestry equivalent to the biomass accumulation models presented in the ecological literature. The important difference is that growth and yield models provide outputs in terms of commercial forest products, whereas biomass accumulation looks at total productivity without regard to commercial value.

The dynamics of most central hardwood stands appear to fit the **even-aged**, **single-cohort** case described by Oliver and Larson (1996). That is, stands

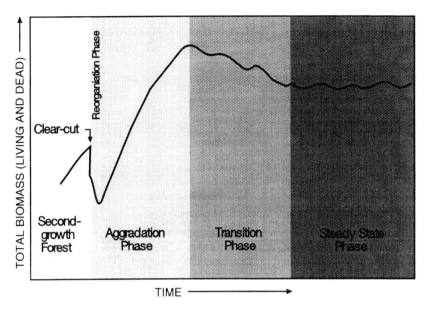

FIGURE 87   Proposed phases of ecosystem development after clearcutting of a second-growth northern hardwood forest. Phases are delimited by changes in total biomass accumulation (living biomass and organic matter in dead wood, the forest floor, and the mineral soil). It is assumed that no exogenous disturbance occurs after clearcutting (redrawn from Bormann and Likens 1979).

normally develop as a result of a single disturbance event, such as the logging/fire scenario that occurred around the turn of the century. Often the regeneration of present-day stands developed synchronously after this disturbance event with different species assuming dominance through time. This fits the "initial floristics" model proposed by Egler (1954) (Fig. 88). In this pattern of vegetation development, most of the species regenerate immediately or soon after the disturbance event, but some short-lived pioneer species dominate early and drop out quickly (**Phase I**). Species such as goldenrod, fire weed, brambles, and a variety of herbaceous species typify this phase. Woody perennials, such as sumacs, sassafras, aspens, pin cherry, and hawthorns, assume dominance after the herbaceous stage. These **Phase II** species grow rapidly at first, but slow down later and are overtaken by others. Many of the Phase II species are either gone or in the process of dropping out of today's central hardwood stands (Fig. 89). Among the long-lived tree species are some that often assume an early dominance (**Phase III**) and are slowly replaced by the longest-lived species in Phase IV. Examples of Phase III species are sweet birch, black cherry, scarlet oak, sweetgum, hickories, and yellow-poplar. Species that fit the **Phase IV** designation include maples, most red and white oaks, and American beech. Oliver

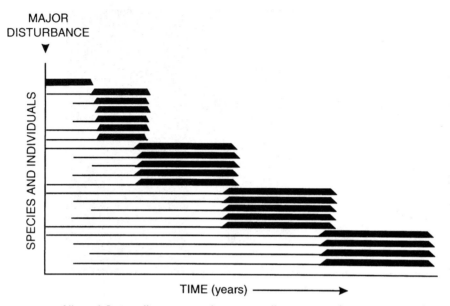

FIGURE 88   Diagram of "initial floristics" pattern. In this pattern all species invade at approximately the same time after a disturbance but assert dominance at different times (from Egler 1954).

(1981) diagrammed the stand development process (Fig. 90) and included four stages (stand initiation, stem exclusion, understory reinitiation, and old growth). These four stages are generally analogous with Phases I–IV. Most of the central hardwood forests are still in Phase III and transitional into Phase IV, which correlates to the Understory Reinitiation Stage of Oliver (1981).

As indicated previously, central hardwood stands are typically even-aged, single-cohort stands. But due to the variety of species that occur, with their different reproductive strategies and silvical characteristics, the development of stands is much more complex than that which takes place in single-cohort stands involving a single or a few species, such as boreal spruce/fir forests and lodgepole pine forests. The forests of the central hardwood region represent an exceedingly complex interactive system where site characteristics, silvical characteristics of species, and random processes combine to produce a forest mosaic across the landscape. This complexity presents challenging management and silvicultural problems and opportunities.

Silviculture is "applied ecology," thus, the silvicultural implications for central hardwood stands are as complex as the stands themselves, but a few generalities can be advanced. First, most central hardwood stands are even-aged (with the exception of stands in the Missouri Ozarks) and are second-growth stands between 60 and 100 years of age. Although most are composed of deciduous hardwoods, there is a considerable variety of species and stand types

FIGURE 89   Dead sassafras trees are a common sight in central hardwood stands. This short-lived species was a pioneer in many second-growth forests in the central hardwood region.

related to site, historical development, and random processes. Most of the present-day stands regenerated from major disturbance events, such as logging, fires, and agricultural abandonment. Because central hardwood stands are generally composed of a mixture of species, the various species groups, although even-aged, are at different stages of maturity and have different requirements and potentials. Thus, in any given stand, some species may be mature or approaching maturity and others may still be biologically, if not economically, immature. Traditional markets for high-value sawtimber and veneer products make it economically feasible to carry central hardwood stands to longer

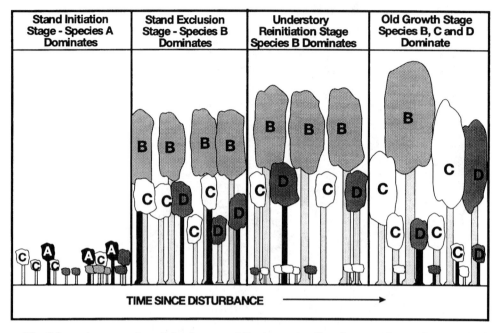

| Stand Initiation Stage - Species A Dominates | Stand Exclusion Stage - Species B Dominates | Understory Reinitiation Stage Species B Dominates | Old Growth Stage Species B, C and D Dominate |

TIME SINCE DISTURBANCE ⟶

FIGURE 90   Schematic stages of stand development following major disturbances. All trees forming the forest start soon after the disturbance; however, the dominant tree type changes as stem number decreases and vertical stratification of species progresses. The height attained and the time elapsed during each stage vary with species, disturbance, and site. (Reprinted from *Forest Ecology and Management*, vol. 3, C. D. Oliver, Forest development in North America following major disturbances, pp. 153–168, 1981, with kind permission of Elsevier Science-NL, Sara Burgerhartstraat 25, 1055 KV Amsterdam, The Netherlands).

rotations, although a recent trend in industrial development of processes that produce composite products is creating new markets for smaller diameter trees. This development could provide economic incentives for silvicultural operations or could lead to overcutting of small diameter products that will ultimately result in a deficit of larger trees.

In general, most central hardwood stands are *not* ready for a final harvest. Thus, the appropriate silvicultural treatments for most stands are **intermediate treatments** and often a combination of cuttings performed simultaneously, owing to the complexity of mixed species and site factors. The threat of pests, such as the gypsy moth, also creates a need for certain silvicultural treatments, focusing on the susceptible and vulnerable component of mixed stands (Gottschalk 1987). Although regeneration may not be the primary concern in many present-day stands, what occurs now will most certainly affect regeneration in the future. For example, partial cuttings (improvement, release, thinning, etc.) stimulate understory development, and, depending on the severity of cutting, they often encourage relatively shade tolerant species to grow. With the exception of the Ozark Plateau, red maple appears to be a major component of midstories and understories throughout the central hardwood region (Perkey and Powell 1988), and oak regeneration is generally viewed as difficult on all but poorer sites (Smith 1992).

In addition to the age and current composition of forest stands in the central hardwood region, the three ecological factors that will most likely have a significant impact on future stands are (1) the **lack of fire** as a disturbance factor in most stands, (2) the invasion of **gypsy moth** and its host preference for oak species, and (3) the **impact of deer** on forest regeneration (Marquis 1981a; Jones et al. 1993). Add these factors to the fact that partial cutting and high-grading will probably continue to be practiced on many privately owned forests in the central hardwood region, and the result will most likely be changes in the composition of future stands, generally leading to forests with a lower oak component and, unless good silviculture is practiced, lower quality stands in general.

ECOLOGICAL RELATIONSHIPS, SUMMARY

Climatic and site conditions generally drive the process of succession, and theoretically there is a climax association that would naturally occur on a given site in the absence of interruption of the successional process. However, the forests that currently exist in the central hardwood region are a product of historical disturbance events, such as logging, fire, disease, and agriculture. These stands are generally even-aged, and most of the trees present in today's stands regenerated at the same time following disturbance (single cohort). The trees that repopulated our forests were those that were adapted to the conditions that existed and were successful because of their silvical characteristics. The individuals that survive today are those that were able to tolerate the site conditions and were efficient in resource utilization so that they could compete successfully with other trees in the stand. Thus, each species' ecological role is a product of its inherent genetic makeup and interactions with environment and other species. Survival, growth, and regeneration are all a function of these adaptations and interactions. In the succeeding chapters, these **silvical characteristics** of important species in the central hardwood region are enumerated and applications of ecological principles to facilitate the regeneration and tending of stands (silviculture) are discussed.

The ecological factors that have produced today's stands are changing. For example, fire is less of a disturbance factor in central hardwood stands than it once was. Gypsy moth threatens the oak resource throughout the region, and high deer populations have a great impact on development of regeneration. Although the central hardwood forests are becoming economically valuable, many of the species are not yet biologically mature. Silvicultural treatments that should be applied to central hardwood stands at this critical juncture are those that are ecologically sound, environmentally safe, and economically viable.

4

Silvical Characteristics of the Major Central Hardwood Species

INTRODUCTION TO SILVICAL CHARACTERISTICS

The term *silvical characteristics* literally means "tree characteristics," but it is most often used to describe the characteristics that define a particular species' ecological role, for example, shade tolerance, growth rate, site requirements, regeneration strategies, and injurious agents, such as insects and diseases. The obvious value of knowledge of silvical characteristics is in making management and silvicultural decisions that are compatible with a species' natural role in the ecosystem. Spurr and Barnes (1980) quote Lutz (1959) as follows:

In conclusion, it seems to me that the silviculturist should seek to understand ecological principles and natural tendencies as they relate to trees and forest communities with which he works. I do not infer that he is bound to follow them blindly but neither do I think that he can safely ignore them. Between the two extremes of passively following nature on the one hand and open revolt against her on the other is a wide area for applying the basic philosophy of working in harmony with natural tendencies.

Most contemporary silviculturists have adopted the philosophy that silviculture must work in concert with the natural ecological processes. Therefore, a thorough understanding of the natural processes, such as succession, competition, disturbance, and reproductive strategy, is critical to the application of silviculture. This is particularly true in the central hardwood region for several reasons. For example, unlike the simpler southern pine, boreal, and western coniferous ecosystems, deciduous hardwood stands are often complex mixtures of species, each with its own silvical characteristics. The central hardwood region, for the most part, consists of hilly or mountainous terrain, making many mechanized operations, such as site preparation and machine planting, difficult or impossible. Natural regeneration of commercial tree species seems to generally occur in adequate quantity throughout the region, therefore, natural regeneration is the most common and practical way to regenerate central hardwood stands. Furthermore, most forestland in the central hardwood region belongs to small private landowners who lack the economic means and motivation to practice intensive silviculture. Thus, silviculture in the central hardwood region is, of necessity, an emulation of natural processes. In managed forests using natural regeneration, the forester's role is to attempt to create conditions that favor regeneration of the desired type and quantity through manipulating the canopy by cutting. But whether or not the desired result is achieved, almost invariably, regeneration by some commercial species will occur. Intermediate cuttings also emulate natural processes, such as competition, suppression, and natural thinning. But here again, the forester's role becomes one of manipulating the stand such that the growing space is occupied by the trees that possess the greatest potential for the intended use. Thus, it is imperative that a person engaged in forest management be familiar with the silvical characteristics of the species.

Silvical characteristics define a species' **ecological role** and are therefore key to managing the species. Important information regarding the management of a species is gained by observing the natural conditions under which it grows best. Factors like geographic distribution, soils, topographic conditions, and associated species all provide valuable clues for proper management.

Bormann and Likens (1979) discussed growth and reproductive strategies of species that colonize an area following disturbance (secondary successional species). In terms of growth and **reproductive strategies**, some species rely primarily on vegetative propagation to spread and colonize an area. Many ferns, for example, utilize this "outgrowth" strategy, and they tend to spread in the horizontal dimension by vegetative propagation and occupy the site quickly after disturbance. These species often give way to trees that employ an "upgrowth" strategy. Ferns and herbs can dominate an area longer when some environmental imbalance, such as a high deer population, exists. Taller-growing trees become established by advance reproduction (seedlings already present, usually of tolerant species) or new individuals. The latter can be from vegetative sources (root or stump sprouts) or from seed. Seed may be residual seed, either from buried or stored seed or from dispersion from surrounding trees. Vectors of seed dispersion include gravity, wind, water, and animals. Different species have evolved different mechanisms for dispersal, which is an important element in the species' ecological role. All the above reproductive and growth strategies are represented by different species in the central hardwood region.

In addition to reproductive and growth strategies, **shade tolerance** is a silvical characteristic that is important in defining the ecological role of a species. The stereotypic "pioneer" community would be dominated by fast-growing intolerant species, which is frequently the case. But, depending on the type of disturbance that initiates a secondary succession, many representatives of shade tolerant species may be present as well. A hypothetical sequence of dominance by various types of species relative to shade tolerance is illustrated by Figure 91.

Bormann and Likens (1979) group species into two categories (**exploitive** and **conservative**) with respect to their ecological role. Exploitive species are those that are capable of rapidly taking advantage of temporary enrichment of a site. Exploitive species are analogous to pioneer species in plant succession terms. Such species characteristically have certain attributes (Table 9). For example, they have mechanisms that disseminate seed widely (light weight, etc.); they are shade intolerant, grow fast, use high-intensity light efficiently, and have indeterminate growth and a lower root/shoot ratio. The conservative strategy is typical of species that occur later in the successional sequence, and such species generally are shade tolerant, are slower growing, have heavier seed, are more efficient at using low-intensity light, and have determinate growth and a higher root/shoot ratio (Table 9).

The morphologic growth type (**determinate** and **indeterminate**) in woody plants is related to the extension of apical meristems (Marks 1975). With the determinate type, buds that are formed at the end of a growing season contain primordia representing all the tissues that will expand in the next growing

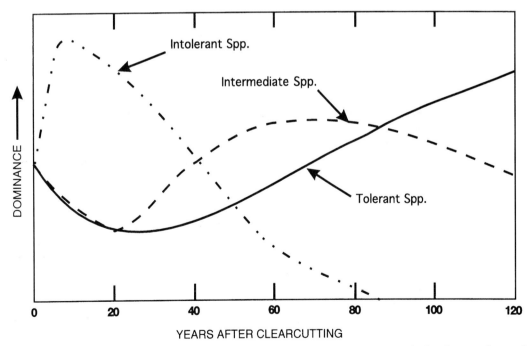

FIGURE 91   A hypothetical sequence showing trends in dominance among different shade tolerance classes of trees after clearcutting (from Nyland 1996).

TABLE 9 Relative Characteristics of Species Exhibiting Exploitive and Conservative Growth Strategies

	STRATEGY	
CHARACTERISTIC	EXPLOITIVE	CONSERVATIVE
I. Longevity	Shorter	Longer
II. Apical growth		
Morphologic type	Indeterminate	Determinate
Timing during season	Throughout	Early
III. Crown geometry		
Width	Narrower	Wider
Depth	Shallower	Deeper
IV. Root morphology		
Root/shoot ratio	Lower	Higher
Depth	Shallower	Deeper
V. Light saturation of leaves	At high intensity	Lower intensity
VI. Efficiency of roots in water and nutrient uptake	Lower	Higher
VII. Reproductive strategy		
Age at sexual maturity	Younger	Older
Seed weight	Lighter	Heavier
Advance regeneration	No	Yes
Buried seed strategy	Yes	No

Source: Borman and Likens 1979.

season. In other words, the enclosed winter bud contains a miniature of the next year's stem and leaf tissue, which expands the following spring. With the indeterminate morphologic type, primordia enclosed in the winter bud represent only a portion of the potential leaf and stem tissue that can be produced the following year. Leaves found in the winter bud of indeterminate species are morphologically distinct from those produced later in the growing season (Critchfield 1960). The determinate species base their next year's growth on the current year's conditions, a more conservative strategy, whereas the indeterminate species have the capacity to exploit temporary episodes of abundant resources. For example, a sugar maple (determinate species), once the leaf and twig tissue is fully expanded in the spring, cannot extend any more during that growing season, but a bigtooth aspen (indeterminate) may produce several flushes of extension growth during a year when resources, such as water, are abundant. If a dry summer occurs, the indeterminate species can cease growing in the early summer.

Species utilizing an exploitive strategy are capable of taking advantage of temporary enrichment, but they sacrifice the ability to compete with conservative species when resources are limited—a condition that often develops later in the life of a community. Kolb et al. (1990), using controlled environmental conditions, observed that under optimal conditions of light, water, and fertility, yellow-poplar seedlings produced 36 percent more dry weight than red oak seedlings, but when at least one resource was limited, red oak grew 38–126 percent more dry weight than yellow-poplar. These results are consistent with a more conservative strategy for red oak as compared with yellow-poplar, which is typical of an exploitive species. Thus, the morphological and physiological characteristics of species seem to predispose them to unique roles in the ecosystem (Grime 1979), and, therefore, silvicultural prescriptions must take these factors into account in order to be effective.

Knowledge of agents that cause damage to trees is important when making management decisions. Insects, diseases, fire, drought, deer browsing, and air pollution have all been implicated as causes of poor forest health. In many situations in the central hardwood region, **forest health** problems take the form of so-called **declines**. A decline is a syndrome involving an accumulation of stress by the tree. As stress accumulates, different secondary organisms become involved. The decline process is illustrated by the diagram of tree health/vigor as it relates to genetic potential, time, and stress factors (Fig. 92).

The following discussion of silvical characteristics of 34 important species in the central hardwood region is intended to preface the subsequent discussion of cultural methods. Discussion of silvical characteristics is organized alphabetically by genus and is in the following order:

1. *Acer rubrum* (red maple)
2. *Acer saccharum* (sugar maple)
3. *Aesculus octandra* (yellow buckeye)
4. *Betula allegheniensis* (yellow birch)
5. *Betula lenta* (sweet birch)
6. *Betula nigra* (river birch)
7. *Carya cordiformis* (bitternut hickory)
8. *Carya glabra* (pignut hickory)
9. *Carya ovata* (shagbark hickory)
10. *Carya tomentosa* (mockernut hickory)
11. *Fagus grandifolia* (American beech)
12. *Fraxinus americana* (white ash)
13. *Fraxinus pennsylvanica* (green ash)
14. *Juglans nigra* (black walnut)
15. *Juniperus virginiana* (eastern redcedar)
16. *Liquidambar styraciflua* (sweetgum)
17. *Liriodendron tulipifera* (yellow-poplar)
18. *Nyssa sylvatica* (blackgum)
19. *Pinus echinata* (shortleaf pine)
20. *Pinus rigida* (pitch pine)

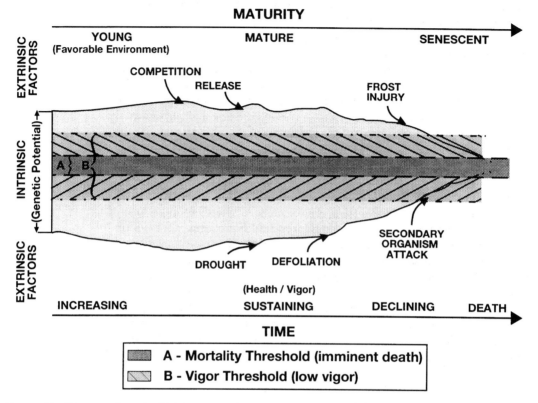

FIGURE 92 Diagram of tree health/vigor as it relates to genetic potential, time, and stress factors. The vertical dimension represents the health and vigor status of the tree while the horizontal line represents time. Trees begin life with a given intrinsic genetic potential represented by the vertical line on the far left. Site conditions may favor expanding vigor or stress factors may reduce it. Once a plant crosses a vigor threshold, either due to stress or senescence, it may become vulnerable to secondary organisms. As vigor declines, the tree enters a threshold of "imminent death" represented by the inner band.

21. *Pinus strobus* (eastern white pine)
22. *Pinus virginiana* (Virginia pine)
23. *Platanus occidentalis* (American sycamore)
24. *Prunus serotina* (black cherry)
25. *Quercus alba* (white oak)
26. *Quercus coccinea* (scarlet oak)
27. *Quercus falcata* (southern red oak)
28. *Quercus prinus* (chestnut oak)
29. *Quercus rubra* (northern red oak)
30. *Quercus stellata* (post oak)

31. *Quercus velutina* (black oak)
32. *Robinia pseudoacacia* (black locust)
33. *Tilia americana* (American basswood)
34. *Tsuga canadensis* (eastern hemlock)

For each species, the Ecoregion (Province) where the species predominantly occurs in the central hardwood region is given (Fig. 45). Table 10 provides a summary of silvical characteristics for the 34 species listed above.

SILVICAL CHARACTERISTICS OF SPECIES

1. Acer rubrum (Red Maple)

General. Red maple is a species that is adaptable to a broad spectrum of environmental conditions and, in the central hardwood region, is emerging as a prominent component of many stands. Red maple grows on a wide variety of sites and can function as a pioneer or a climax species. It grows at a moderate rate and produces trees of moderate value for sawtimber products but is adaptable for use in composite products. Red maple is not particularly valuable for wildlife habitat, but its brilliant autumn coloration makes it valuable as an amenity species.

Distribution and Associates. Red maple is among a handful of species in the eastern United States that is found from New England to Florida and from the Atlantic coast to the Peninsular Great Plains. Within this wide geographic area, red maple is also very abundant in many stands. Walters and Yawney (1990) indicate that red maple can tolerate a wider range of soil and climatic conditions than any forest species in North America. Red maple occurs in the following provinces in the central hardwood region (Bailey 1994): 212, M212, 221, M221, 222, 231, and M231 (Fig. 45). Because of its broad range of tolerance, red maple can exist on a wide range of soils and has no distinct preference for aspect (Stephenson 1974; Hicks and Frank 1984). It grows on sites ranging from dry ridges to swamps and occurs on Entisols, Inceptisols, Ultisols, Alfisols, Spodosols, and Histosols soil orders, derived from sedimentary as well as igneous rock types. Red maple, therefore, is associated with a great diversity of species ranging from bottomland species (especially in the South) to upland types.

Ecological Role. Red maple is capable of functioning in a variety of ecological roles. It is relatively shade tolerant with a moderate growth rate (Trimble 1975). But in spite of those characteristics that are typical of conservative or climax species, red maple is frequently a pioneer species, especially on old fields, where it can form almost pure stands. Red maple matures in 70–80 years with a maximum longevity of about 150 years. Red maple is shade tolerant enough to persist in an understory role. It may dominate a site early, giving way to oaks

TABLE 10 Summary of Some Silvical Characteristics of Important Species in the Central Hardwood Region

Species	Seed Dissemination			Ecological Strategy		Shade Tolerance			Growth Rate			Longevity			Injurious Agents				Sprouting	
	Gravity	Animals	Wind	Exploitive	Conservative	Intol.	Intermed.	Tol.	Slow	Med.	Fast	<100	100–200	>200	Fire	Gyp. Moth	Deer	Other	Yes	No
Red maple			✓	✓	✓			✓		✓			✓		✓		✓	✓	✓	
Sugar maple			✓		✓			✓		✓			✓		✓		✓	✓	✓	
Yellow buckeye		✓			✓			✓		✓				✓						✓
Yellow birch			✓	✓	✓		✓				✓			✓	✓		✓	✓		✓
Sweet birch			✓	✓			✓			✓			✓		✓			✓		✓
River birch			✓	✓		✓					✓	✓			✓	✓			✓	
Bitternut hickory	✓	✓			✓		✓			✓			✓		✓			✓	✓	
Pignut hickory	✓	✓			✓		✓			✓				✓	✓			✓	✓	
Shagbark hickory	✓	✓			✓		✓			✓				✓	✓			✓	✓	
Mockernut hickory	✓	✓			✓		✓			✓				✓	✓			✓	✓	
American beech	✓	✓			✓			✓	✓					✓	✓			✓	✓	
White ash			✓	✓		✓					✓		✓				✓	✓	✓	
Green ash			✓	✓		✓					✓		✓				✓	✓	✓	
Black walnut	✓	✓		✓	✓	✓					✓		✓		✓		✓	✓		✓
Eastern redcedar		✓		✓		✓			✓				✓		✓		✓			
Sweetgum			✓	✓		✓					✓			✓	✓	✓			✓	
Yellow-poplar			✓	✓		✓					✓			✓	✓				✓	

138

Species													
Blackgum	✓	✓		✓	✓	✓	✓	✓	✓		✓	✓	
Shortleaf pine	✓	✓	✓	✓	✓			✓				✓	✓
Pitch pine	✓	✓	✓	✓	✓		✓		✓			✓	✓
Eastern white pine	✓	✓	✓		✓	✓		✓		✓			✓
Virginia pine	✓	✓	✓	✓		✓			✓			✓	✓
American sycamore	✓	✓	✓			✓		✓				✓	✓
Black cherry	✓	✓	✓	✓	✓	✓		✓	✓		✓	✓	✓
White oak	✓	✓		✓	✓	✓	✓		✓		✓	✓	✓
Scarlet oak	✓	✓	✓	✓	✓		✓		✓		✓	✓	✓
Southern red oak	✓	✓	✓		✓			✓	✓		✓	✓	✓
Chestnut oak	✓	✓	✓	✓	✓			✓			✓	✓	✓
Northern red oak	✓	✓	✓		✓		✓		✓		✓	✓	✓
Post oak	✓	✓	✓	✓		✓			✓		✓	✓	✓
Black oak	✓	✓	✓	✓			✓				✓	✓	✓
Black locust	✓	✓	✓	✓	✓			✓	✓		✓	✓	✓
American basswood	✓		✓		✓	✓		✓			✓	✓	✓
Eastern hemlock		✓	✓	✓	✓	✓		✓			✓	✓	✓

139

or sugar maple and forming a midstory as these mixed-species forests mature.

Red maple is a prolific sprouter and produces good seed crops almost annually, with bumper crops every other year. The fruit is a double samara that is relatively light and can travel a great distance in the wind. Seed are capable of germination soon after dispersion and can germinate on mineral soil or a thin layer of litter.

Injurious Agents. Sprout-origin red maples are susceptible to decay and mechanical wounding initiates decay that spreads rapidly causing wood discoloration and weakening. Red maple is especially susceptible to fire damage and is a preferred species for deer browsing. It is host to a number of borers and gall-forming insects, none of which usually kill healthy trees (Walters and Yawney 1990). Red maple is moderately susceptible to defoliators, such as the looper complex and gypsy moth, but in mixed stands with oaks, the insects' preference for oaks often provide a competitive advantage for red maple.

2. Acer saccharum (Sugar Maple)

General. Sugar maple is a species that is more important in the northern portion of the central hardwood region but is distributed through most of the region. It is a relatively slow-growing species that is a key component of the so-called **northern hardwoods** or beech-maple type. Especially in the northern part of its range, sugar maple is an important timber species. It is also used extensively for maple syrup production, again more so in the northern portion of its range. Like red maple, sugar maple is of limited wildlife habitat value but is among the most desirable species for aesthetic value due to its brilliant yellow, orange, and red fall foliage. Sugar maple wood is especially hard, being known by lumber manufacturers as "hard maple," and is used extensively for flooring and furniture.

Distribution and Associates. Sugar maple is distributed from northern Georgia to the Maritime Provinces of Canada and from the Atlantic coast of New England down the Appalachians and west to northern Arkansas. But in the southern portion of its range, it is usually found only as scattered individuals, except at high elevations in the southern Appalachians where climatic conditions correspond to more northern latitudes. Sugar maple is found in the following central hardwood provinces: 212, M212, 221, M221, 222, M222, and M231 (Fig. 45).

Sugar maple can grow on a wide variety of sites but grows best on sandy or loamy soils. It occurs on the following soil orders: Spodosols, Alfisols, and Mollisols (Godman et al. 1990). Sugar maple may be associated with many different species including hemlocks, basswoods, black cherry, and northern red oak, but its classical associates in the so-called northern hardwoods cover type are American beech and yellow birch.

Ecological Role. Sugar maple is a good example of a species that employs a conservative ecological strategy. The northern hardwoods are considered a climax over much of New England and on better sites in the Lake States, and sugar maple is among the most shade tolerant of native American species. Sugar maple grows at a medium rate in diameter (Trimble 1975). Young seedlings can persist with virtually no growth under heavy shade but often outlive their competitors and respond well to release. Sugar maple is relatively long-lived, reaching biological maturity at age 300–400 years.

Sugar maple is capable of stump sprouting, but the principal source of regeneration is seed. Like red maple, the sugar maple fruit is a double samara that is somewhat larger in size than that of red maple. Samaras ripen and are shed from the tree in early fall. Good seed crops occur almost every other year, but trees must be at least 70–100 years old to produce heavy seed crops.

Injurious Agents. Sugar maple seems especially sensitive to declines that are stress-related syndromes, often involving site, climatic factors, and attacks by insects and diseases. Lately, atmospheric pollutants, such as acidic deposition, nitrate saturation, and ozone, are being implicated as part of the contributory stress in declines (Smith 1987). For example, maple decline in the Allegheny section of northern Pennsylvania is more frequent on glacial soils and is also associated with outbreaks of defoliating insects and occurrence of drought. Liebhold et al. (1995) list sugar maple as intermediate in susceptibility to gypsy moth, and the species is very sensitive to fire injury, which causes basal wounding and ultimately leads to decay. Fungal root diseases, such as Armillaria root disease, seem to be the organisms that ultimately kill trees. Because sugar maple is only locally abundant in the central hardwood region, the impact of maple decline is significant only in such localities. Like red maple, sugar maple is preferred by white-tailed deer and in areas with high deer density it may be eliminated from the understory.

3. *Aesculus octandra (Yellow Buckeye)*

General. Yellow buckeye is the largest native American species of the genus *Aesculus* but has the softest and lightest weight wood of any hardwood species in the central hardwood region (Williams 1990a). As a result, it is not a good species for lumber, although it is sometimes used for pulpwood.

Distribution and Associates. Yellow buckeye ranges from southwestern Pennsylvania through West Virginia, eastern Kentucky, and Tennessee into western North Carolina, where it makes its best development in the southern Appalachians. Another species, Ohio buckeye (*A. glabra*), grows in a more western range from Ohio to Oklahoma and is similar to, although smaller than, yellow buckeye. In the central hardwood region it is found in the following provinces: 221 and M221 (Fig. 45).

Yellow buckeye is often a part of a mixed mesophytic forest community in the central and southern Appalachians and Appalachian Plateaus. It grows on a variety of soils (generally Alfisols and Entisols), making its best growth on deep alluvial soils. Yellow buckeye is usually found as a scattered tree in mixed stands, often associating with mesophytic hardwoods, such as yellow-poplar, northern red oak, maples, and ashes.

Ecological Role. Yellow buckeye has many of the characteristics of species that employ a conservative ecological strategy. It produces large heavy seed (about 30/lb.) that are distributed primarily by animals and gravity. Buckeyes are considered shade tolerant but utilize a strategy of leafing out ahead of most of their associates, thereby obtaining light through the leafless canopy in the early spring. Buckeye is intermediate in growth rate and long lived. Thus, buckeyes are part of a complex mix of species that form the climax community in the mixed mesophytic forest of the central hardwood region.

Injurious Agents. There are several defoliators (the buckeye lacebug, walnut scale, etc.) that attack buckeyes, but none cause catastrophic losses. Yellow buckeye is listed as resistant to gypsy moth defoliation by Liebhold et al. (1995). Likewise, buckeye is relatively free from major disease pests, occasionally being infected by powdery mildew and leaf spot diseases (Carmean 1965).

4. *Betula allegheniensis (Yellow Birch)*

General. Yellow birch is the most valuable of the native birch species for timber production (Erdmann 1990). It is relatively fast growing, produces lumber of good quality, and provides browse and food for several species of wildlife.

Distribution and Associates. Yellow birch ranges across the northern tier of states in the eastern United States and southern Canada and down the highest spine of the Appalachians to North Carolina. In the southern portion of its range, yellow birch is found at higher elevations where it is typically associated with the so-called northern hardwoods (beech, birch, maple). It may often mix with red spruce at higher elevations or more northern latitudes and sometimes associates with mesophytic hardwood species, such as yellow-poplar and northern red oak, especially in cove situations. Yellow birch performs best on deep soils with good drainage. It is most often found on soils belonging to the Spodosol, Inceptisol, and Alfisol orders. Although it does not prefer wet sites, it is tolerant of them and can be found on such sites due to the reduced competition that these sites afford. Yellow birch occurs in the following provinces of Bailey (1994): 212, M212, 221, and M221 (Fig. 45).

Ecological Role. Yellow birch can function as both a conservative and an exploitive species. It is considered intermediate in shade tolerance, but its seedlings need some opening in the forest canopy to become established

(Winget and Kozlowski 1965). Yellow birch trees begin producing seed at about age 40, and good seed crops are produced at 2–4 year intervals. Seed are light in weight, and after maturing in the fall, they are spread by wind and can readily colonize newly disturbed habitats. Yellow birch can become established after forest fire, and a disturbed soil seedbed seems to benefit regeneration. Birch seed often germinate on top of old logs or stumps. Roots grow quickly down into the soil, and after the log or stump decays it leaves the characteristic "stilt-rooted birch." Yellow birch is a poor stump sprouter, and only stumps from young saplings produce vigorous sprouts.

Yellow birch is a relatively fast grower but requires full overhead light to make good growth. Trees respond well to release, and if grown in closed stands, yellow birch trees have good natural pruning of lower limbs. Care should be taken when opening up yellow birch stands to avoid thinning that is too heavy, since birch trees can be damaged by sudden increases in light. Rapid stand opening can also contribute to development of epicormic branching in yellow birch. Yellow birch trees can live to an age of 300 years but probably reach their physiological maturity in about half that time. Yellow birch competes with other northern hardwood species by exploiting forest openings that occur due to disturbances and by growing fast enough to stay above its competitors.

Injurious Agents. Yellow birch, although it possesses several reproductive and growth characteristics that enable it to disseminate and become established on a variety of sites, is vulnerable to a number of damaging agents. It has thin bark and is, therefore, very susceptible to fire. With fine branches and somewhat brittle wood, yellow birch is subject to snow and ice damage. Logging damage can introduce decay fungi, and decay spreads rapidly in birch trees. Birch trees have undergone episodic bouts with so-called *dieback* or *decline* disease. Symptoms begin with small yellowish foliage and dieback of smaller limbs. As the dieback progresses downward, secondary organisms, such as stem borers and root rots, may ultimately kill the tree. These declines are typically associated with stressing factors, such as insect defoliation and drought. The sensitivity of birch to wounding and environmental stress makes it a species that must be managed carefully. Yellow birch is also a preferred browse species for white-tailed deer. Therefore, in areas with high deer populations and in the absence of canopy gaps, yellow birch may be almost completely excluded from regenerating. Yellow birch is listed as resistant to gypsy moth (Liebhold et al. 1995).

5. *Betula lenta* (Sweet Birch)

General. Sweet (or black) birch is a relatively aggressive, but short-lived species that is prone to disease and decay. Trimble (1975) classes sweet birch as slow growing in diameter. For the above reasons, sweet birch has limited commercial potential, but the species is an aggressive invader of forest canopy gaps, and as such it competes with more desirable species. Sweet birch is one of the most common species that regenerates in oak stands following heavy gypsy moth-induced tree mortality in the Ridge and Valley and Appalachian Plateau of Pennsylvania, western Maryland, and West Virginia (Hix et al. 1990).

Distribution and Associates. Sweet birch occurs from southern Maine through New England, New York, and Pennsylvania, down to the southern Appalachians, with isolated populations as far south as northern Alabama. Sweet birch grows on a variety of topographic features and over a wide range of altitudes from near sea level in the north to 4500 ft. in the southern Appalachians (Lamson 1990). Sweet birch grows best on moist, well-drained, protected sites on northern and eastern exposures, although it has a fairly broad tolerance of site conditions. It is found primarily on Spodosols, Inceptisols, and Ultisols. Sweet birch is often associated with species of the northern hardwoods and me-sophytic hardwoods, but, as stated earlier, it appears to be very successful in regenerating under oak stands that have suffered heavy mortality from gypsy moth. Sweet birch occurs in the following provinces of Bailey (1994): 212, M212, 221, and M221 (Fig. 45).

Ecological Role. Sweet birch is somewhat like yellow birch in that it combines attributes typical of conservative and exploitive species. Perhaps it is best described as "opportunistic." It is a prolific seed producer, beginning before age 40. Abundant crops of seed are produced every year to alternate years (Lamson 1990) and are disseminated by wind in the fall, germinating the following spring. Sweet birch seed express little or no ability to remain viable in the litter beyond the first year (Trimble 1975). Stump sprouts can develop from small stumps, but sprouting is not as common in sweet birch as with many of its competitors.

Sweet birch was listed as intermediate in shade tolerance by Trimble (1975) and intolerant by Lamson (1990). Although definitely more shade tolerant than yellow-poplar and black cherry, sweet birch seems to respond positively to light. For example, on the Kane Experimental Forest in Pennsylvania, Allegheny hardwood stands thinned from below, where a continuous canopy was main-tained, developed a sweet birch understory, apparently resulting from the greater amount of diffuse light entering below the "high shade." It is also well known that sweet birch quickly regenerates in canopy gaps.

Growth rate of sweet birch is listed as slow in diameter by Trimble (1975). Although Lamson (1990) indicates that saplings grow rapidly in height on good sites, sweet birch seems to peak out at a fairly young age. In 50-year-old stands, many trees become overtopped, and by age 60–70 mortality is high. A few trees of about 200 years of age have been reported, but, by-and-large, sweet birch seldom lives beyond 120 years.

Underscoring sweet birch's role as an opportunist, it appears to compete very well with other species in the first stages of succession (either in open areas or canopy gaps) but is often overtaken later.

Injurious Agents. Glaze storms, logging, and fire are frequent causes of damage to sweet birch, and damaged trees are highly susceptible to several decay fungi. It is listed as intermediate in susceptibility to gypsy moth (Liebhold et al. 1995) and is intermediate in deer browse preference.

6. *Betula nigra (River Birch)*

General. River birch is a small tree that generally grows in riparian habitats and in bottomlands. It is of limited use for wood products, due to its small size, but has become an important ornamental species and has many of the aesthetic qualities of gray or paper birch without the disease problems.

Distribution and Associates. River birch is distributed widely, from New England to Texas. It is found mostly along streams and rivers throughout the central hardwood region, excluding most of the higher Appalachians and the Nashville Basin and Bluegrass Regions of Tennessee and Kentucky. River birch, although generally found on alluvial soils, is capable of growing on a variety of sites and cannot tolerate extended periods of inundation (Grelen 1990). It is frequently associated with other bottomland species, such as American sycamore, silver maple, black willow, hazel alder, and American hornbeam, but is also sometimes associated with yellow-poplar, red maple, and black cherry in the ecotone between cove hardwoods and riparian zones. River birch is found in Bailey's (1994) provinces 221, M221, 222, M222, 231, and M231 (Fig. 45).

Ecological Role. River birch functions in an exploitive ecological role, although it has a narrow range of site specificity, being generally associated with riparian or floodplain situations. It is a good seed producer, and trees begin bearing seed at an early age (approximately 15–20 years) and produce good crops annually. Seed are light and wind disseminated. River birch seed are mature in late spring/early summer (the only birch not maturing in the fall). Seed require no cold treatment or stratification and germinate immediately. Seedlings grow rapidly but require a moist seedbed. Floodplains that are resculpted and silted after spring floods provide an ideal environment for river birch seedling establishment. No information was available on vegetative regeneration although the abundance of clumped trees indicate that stump sprouting probably occurs. Dosser and Hicks (1975) found that river birch cuttings could be rooted in mist bed conditions.

River birch is classed as shade intolerant, and seedlings and saplings grow fast. However, river birch trees lack a strong apical dominance and, therefore, begin branching at 15–20 ft. above ground. The species may dominate a site for up to 40 years but, in the absence of disturbance, will be overtopped by taller-growing competitors. No information on longevity was found, but it is probably short-lived (under 100 years).

Injurious Agents. River birch can be damaged by extended flooding but has few serious insect or disease pests, which makes it a particularly good ornamental species. River birch is classed as susceptible to gypsy moth defoliation (Liebhold et al. 1995).

7. *Carya cordiformis (Bitternut Hickory)*

General. Bitternut hickory belongs to the "pecan hickory" subgenus, and it is among the widest in distribution of all the hickories, occurring throughout the central hardwood region. Its wood is hard and shock resistant, making it an excellent species for tool handles. Bitternut hickory produces nuts that are valuable as mast for a variety of wildlife, including several species of rodents, but they contain more tannins than most other hickories and are not palatable to humans (Hicks and Stephenson 1978).

Distribution and Associates. Bitternut hickory is widely distributed from southern Ontario to the Florida Panhandle. It occurs from the Coastal Plains of Virginia to the edge of the Great Plains in Kansas. It grows on a wide variety of sites. In the northern part of its range, it can be found on moist mountain slopes to drier sites, whereas it is more likely to be found on moist bottomlands in the southern portion of its range. It is found primarily on Ultisols but is also found on Inceptisols, Mollisols, and Alfisol soils. Bitternut hickory occurs in the provinces (Bailey 1994) 212, M212, 222, M222, 221, M221, 231, and M231 (Fig. 45).

Bitternut hickory seldom grows in pure stands but is a scattered component of many stands (Smith 1990a). It is present with other hickories (shagbark, mockernut) as part of the oak-hickory association. Upland oaks of the central hardwoods commonly associated with bitternut hickory include white oak and northern red oak. On mesic sites, bitternut hickory may be associated with yellow-poplar, maples, green ash, elms, and black cherry.

Ecological Role. Bitternut hickory, like other hickories, employs primarily a conservative strategy. It produces few, but large, seed that are spread primarily by animals and gravity. Optimum seed production age is from 50 to 125 years old (Smith 1990a) with good crops every 3–5 years. Nuts ripen and are dropped from the tree in the fall (September and October), and germination requires stratification; thus, seed overwinter and germinate the following spring. Seed do not remain viable for more than one year (Trimble 1975). Bitternut hickory, like other pecan hickories, is capable of vigorous sprouting and most sprouts arise from the root collar.

Bitternut hickory is classed as intermediate to intolerant of shade but is usually associated with species of a similar tolerance, and seedlings seem to be able to survive in shaded understories for several years. Natural pruning of bitternut hickory occurs readily due to the relative intolerance of the species.

Bitternut hickory is intermediate in growth rate, growing slower in diameter than northern red oak, yellow-poplar, and black cherry on good sites. Hickories grow at a rate similar to beech, chestnut oak, white oak, and sweet birch (Smith 1990a). Bitternut hickory is shortest-lived of the hickories, living to a maximum age of about 200 years, but it sustains a fairly consistent rate of growth over many years, managing to keep up with its competitors on all but the best of sites.

Injurious Agents. Hickory trees are relatively resistant to decay, although fire can damage and kill trees, especially young saplings. There are a number of leaf diseases, as well as defoliating insects, such as leaf rollers and caterpillars in the looper complex, that affect hickories. Bitternut hickory is classed as intermediate in susceptibility to gypsy moth (Liebhold et al. 1995). There are several bark beetles and twig girdlers that affect bitternut hickory. Hicks and Mudrick (1994) reported that a higher-than-normal rate of mortality had been reported for hickories in several West Virginia counties in 1985. It was believed to be a complex "decline" phenomenon involving drought, insects, and disease. Hickories are not listed by Oefinger and Halls (1974) among species preferred by white-tailed deer.

8. *Carya glabra (Pignut Hickory)*

General. Pignut hickory is a widely distributed species in the "true" hickory subgenus. Wood is used for tool handles, fuel wood, and other products, while nuts are an important source of wildlife food.

Distribution and Associates. Pignut hickory is distributed over approximately the same area as bitternut hickory, only slightly more southerly. It occurs mostly as scattered individuals and is found on dry to moist slopes (Smalley 1990). It grows on soils of the Alfisol, Mollisol, and Ultisol orders. Pignut hickory is frequently associated with oaks and other hickories in the broad forest association called the oak-hickory type. Common associates are white oak, chestnut oak, northern red oak, scarlet oak, black oak, and other hickories. Like other hickories, pignut hickory is widely distributed in the central hardwood region, occurring in the following provinces (Bailey 1994): 212, M212, 221, M221, 222, M222, 231, and M231 (Fig. 45).

Ecological Role. Pignut hickory and other hickories are relatively conservative in their ecological strategy. They produce large seed that are disseminated by animals and gravity. Pignut hickory trees may bear seed at 30 years of age, but optimum production occurs after age 70, to about age 200 (Nelson 1965). Good seed crops occur in alternate years with lesser amounts in between. Pignut hickory regenerates readily by root and stump sprouting.

Pignut hickory is classed as intermediate in shade tolerance, although it seems to be more tolerant in the southern portion of its range. This may reflect pignut hickory's tolerance relative to its competitors rather than a difference within the species itself. Pignut hickory is moderately long-lived, living to an age exceeding 200 years and attaining a relatively large size (120 ft. tall and 48 in. in dbh). Pignut hickory grows at a moderate rate, generally able to maintain its position in mixtures with oaks.

Injurious Agents. Fire can cause serious damage to pignut hickory, which has fairly thin bark. Fire scars and basal wounds from logging can introduce decay-causing fungi into pignut hickory trees, resulting in defective wood. Leaf diseases, such as anthracnose and mildews, attack pignut hickory. Droughts apparently induce attacks of hickories by bark borers, and these, in turn, lead to entry of diseases. Such decline syndromes have apparently affected hickories in West Virginia during the 1980s and 1990s (Hicks and Mudrick 1994). Pignut hickory is susceptible to several defoliating insects and is listed by Liebhold et al. (1995) as intermediate in susceptibility to gypsy moth.

9. *Carya ovata (Shagbark Hickory)*

General. Shagbark hickory is a distinctive species of "true" hickory with shaggy bark. The nuts are large and edible and the wood is used for tool handles where high shock resistance is required.

Distribution and Associates. Like bitternut and pignut hickories, shagbark hickory is widely distributed in the eastern United States, including most of the coastal plain in the Atlantic and Eastern Gulf Regions. It is found throughout the central hardwood region and occurs in Bailey's (1994) provinces 212, M212, 222, M222, 231, M231 (Fig. 45). Shagbark hickory is normally an upland species found on good sites but often occurring on bottomlands in the southwestern portion of its range. It grows on soils ranging from Alfisols and Mollisols to Ultisols (Graney 1990). It is part of the broad oak-hickory forest association, occurring with northern red oak, white oak, black oak, and chestnut oak, as well as other hickories.

Ecological Role. Like other hickories, shagbark hickory is a conservative species in its ecological strategy. It has large edible seed that are dispersed by gravity and animals. Trees begin producing seed at about 40 years of age and produce maximum seed crops from age 60–200 years. Good crops occur at intervals from 1 to 3 years. Trees with large open-grown crowns are heavier seed producers than those with smaller crowns. In addition to regenerating by seed, shagbark hickory is a prolific sprouter with best sprouting potential from smaller stumps.

Shagbark hickory is classed as intermediate in shade tolerance, with seedlings and saplings being more tolerant than mature trees. Shagbark hickory, like other hickories, is somewhat more slow growing than some of its competitors, such as yellow-poplar and northern red and scarlet oaks; thus, it is at a competitive disadvantage when growing with these species. On sites where it competes with white and chestnut oaks it maintains its position. Shagbark hickory is long-lived and a species that can be managed on rotations of up to 200 years.

Injurious Agents. Hickories are all susceptible to fire. Many wood-rotting fungi are known to cause decay and rot in hickory species and are frequently introduced by fire or mechanical wounding. Leaf and twig diseases, such as anthracnose, mildew, and viral "bunch" diseases, affect leaves and twigs. Defoliators, including loopers and gypsy moth, will infest hickories, although Liebhold et al. (1995) list shagbark hickory as low in gypsy moth susceptibility. Hicks and Mudrick (1994) noted a decline syndrome of hickories in West Virginia, apparently involving drought, bark beetles, and disease.

10. *Carya tomentosa (Mockernut Hickory)*

General. Mockernut hickory is a "true" hickory that is widely distributed in the eastern United States and generally the most abundant of the hickories (Smith 1990b). Its wood is very hard and shock resistant, and the large nuts are edible by animals as well as humans.

Distribution and Associates. Mockernut hickory is distributed across the entire eastern United States and throughout the central hardwood region, in the following provinces of Bailey (1994): 212, M212, 221, M221, 222, M222 231, and M231 (Fig. 45). Its northern extent is more or less consistent with the southern extent of the Wisconsin Ice Sheet. Mockernut hickory is very abundant through Virginia, the Carolinas, and into northern Florida. It grows largest in the lower Ohio River Basin into Arkansas (Nelson 1959). It is mostly found growing on Ultisols, with lesser amounts found on Inceptisols, Mollisols, and Alfisols (Smith 1990b).

Mockernut hickory, like the other hickories, is a major component of the oak-hickory association and is found growing in proximity to white oak, post oak, blackjack oak, chestnut oak, northern red oak, and southern red oak, as well as other hickories and several species of pines.

Ecological Role. Mockernut, like the other hickories, employs a conservative ecological strategy. It produces large, heavy, and edible seed that occur in good crops every 2 3 years. Seed are dispersed by animals and gravity. Trees begin producing seed after 25 years of age but reach optimal seed-producing age between 40 and 125 years of age. Mockernut hickory, like other true hickories, is capable of vigorous stump sprouting.

Mockernut hickory is classed as tolerant in the seedling/sapling stage and intolerant as a mature tree (Baker 1949; Trimble 1975). Thus, advance regeneration can persist in the understory and responds well to release. Mockernut hickory grows at a moderate rate, slower than competitors, such as northern red oak, but about equal to white oak. Mockernut hickory is fairly long-lived (up to 200 years or more), and it competes with white, chestnut, and southern red oaks and can be carried to very long rotations with these species.

Injurious Agents. Mockernut hickory is very sensitive to fire, which may cause tree mortality or serve as a precursor to the introduction for decay into the bole. Foliage diseases and "witches brooms" may cause damage to mockernut hickory. The species is also susceptible to insect defoliation by loopers and is listed by Liebhold et al. (1995) as intermediate in gypsy moth preference. A variety of bark borers, twig girdlers, and spiral borers attack mockernut hickory, and these attacks seem to intensify during or following droughts (Smith 1990b).

11. *Fagus grandifolia (American Beech)*

General. American beech is a widely distributed species that grows to large size and produces wood that is especially usable in turned and bentwood products. Nuts of beech are eaten by a wide variety of wildlife from wild turkey to white-tailed deer, and hollow trees provide excellent denning cavities.

Distribution and Associates. American beech occurs throughout most of the eastern states and is found through all of the central hardwood region with the exception of the Ozark Plateau, where it only occurs in a few isolated populations. Beech generally occurs on lateritic (Acrorthox) soils and gray-brown podzols (Hapludalfs) in the order Alfisol, Oxisol, and Spodosol (Tubbs and Houston 1990). It is found at low elevations in the north and at high elevations in the southern Appalachians and may grow on alluvial or protected cove soils in the southwestern portion of its range. It occurs in the central hardwood region in the following provinces (Bailey 1994): 212, M212, 221, M221, 222, M222, 231, and M231 (Fig. 45).

Beech is associated with a wide array of species including upland oaks and hickories in the oak-hickory, western mesophytic, and mixed mesophytic regions. Along the northern tier of the central hardwood region and at higher elevations in the Appalachians, beech is commonly associated with eastern white pine, maples, and birches.

Ecological Role. American beech is perhaps the best example of a species that utilizes a conservative ecological strategy. It reproduces by nut-like seed that are moderately heavy and are carried by gravity and animals including birds. Beech is a fairly good seed producer, beginning at about age 40 and producing large quantities by age 60. Grisez (1975) described beech in the Allegheny region of Pennsylvania as being a consistent seed producer with some seed produced almost every year. Tubbs and Houston (1990) indicated that good crops were produced at 2 to 8-year intervals. Beech sprouts well from cut stumps, but sprouting ability is greatest when trees are less than 4 in. dbh (Tubbs and Houston 1990). Root suckering is also common in beech, and for older and larger trees this may be an important source of regeneration.

American beech is classed as very shade tolerant by Trimble (1975), and this appears to be a consensus among foresters who have worked with the species. Beech seedlings, sprouts, and suckers can persist under a completely closed canopy, even under other very tolerant species, such as sugar maple and eastern hemlock. Beech is listed as being slow growing by Trimble (1975), but beech seedlings and saplings that have been suppressed for many years are capable of responding to release. In spite of its relative shade tolerance, beech in closed stands prunes itself well.

Capable of living more than 300 years and often outliving its competitors, beech is one of the longest-lived species in the central hardwood region. Beech probably represents the prototype conservative species with heavy seed, shade tolerance, slow growth, and great longevity.

Injurious Agents. Beech seems to be highly prone to damage from a number of agents. It is highly susceptible to fire and logging damage due to its thin bark, and over 70 species of fungi are known to cause decay in beech. Older beech trees are almost invariably hollow, especially in stands with a history of fire and logging. Beech is listed as intermediate in susceptibility to gypsy moth by Liebhold et al. (1995). Beech is intermediate in deer browse preference. The latest, and potentially most serious, pest to affect beech is the "beech bark disease" complex, which was introduced from Europe to Nova Scotia in 1890. The leading edge of beech bark disease is currently into West Virginia (Hicks and Mudrick 1994) and will probably spread through much of the natural range of the species. This complex is initiated by a scale insect (beech scale), which creates entry ports for the fungus *Nectria coccinea* var. *faginata*. Trees that are infected usually die within a year or two. Symptoms are yellowing and thinning foliage. Bark on infected trees takes on a warty or pustular appearance.

12. *Fraxinus americana (White Ash)*

General. White ash is the most common species of ash in North America, and its tough wood is very usable for handles, baseball bats, and furniture.

Distribution and Associates. White ash is distributed throughout the eastern United States, including the central hardwood region, but is a site-demanding species that requires moist and fertile conditions (Schlesinger 1990). White ash occurs on Alfisol, Spodosol, and Inceptisol soils and does best on moderately well-drained soils with high nitrogen and calcium content. White ashes grow on a variety of topographic features, from the Atlantic Coastal Plain to the mountains of the Appalachian region. White ash occurs in the following provinces of the central hardwood region (Bailey 1994): 212, M212, 221, M221, 222, M222, 231, and M231 (Fig. 45).

White ash, because of its high site requirements, is frequently associated with other species having similar requirements. For example, in the central hardwood region, white ash is often associated with yellow-poplar, black cherry, and northern red oak. It also associates frequently with red maple throughout its range and yellow birch in the northern parts.

Ecological Role. White ash generally functions as an exploitive species, taking advantage of sites that are temporarily enriched due to disturbance. Seed are winged and light in weight and spread up to 460 ft. from the parent tree (Schlesinger 1990). Trees begin seed production at an early age (about 20 years). Grisez (1975) found a correlation between abundant seed crops in white ash and sugar maple, indicating that the two species respond to the same environmental cues. Following seed production over 6 years, Grisez observed one bumper seed crop, one fair seed crop, and one year with no seed production. This suggests a 2 to 3-year periodicity to good seed production in white ash. As is often the case, the heavy seed crop was followed by the year with no seed production. White ash seedlings and saplings sprout prolifically from cut stumps.

White ash is intolerant of shade and a relatively fast-growing species with an excurrent growth form. Even open-grown trees maintain a single stem to a height of 30–40 ft. Seedlings of white ash may take up to 15 years to reach a height of 5 ft., but once the root system is well established, white ash trees can generally outgrow their competitors. White ash is shorter-lived than oaks or maples, similar to black cherry, and longer lived than aspens, sweet birch, and sassafras. Therefore, white ash should not be managed on rotations greatly exceeding 100 years.

Injurious Agents. White ash is susceptible to a syndrome called ash decline, especially north of a latitude of 39°–45°. The disease associated with ash decline (ash yellows) is caused by mycoplasma-like organisms (MLO). The decline appears to be triggered by stresses, such as drought. Seedlings of white ash are preferred by deer and rabbits, and extensive damage can result from browsing. Ashes are considered resistant to defoliation by gypsy moth (Liebhold et al. 1995).

13. *Fraxinus pennsylvanica (Green Ash)*

General. Green ash has the widest distribution of native ash species in North America. It tends to grow on wetter sites than white ash, but the wood of the two species is very similar and used for the same purposes (Kennedy 1990).

Distribution and Associates. Green ash has an extensive geographic range, from southcentral Canada to Texas and from extreme eastern Colorado to the coastal plains of the Atlantic states. Green ash is found throughout the central hardwood region, except for an area of western Pennsylvania into southern New York. Green ash occurs primarily on bottomlands and floodplains of rivers and streams throughout its natural range but will grow when planted on other sites. Like white ash, green ash is very site demanding and is a good indicator species for excellent growing sites. Green ash is tolerant of flooding and can withstand inundation up to 40 percent of the growing season (Kennedy 1990). Soil orders that generally support green ash stands are Inceptisols and Entisols. Green ash can be found in the central hardwood region in Bailey's (1994) provinces: 212, M212, 221, M221, 222, M222, 231, and M231 (Fig. 45).

Green ash is generally associated with other species that are found in bottomlands, such as silver maple, cottonwood, pin oak, sweetgum, black willow, and American sycamore. It may also be associated with species with a wide site tolerance, such as red maple, American elm, and boxelder.

Ecological Role. Green ash, like white ash, is functionally an exploitive species, capable of responding to enriched site conditions. Green ash is a prolific seed producer, and the winged seed are primarily disseminated by wind and water. Seed begin to ripen in the fall and are shed throughout the winter. Seed require a stratification treatment and may also remain dormant in the litter as stored seed for several years. Green ash sprouts vigorously from cut stumps, especially when trees are in the pole-size class or less.

Green ash is classed as shade intolerant to moderately tolerant in the northern part of its range and tolerant to moderately tolerant in the southern part. Open-grown trees grow fast on good sites, averaging more than 2 ft. in height per year for the first 20 years. In the southern portion of its range, green ash develops a spreading root system, which is probably an adaptation to flooding.

Although Kennedy (1990) did not report any information on the longevity of green ash, it is probably similar to white ash with a life span exceeding 100 years, but probably not 200 years. Suppressed seedlings can persist in the understory for up to 15 years and respond well to release.

Injurious Agents. Several defoliators, scale insects, and borers feed on green ash. Green ash is susceptible to ash yellows disease, which could become a serious problem to the species. Green ash is classed as resistant to gypsy moth defoliation (Liebhold et al. 1995) but is highly favored as a browse species for white-tailed deer and rabbits.

14. *Juglans nigra (Black Walnut)*

General. Black walnut is the most valuable of the native hardwoods in North America. Its dark-colored wood is valued for furniture and gun stocks, and the nuts are used in baking as well as providing food for wildlife. However, trees yielding good quality sawtimber are scarce, and not abundant anywhere.

Distribution and Associates. Black walnut is widely distributed throughout the eastern United States and occurs everywhere in the central hardwood region except for the northeastern portion, down through the higher mountains of Pennsylvania and into West Virginia. Black walnut is site demanding and requires deep, well-drained soils of near neutral pH for best growth (Williams 1990b). Black walnut seems to make its best development on alluvial soils and on north and east aspects. It performs particularly well along the well-drained bottomlands of the Ohio and Mississippi Rivers in southern Illinois, Kentucky, and Indiana. The primary soils with which black walnut is associated are Alfisols and Entisols, and it occurs in the following provinces of Bailey (1994): 221, M221, and 222 (Fig. 45).

Black walnut is usually found as scattered individuals throughout the mesophytic forests. It is often associated with other mesophytic species, such as yellow-poplar, northern red oak, white ash, black cherry, maples, boxelder, and American elm.

Ecological Role. Black walnut has attributes of both conservative and exploitive strategies. For example, it produces few large and heavy seed (a conservative trait) but is classed as shade intolerant, which is an exploitive trait.

Black walnut generally reproduces by seed, although young trees are capable of sprouting. Seed are large and heavy and are spread primarily by gravity and animals. Trees can begin seed production as young as 10 years old and continue to yield seed for about 100 years (Williams 1990b).

In shade tolerance, black walnut is classed as intolerant and must maintain a dominant or codominant crown position to survive. Young trees can grow rapidly in height (3–4 ft./yr.) under ideal conditions, which helps trees to maintain a competitive advantage. Black walnut also grows rapidly in diameter, and it is possible to produce sawlogs in 35 years and veneer logs in 50 years on the best sites (Schlesinger and Funk 1977). In spite of its ability to grow rapidly, black walnut trees are moderately long-lived, with a longevity exceeding 150 years.

Black walnut uses a unique method of interacting with its competitors. A toxic substance (juglone) occurs in the leaves, bark, nut husks, and roots that is antagonistic to competing vegetation. This process, called **allelopathy**, enables walnut trees, once established, to control competitive interactions with many other species.

Injurious Agents. Black walnut is susceptible to a number of defoliating insects, such as the walnut caterpillar and fall webworm. It is listed as intermediate in susceptibility to gypsy moth (Liebhold et al. 1995). It is susceptible to *Fusarium* and *Nectria* canker diseases, both of which are capable of causing severe damage to the tree and to its value. Black walnut is also browsed by deer and rabbits, and trees can be damaged severely by sapsuckers.

15. *Juniperus virginiana (Eastern Redcedar)*

General. Eastern redcedar is a small tree that often occurs as an old-field invader on limestone sites. The decay-resistant wood is used as fence posts, and lumber is used in producing closet liners and cedar chests. Redcedar is also important as wildlife food and habitat.

Distribution and Associates. Eastern redcedar is widely distributed throughout the eastern United States and the central hardwood region, with the exception of northcentral Pennsylvania. It is found on a variety of sites, from dry rocky outcrops to swampy areas. Although redcedar can grow on soils ranging in pH from 4.7 to 7.8 (Lawson 1990a), it apparently prefers sites with a high base cation content, such as occurs in soils formed from dolomite or calcareous shales. The most common soil orders for redcedar occurrence are Mollisols and Ultisols. It occurs, within the central hardwood region, in the provinces 221, M221, 222, M222, 231, and M231 (Bailey 1994) (Fig. 45).

Eastern redcedar often occurs as pure stands, developing from invasion of old fields by seedlings. But the species is also commonly associated with shortleaf and loblolly pines, upland oaks, and hickories.

Ecological Role. Eastern redcedar utilizes an exploitive ecological strategy. It is often an early invader of old fields or other open sites. Seed of eastern redcedar are produced in some quantity nearly every year, with good crops occurring at 2- to 3-year intervals. Birds and animals are very important in distributing the seed, which are contained in small, green, berry-like conelets.

Eastern redcedar is intolerant to shade, but trees may exist below hardwood or pine canopies, especially on lower quality sites where the overstory is less dense.

Redcedar is a relatively slow-growing species (about 1 ft. per year in height). Although overtopping by trees of other species often limits the longevity of redcedar, open-grown trees can live for more than 100 years and achieve sizes of up to 120 ft. tall and 48 in. in dbh (Lawson 1990a). As a competitor, redcedar performs well on open sites until it is overtopped by trees of other species. It can maintain a presence in stands on poor sites longer than on good sites. In certain cases where thin and rocky soils occur, especially of limestone origin, pure stands of redcedar may form a more or less permanent cover, such as the "barrens" that occur in the Nashville Basin.

Injurious Agents. Eastern redcedar is particularly susceptible to damage from fire. It has thin bark and resinous foliage that burns readily. Several insects, including defoliators, such as bagworms and thrips, feed on redcedar, and several borers and weevils feed on the inner bark and twigs. Redcedar is considered resistant to gypsy moth (Liebhold et al. 1995). Several root rot diseases and rust fungi also attack redcedar. It is the alternate host for cedar-apple rust, which produces orange fruiting bodies on cedar foliage in the spring. Deer will browse eastern redcedar and can destroy regeneration crops.

16. Liquidambar styraciflua (Sweetgum)

General. Sweetgum is a large, fast-growing tree of bottomlands and uplands of the southeastern states. The wood is used for lumber and veneer, seed are eaten by several species of wildlife, and it is widely planted as an ornamental.

Distribution and Associates. Sweetgum is generally a tree of the southeastern United States, but it occurs naturally in the Atlantic Coastal Plain from southern Connecticut to central Florida, and from eastern Texas to North Carolina. In the central hardwood region, sweetgum is absent in most of the northeastern portion, the Ozark Plateau, and from the Ridge and Valley and Blue Ridge physiographic provinces. It occurs in Bailey's (1994) provinces 221, M221, 231, and M231 (Fig. 45). Sweetgum is much like red maple in its adaptability to a variety of sites. It grows best on the moist alluvial soils of bottomlands (Kormanik 1990) but can be found on sites ranging from swamps to uplands. Sweetgum is often found on soils of the Alfisol order.

Because of its broad site tolerance, sweetgum is associated with a wide array of species, including bottomland hardwoods and southern pines, as well as a variety of upland hardwoods, such as red maple, hickories, river birch, and southern red oak.

Ecological Role. Sweetgum employs an exploitive strategy, generally taking advantage of canopy openings and using its rapid excurrent early growth to stay ahead of its competitors. Sweetgum begins producing seed at about age 20–30 and remains productive until about age 150 (Kormanik 1990). Seed are produced in some quantity every year, with bumper crops occurring about every 3 years. The winged seed are borne in a spherical fruit formed from fusion of up to 60 capsules. Seed capsules open in the fall (Hicks and Reines 1967), and seed are distributed by wind. Germination takes place the following spring, and seedling survival is dependent on good soil moisture conditions. Sweetgum also apparently regenerates vigorously from root sprouts (Hook et al. 1970).

Sweetgum is classed as shade intolerant with seedlings requiring full sunlight to survive. Trees respond well to release when young, but older trees often lose their capacity to respond. Sweetgum is prone to develop epicormic branches when heavy thinning is conducted. The species is relatively fast growing in both height and diameter. Through the sapling and pole stages, sweetgum maintains a strong excurrent growth form, which allows trees to grow up and stay ahead of their competitors. Older trees develop a more spreading crown and almost cease height growth entirely. Sweetgum is moderately long-lived and can survive 200 years or more, but in terms of commercial timber production, growth severely declines after trees reach their full height, and mature trees often become defective. Thus, a rotation length of less than 100 years, 50 years on excellent sites, is optimal.

Sweetgum is a vigorous competitor on good sites where light, moisture, and nutrient resources are abundant. It often seeds into openings and may form pure stands. In the southern pine region, sweetgum is becoming an important species for planting on bottomland sites (Kormanik 1990).

Injurious Agents. Sweetgum has few major insect and disease pests. Perhaps fire during the growing season is the major damaging agent to sweetgum. Basal wounds from fire or logging can serve as entry ports for decay fungi. Browsing by animals is another form of damage that can be locally important. Sweetgum is listed as susceptible to gypsy moth by Liebhold et al. (1995), but the leading edge of gypsy moth spread has only recently entered the range of sweetgum, and it remains to be seen as to how serious the damage will be.

17. *Liriodendron tulipifera (Yellow-Poplar)*

General. Throughout the mesophytic regions, and to a lesser extent in the oak-chestnut region of Braun (1950), yellow-poplar is the most significant individual species. In terms of timber volume, yellow-poplar is the single species with the greatest volume in these regions. Although volume for all oaks combined exceeds it, no single oak species is greater. Yellow-poplar is a fast-growing species generally found on the best sites and produces wood that is usable for a wide variety of products from composites and structural timber to furniture. Seed heads are consumed by wildlife, and the large flowers are an important source of nectar for bees (Beck 1990).

Distribution and Associates. Yellow-poplar is found throughout the United States east of the Mississippi River, except for New York, New England, and southern Florida. It occurs everywhere in the central hardwood region except the Ozark/Ouachita Region of Missouri and Arkansas and occurs in Bailey's (1994) provinces 212, M212, 221, M221, and 222 (Fig. 45). Yellow-poplar seems to be most abundant and grows best on the deep, well-drained sites of the region described by Braun (1950) as the "mixed mesophytic association." Yellow-poplar is very site demanding, and although it grows on a wide variety of soils, it has a decided preference for the best sites. Hicks and Frank (1984) found aspect to be the key site factor controlling the importance of yellow-poplar in northcentral West Virginia, with yellow-poplar predominating on northeast-facing slopes. Most of the soils within its range supporting stands of yellow-poplar are in the orders Inceptisols and Ultisols.

As an important component of the mesophytic hardwoods, yellow-poplar typically associates with some 30 species, including maples, northern red oak, sweetgum, black cherry, white ash, and sweet birch.

Ecological Role. Yellow-poplar is an excellent example of a species that employs an exploitive ecological strategy. It often functions as an invader of disturbed sites, has light, windborne seed with the ability to remain stored in the litter for several years, displays rapid excurrent height growth, and often develops in pure even-aged stands.

Seed of yellow-poplar are light in weight and winged and occur in cone-like aggregates. They ripen in the fall with good seed crops being produced almost every year. Seed can retain their viability in the leaf litter for up to 11 years (Clark and Boyce 1964). Yellow-poplar is a prolific sprouter, and sprouts usually exceed seedlings in growth rate. Sprouts arising from near the root collar are less prone to decay than those arising from higher on the stump.

Yellow-poplar is classed as shade intolerant and needs nearly full sunlight to grow well. As a result of its intolerance, yellow-poplar trees tend to lose their lower branches in closed stands with the live crown occurring in the upper portion of the bole.

Yellow-poplar is a very fast-growing species, especially on the best sites. On good sites, yellow-poplar can sustain a height growth of almost 1.5 ft. per year to an age of 100 years and a diameter growth of about 0.25 in. per year in unthinned stands (Beck and Della-Bianca 1970). Although yellow-poplar is not generally regarded among the truly long-lived species of the central hardwood region, on some sites, yellow-poplar trees can live up to 300 years. The age of natural senescence is usually around 200–250 years, but trees slow in growth long before this age and for maximum wood production should not be managed beyond 100–120 years.

Yellow-poplar is faster growing than almost any of its competitors, except eastern white pine, and is capable of maintaining dominance in the canopy of mixed stands for at least 100 years. However, as previously indicated, yellow-poplar is selective for the best sites, and when growing on lower quality sites, it is shorter-lived and can be overtaken or outgrown by northern red oak. Kolb et al. (1990) demonstrated this in an experiment comparing growth of yellow-poplar and red oak seedlings using limited and adequate environments. Yellow-poplar is a species that lends itself well to intensive culture in pure stands. Trees respond well to thinning by sustaining higher rates of diameter growth and higher yield of high-value products (Beck 1990).

Injurious Agents. Yellow-poplar is one of the least prone of the hardwoods to pests. It is prone to fire damage, although the pure yellow-poplar type is not as likely to burn as oak forests due to the difference in type of litter and moisture conditions between the two types. Root collar borers may attack yellow-poplar, and these injuries plus fire scars and logging damage can serve as entry ports for decay, often resulting from shoestring fungus (*Armillaria*). The yellow-poplar weevil causes episodic damage to foliage of yellow-poplar. In West Virginia, large outbreaks of this insect were reported for 5 of the past 20 years (Hicks and Mudrick 1994). Yellow-poplar is sensitive to summer droughts and responds to extended periods of low rainfall by premature leaf shed. Liebhold et al. (1995) list yellow-poplar as among the most resistant of hardwoods to gypsy moth, and green trees can often be seen in the midst of defoliated oaks during outbreaks. Deer will browse yellow-poplar seedlings, but it is intermediate in browse preference.

18. *Nyssa sylvatica (Blackgum)*

General. Blackgum or black tupelo is a widely distributed species. The upland or typical form (var. *sylvatica*) is found as a scattered tree in hardwood stands. The species provides wildlife food in the form of soft mast, and wood is used as a construction grade material. It has good potential for a landscaping tree, with attractive shiny foliage and bright red fall coloration.

Distribution and Associates. Blackgum grows throughout the central hardwood region, in Bailey's (1994) provinces 212, M212, 221, M221, 222, M222, 231, and M231 (Fig. 45), and can grow on sites ranging from creek bottoms and lower slopes to upper slopes. It makes its best growth on loamy alluvial soils. The species is predominantly found on soils of the Ultisol order (McGee 1990). Because of its wide range and broad site tolerance, blackgum is associated with a wide array of species but is seldom a large component of any stand. It is associated with hickories, oaks, red maple, sweetgum, yellow-poplar, sweet birch, and pines.

Ecological Role. Blackgum appears to be somewhat conservative in its ecological strategy, although the term "opportunistic" might better describe it. Blackgum seed are contained within small black drupes, and seed production is highly variable from year to year. Seed are dispersed by gravity and probably by animals and birds as well. Sprouting of cut stumps also occurs but less so as the stump becomes larger.

Blackgum is classed as tolerant of shade, and it usually occurs in mixtures with species that are capable of overtopping it. Blackgum can live to an age about equal to many of its associates and is capable of responding to release when the overstory is removed. It has a moderate growth rate and is usually overtopped by species such as northern red oak, hickories, and yellow-poplar.

Injurious Agents. Blackgum is susceptible to fire damage, and basal wounds caused by fire provide infection courts for decay. Blackgum is listed as resistant to gypsy moth defoliation (Liebhold et al. 1995).

19. *Pinus echinata (Shortleaf Pine)*

General. Shortleaf pine is one of the four most important commercial southern pines in the southeast (Lawson 1990b). It has a wider geographic distribution than the other three and is used for paper pulp, lumber, and plywood.

Distribution and Associates. Shortleaf pine ranges both east and west of the Mississippi River and is the only pine species that is a significant component of hardwood stands throughout the central hardwood region. It occurs from the Ridge and Valley of central Pennsylvania to southeastern Texas. It is a particularly important timber species in the Ozark/Ouachita region, where it often makes up a significant component of stands. It can be found in Bailey's (1994) provinces 221, M221, 222, M222, 231, and M231 (Fig. 45). Shortleaf pine can grow on a variety of soils and sites. It occurs on sandy soils of the Upper Coastal Plains, where it makes its best growth, and also on eroded clays of the Piedmont as well as on thin, rocky soils of the mountainous regions of Arkansas, Missouri, and the Appalachian Ridge and Valley Province, where it makes its poorest growth. Most of the soils that support shortleaf pine stands are classed as Ultisols.

Because of its broad geographic range and wide tolerance to site, shortleaf pine is associated with a large variety of species. These range from eastern white and Virginia pines to upland oaks, such as southern red oak, white oak, chestnut oak, and black oak. Shortleaf pine is often associated with red maple, sweetgum, hickories, eastern redcedar, and, on some sites, yellow-poplar. In addition, shortleaf pine sometimes grows in pure stands, often as a result of old-field abandonment.

Ecological Role. Shortleaf pine is a good example of a species that employs an exploitive ecological strategy. It is shade intolerant, relatively fast growing with an excellent growth form, capable of responding to good growing seasons by putting on additional flushes of height growth, and has light, windborne seed.

Shortleaf pine is a relatively prolific seed producer, beginning production as early as 10–20 years of age. Seed occur in cones, about 25–40 per cone, and are released by opening of cone scales from about October to November. Winged seed are capable of traveling up to 130 ft. from the parent tree. Seed production occurs almost every year, with good to excellent crops occurring every 3–10 years. In some portions of its range, shortleaf pine naturally hybridizes with loblolly pine, and hybrids are difficult to identify. Shortleaf pine is one of the few pines capable of stump sprouting, but only stumps from young seedling or sapling-size trees will sprout.

Shortleaf pine is shade intolerant. It commonly seeds into open areas with a mineral soil seedbed, and in the absence of weedy competition, seedlings become established and grow rapidly. Using an excurrent growth habit, shortleaf pine seedlings, once established, can usually maintain their height advantage over their competitors. In pure stands, competition leads to expression of dominance and natural thinning. As shortleaf pine trees mature, they develop a straight, clear, naturally pruned bole, and the high shade that develops as the stand matures is favorable to the development of more tolerant species in the understory. Red maple, oaks, and hickories are common in understories of shortleaf pine.

Shortleaf pine is capable of exceeding 100 ft. in height and 40 in. in diameter on good sites and can live for 150–200 years. However, the period of maximum growth occurs from about age 15 to age 60, depending on the site. Shortleaf pine can maintain a position of dominance over hardwoods in the understory for many years, but in the absence of disturbance, such as logging or fire, the longer-lived hardwoods will ultimately take over the site.

Injurious Agents. Shortleaf pine has several serious pests, the worst being southern pine beetle and littleleaf disease. Southern pine beetle is more likely to kill shortleaf pines growing in overcrowded stands (Hicks 1981). Shortleaf pine is ranked as intermediate in susceptibility to gypsy moth (Liebhold et al. 1995). Littleleaf disease is a fungal root disease that is associated with certain site factors, notably soil moisture deficits, poor aeration, low fertility, and nematode damage. It is particularly prevalent on eroded clay uplands. Shortleaf pine is fire-resistant, although very hot burns can kill trees. Shortleaf pine is sensitive to acid deposition (Schier 1987), and droughts can compound the stress loads of trees, leading to attack by some of the organisms discussed above.

20. *Pinus rigida (Pitch Pine)*

General. Pitch pine is a somewhat scrubby pine that often grows on poorer sites. It can be used for lumber and pulpwood where trees reach adequate size. A hybrid between pitch and loblolly pine is being planted widely in West Virginia and shows considerable promise for pulpwood production.

Distribution and Associates. Pitch pine is generally associated with fire-dominated ecosystems from southern Maine, down the Appalachians to North Carolina. It occurs only in the northeastern portion of the central hardwood region, essentially the mixed mesophytic and oak-chestnut associations of Braun (1950). Where it occurs on sandy or gravelly till, from New England into New Jersey, pitch pine often exists in so-called pine barrens. These areas often have poor, droughty soils, although pitch pine is capable of growing on poorly drained sandy and gravelly soils as well. In such areas, a combination of nutrient-poor soils and fire maintain pitch pine, either in pure stands or in mixtures with scrub oaks and other fire-resistant species. In the central hardwood region, pitch pine often grows in steep mountainous terrain and usually occurs on ridgetops or south-facing ridges, usually on thin and rocky soils, and is found in the provinces 212, M212, 221, and M221 (Bailey 1994) (Fig. 45). Pitch pine is most abundant on Spodosols, Alfisols, Entisols, and Ultisols.

Associates of pitch pine are other species that are capable of tolerating poor and adverse site conditions. Examples include bear oak (*Quercus ilicifolia* Wangenh.), chestnut oak, scarlet oak, black oak, hickories, red maple, and Virginia pine.

Ecological Role. Pitch pine is an example of a species that utilizes an exploitive ecological strategy. It has light seed and is intolerant of shade. However, unlike some exploitive species, pitch pine is not a particularly fast-growing species.

Rather, it succeeds by being able to tolerate fire and to thrive on sites that are too harsh for most other species.

Pitch pine begins seed production at a very early age (Little and Garrett 1990), with trees as young as 2 years old reported to produce cones (Andresen 1957). Good seed crops are produced at about 3-year intervals with excellent crops every 4–9 years. Cones may open in the autumn of the year in which they mature, but in areas with a history of wildfire, there is usually a substantial population of trees that have serotinous cones. For these, the cones may remain closed for many years, releasing seed after fire induces them to open. Seed are wind-dispersed and germination usually occurs in the spring. Survival of seedlings is best where litter is removed from the site, as occurs after fire. In addition to regeneration from seed, pitch pine is one of the few conifers that is capable of vigorous sprouting. Trees up to 60 years old are capable of producing basal sprouts after the top has been killed by fire.

Pitch pine is shade intolerant and is unable to grow in the shade of its competitors. It does have an excurrent stem and if given a head start can maintain its canopy above the spreading-formed oaks that typically are associated with it. It is particularly effective at maintaining its position on poor sites. Fire is a key element in maintaining pitch pine, since it helps open the cones, prepares the seedbed, and knocks back the competition.

Pitch pine can live as long as 200 years (Illick and Aughanbaugh 1930), but height growth is relatively slow and declines after age 60, with little additional height growth occurring after age 100. Diameter growth rates of 1 in. in 5 years can be expected on good sites at 20 years of age, dropping in later years.

Injurious Agents. Pitch pine is damaged by deer and rabbits, and ice damage is also a problem with the species. In coastal areas, trees are damaged by salt spray. The most damaging insects are tip moths, several defoliators, and southern pine beetle. Pitch pine is listed as intermediate in susceptibility to gypsy moth (Liebhold et al. 1995). Several diseases, such as needle rusts, twig cankers, and heart rot, also infect pitch pine but usually do not cause widespread damage.

21. *Pinus strobus (Eastern White Pine)*

General. Eastern white pine is a large, fast-growing, "soft pine" that is among the most widely planted in the eastern United States. Its wood is used for lumber, and it is used for Christmas trees and as an ornamental. It has unique wildlife value, where supercanopy trees serve to promote structural diversity of forests and provide nesting and denning sites (Rogers and Lindquist 1992).

Distribution and Associates. Eastern white pine is distributed from the Lake States across southern Canada and down the Appalachians into northern Georgia. In the central hardwood region, it occurs throughout Braun's (1950) oak-chestnut association and through most of the mixed mesophytic association but is generally absent in the western mesophytic and oak-hickory associations. In the central hardwood region, eastern white pine occurs in the following provinces of Bailey (1994): 212, M212, 221, and M221 (Fig. 45). It grows on a variety of topographic situations, ranging from near sea level in New England to higher mountain slopes in the southern Appalachians. The primary soil orders on which eastern white pine grows are Inceptisols, Spodosols, Ultisols, Entisols, and Alfisols (Wendel and Smith 1990). It can be found on soils of igneous or sedimentary origin and in glaciated and unglaciated regions. Eastern white pine can occur in pure stands but is often a component of mixed stands. Common associates in the central hardwood region include eastern hemlock, white oak, chestnut oak, maples, American beech, yellow-poplar, and northern red oak.

Ecological Role. Eastern white pine has characteristics of a species utilizing an exploitive ecological strategy (light seed, rapid excurrent growth, etc.) and other characteristics of more conservative species (intermediate shade tolerance and great longevity).

Eastern white pine trees can produce seed at an early age (5–10 years old), although good seed production usually does not occur before trees are 20 years old or more. Trees produce good seed crops every 3–5 years with few seed being produced in between (Wendel and Smith 1990). Cones mature in the fall, and seed are dispersed primarily by the wind. Germination occurs the following spring, and seedlings are capable of surviving on both disturbed and undisturbed forest litter. Eastern white pine does not regenerate vegetatively under natural conditions.

Once established, young seedlings grow somewhat slowly for the first 5 years or so. By age 10, the rate of growth has increased, and between age 10 and 20, open-grown white pine can grow as much as 54 in. in height in a single year. This growth flush occurs in the early part of the growing season, and the annual height growth is usually completed by July 1 (Rexrode and Carvell 1981). Eastern white pine, unlike the southern yellow pines, puts on a single growth flush each year.

White pine is classed as intermediate in shade tolerance, and advance regeneration can develop and survive under the canopy of species with a low foliage density. Under oaks and maples, advance white pine seedlings seldom reach a position of dominance unless released by disturbance, such as overstory mortality due to gypsy moth. When starting out in even-aged mixed stands, white pine, with its rapid excurrent growth, does well against competitors, such as oaks, that typically start slowly, but in competition with black cherry and yellow-poplar, white pine can be overtopped quickly and may become suppressed and eventually drop out of the stand. In pure even-aged stands, white pines seldom stagnate with the more vigorous trees gaining dominance, which leads to natural thinning. White pine, in spite of being intermediate in shade tolerance, naturally prunes very well in closed stands. Open-grown trees, conversely, have a tendency to retain large lower limbs and carry a deep crown.

Eastern white pine is relatively long-lived. Trees of up to 450 years of age are reported (Wendel and Smith 1990), and trees commonly reach 200 years of age. Dominant trees, once out of the seedling-sapling stage, can sustain growth for a remarkably long time, and white pine trees that gain a position of dominance in mixed stands are usually able to maintain it.

Injurious Agents. Eastern white pine is susceptible to a number of insects and disease. White pine blister rust, white pine weevil, and shoestring fungus (*Armillaria mellea*) can cause serious problems for the species. White pine is listed as intermediate in susceptibility to gypsy moth (Liebhold et al. 1995), but in mixture with oaks, epidemic populations of gypsy moth may feed heavily on white pines. Unlike oaks, white pine is incapable of refoliating after complete defoliation, and the result is usually death (Stephens 1987). White pine seedlings are often heavily browsed by deer, and the species is particularly sensitive to damage from high levels of atmospheric pollution and has been used as an indicator of SO_2 levels.

22. *Pinus virginiana (Virginia Pine)*

General. Virginia pine is a medium-sized yellow pine with wide tolerance for growing conditions. Its appearance is governed, to a larger extent than most species, by its environment. On good sites and in pure even-aged stands it can produce a straight stem of good to fair quality, but on poor sites or in open-grown situations it develops into the typical branchy "field pine." Thus, its uses are dependent on where it grows. Because of its ability to tolerate harsh environments, Virginia pine has been used in mine reclamation throughout the Appalachians.

Distribution and Associates. Virginia pine occurs from Long Island to the hilly sections of northern Alabama (Carter and Snow 1990). It grows throughout the Piedmont and Appalachian provinces but only occurs sparsely west of the Allegheny and Cumberland Plateaus. Virginia pine is not native to the area of the central hardwood region that lies west of the Mississippi River. In the central hardwood region, Virginia pine is distributed in the following provinces of Bailey (1994): 221, M221, and 222 (Fig. 45).

Virginia pine can occur on a wide array of sites and, therefore, is associated with a large number of upland species. These include upland oaks and hickories, eastern redcedar, sweetgum, and red maple. Virginia pine is a vigorous old-field invader, often developing pure stands on these sites.

Ecological Role. Virginia pine utilizes an exploitive ecological strategy with light seed, shade intolerance, rapid early growth, excurrent form, and the capability of multiple growth flushes in a season.

Virginia pine is among the most prolific and reliable seed producers in its range. Trees as young as 5 years of age can produce seed. Some seed are produced almost every year with heavy seed crops at intervals of about 3 years. Seed are light and windborne and can travel at least 100 ft. from the parent tree. Seed germination is greatly enhanced on mineral soil seedbeds, such as occur after logging and fire. Sprouting is uncommon on Virginia pine stumps and is seldom successful in producing a viable plant.

Virginia pine is shade intolerant. When it regenerates in a mixture with hardwoods, such as oaks, hickories, and red maple, the pine may outgrow the competition for 40–50 years, but these more tolerant competitors will eventually dominate the site since Virginia pine slows in growth before these competitors do and single pines do not cast a shade dense enough to suppress the competitors. When in dense pure stands, Virginia pine may dominate the site longer due to the mutual shading of such stands. As stands mature and natural pruning occurs, the higher shade permits entry of understory oaks, hickories, and maples, which will ultimately replace Virginia pine. Due to its intolerance to shade, Virginia pine loses its lower limbs in dense stands, but the resinous branch stubs are slow to shed and produce a knotty wood. Open-grown Virginia pines form branchy, almost shrubby, trees that are often called "field pines" and have little commercial value. Virginia pine is a frequent invader of disturbed sites, such as old fields or cut-over and burned-over land.

Virginia pine growth is fairly rapid when young. On poor sites, it can outgrow shortleaf pine, oaks, and hickories. On good sites species, such as yellow-poplar and sweetgum, can overtop Virginia pine. Longevity of Virginia pine is less than for most of its competitors. Although it can live up to 100 years of age, it usually does not persist much beyond that age. The shallow-rooted characteristic makes Virginia pine especially susceptible to wind throw and snow or ice damage, particularly as trees get older.

Injurious Agents. Virginia pine is susceptible to pitch canker. The most serious insect pest is southern pine beetle. Wind throw is also a serious problem and, for this reason, thinning of mature stands of Virginia pine is not advisable. Liebhold et al. (1995) list Virginia pine as intermediate in susceptibility to gypsy moth. Voles sometimes cause extensive mortality of seedling and pole-size trees.

23. *Platanus occidentalis (American Sycamore)*

General. Sycamore is a large, widely distributed tree that is often found in bottomlands. It has uses for lumber, as a shade tree, and a favored species for short-rotation "biomass" production in the southeastern states (Wells and Schmidtling 1990). A hybrid of American sycamore and *P. orientalis* is used as a street tree in European and North American cities. The hybrid, called London planetree, is thought to be pollution-resistant.

Distribution and Associates. American sycamore is distributed throughout the eastern United States, with the exception of northern New England and the Lake States. It occurs throughout the central hardwood region. Although sycamore is usually found in stream bottoms, it is capable of functioning as a pioneer species on upland sites, such as old fields, or even surface mines. It is found on a variety of soils, most commonly on Entisols, Inceptisols, and Alfisols. Sycamore is tolerant of excessive soil moisture although it makes its best growth on sandy loam soils at the edges of lakes and streams. In the central hardwood region, American sycamore occurs in the following provinces of Bailey (1994): 212, M212, 221, M221, 222, M222, 231, and M231 (Fig. 45).

Sycamore is rarely found in extensive pure stands; therefore, it is often associated with other bottomland and mesic site species, such as sweetgum, cottonwood, green ash, black willow, boxelder, river birch, silver maple, and pin oak.

Ecological Role. Sycamore has characteristics of both exploitive and conservative species. Regarding the former, it has light wind-disseminated seed and fast growth but, on the other hand, is intermediate in shade tolerance and very long-lived.

As a seed producer, plantation-grown sycamore trees are capable of flowering by age 6 or 7 (Wells and Schmidtling 1990) but in dense natural stands begin appreciable production at around age 25. Trees produce good crops almost every year. Ripening in the fall, seed are very light and are shed from their globose heads through the winter. Dissemination is mostly by wind, although water and birds may also contribute to seed dispersal. Sycamore is capable of vigorous stump sprouting, especially from stumps of sapling- or pole-sized trees.

Shade tolerance for sycamore is listed as intermediate, but Wells and Schmidtling (1990) indicate that seedlings require direct light to survive. Sycamore generally is capable of replacing cottonwoods and willows when these species grow together. Due to its great longevity (up to 500 years), sycamore may persist in old-growth forests.

Injurious Agents. Sycamore is fed upon by a number of insects but few cause serious damage. Liebhold et al. (1995) class sycamore as resistant to defoliation by gypsy moth. Diseases, such as anthracnose, commonly attack sycamore. The infestation occurs in the spring, and refoliation takes place in the later summer. Repeated severe attacks may weaken and stress trees, making them susceptible to other organisms. Sycamore is susceptible to damage by late spring frosts, and such damage may be widespread in years with late-occurring frosts.

24. *Prunus serotina (Black Cherry)*

General. Black cherry has a large and varied natural range and produces wood that is in great demand for furniture. The portion of the range producing commercially valuable cherry wood is generally confined to the Allegheny Plateau of Pennsylvania, New York, and West Virginia and scattered stands into the southern Appalachians (Marquis 1990). Elsewhere, cherry trees are smaller and poorly formed and the wood contains resinous deposits called "gumnosis," making them less valuable for lumber.

Distribution and Associates. Black cherry is found throughout the eastern United States and the central hardwood region. Disjunct populations extend down the western mountain ranges from southern Arizona to Central America. In the eastern United States, black cherry grows mostly on fairly acidic mesic sites. Soils of the Inceptisol and Ultisol orders most commonly support black cherry stands, although Alfisols also frequently grow black cherry. Best productivity in the dissected Allegheny Plateau occurs on the north- and east-facing slopes. In the central hardwood region, black cherry is found in the following provinces (Bailey 1994): 212, M212, 221, M221, 222, M222, 231, and M231 (Fig. 45).

Because of its wide geographic range and site tolerance, black cherry is associated with a great array of upland tree species. Some of its most common associates are maples (red and sugar), American beech, black birch, yellow-poplar, northern red oak, eastern hemlock, pin oak, and white oak.

Ecological Role. Black cherry employs an exploitive ecological strategy. It utilizes a buried (stored) seed strategy in addition to animal dispersal of current-year seed and is shade intolerant with a rapid growth rate. Pure stands of cherry may develop in abandoned fields or following forest disturbance.

Seed production in black cherry begins when trees are as young as 10 years of age, although the age of prime seed productivity is between 30 and 100 years of age (Marquis 1990). Some seed are produced almost every year with bumper crops occurring at 3- to 5-year intervals. Seed are dispersed by birds, animals, and gravity. Seed are enclosed in a sweet drupe, which is fed upon by a large variety of birds and animals, from grouse to bears. Seed passing through the digestive systems of such animals are cleaned, and the digestive acid helps scarify the thick seed coat. Seed may lay on the forest floor for 3 years or more (Trimble 1975), creating a seed bank that is capable of responding to stand openings. Black cherry is also capable of vegetative reproduction via stump sprouting, and sprouts grow rapidly.

In shade tolerance, black cherry is classed as intolerant, but seedlings are more tolerant than mature trees. As a result, a standing crop of black cherry regeneration is often present that is in constant turnover. These seedlings are capable of rapid response to canopy gaps. Black cherry, when given adequate light and good site conditions, is capable of very rapid height and diameter growth, especially in the seedling, sapling, and pole stages. In mixed stands, black cherry frequently occupies the dominant canopy position when growing with maples and beech. It can compete with yellow-poplar on average and below-average sites, but not on better sites. The crown geometries of yellow-poplar and black cherry seem complementary, with yellow-poplar having a pyramidal crown when young and cherry having an inverted vase shape. At some stage, the growth rate of black cherry begins to slow, and species, such as yellow-poplar and northern red oak, that sustain their growth longer begin to overtake the cherry and crowd it out. Stand-grown cherry is usually naturally pruned, owing to its shade intolerance.

Black cherry is a species with a relatively long life expectancy. On good sites in the Allegheny Plateau, black cherry can exceed 200 years of age, although its growth rate peaks much earlier. On poorer sites, black cherry may be mature at 100–150 years of age. It is longer-lived than species such as sweet birch and shorter-lived than oaks, maples, and beech.

Injurious Agents. A number of defoliating insects periodically cause damage to black cherry. One of the most noticeable is the eastern tent caterpillar, which builds the conspicuous webs in the crotches of limbs in the spring, and in some years may totally defoliate trees. The cherry scallop shell moth larvae also cause serious damage to black cherry. Liebhold et al. (1995) classified black cherry as intermediate in susceptibility to gypsy moth. Several wood-boring insects attack black cherry and cause damage in the form of gum spots in the wood. Two of these are the Agromyzid cambium miner and the peach bark beetle.

The most significant disease of black cherry is black knot. This fungus attacks stems and twigs, causing a large gall-like swelling that often results in twig breakage. When these lesions occur on the main trunks, they can render the wood valueless. Deer browsing is important in cherry, although cherry is less preferred by deer than maples, oaks, and birches, and high deer populations may actually increase the ratio of cherry regeneration. Cherry is also susceptible to fire damage and is vulnerable to damage resulting from ice or glaze storms.

25. *Quercus alba (White Oak)*

General. White oak is the most common and widely distributed of the North American oaks. It is a true "white" or leucobalanus oak with wood vessels that become plugged after 2–3 years, non-bristle-tipped leaves, and acorns that mature in a single season. White oak is long-lived and capable of growing to large size and producing wood that has many uses, including staves for "tight" barrels, lumber, and veneer. Acorns of white oak are nutritious and are used for food by many wildlife species, including white-tailed deer, black bear, and wild turkey.

Distribution and Associates. White oak is distributed from southern Canada to Florida and from the eastern Great Plains to the Atlantic coast. It is found throughout the central hardwood region and occurs in Bailey's (1994) provinces 212, M212, 221, M221, 222, M222, 231, and M231 (Fig. 45). White oak grows on a wide range of soils, mostly in the orders Alfisols and Ultisols (Rogers 1990), and is productive on all but the poorest of sites. It grows in hilly and mountainous terrain in the central Appalachians to rolling or level terrain in the southeast.

Because of its wide range and great site tolerance, white oak is found with almost all the species found in the central hardwoods region, except for those that are exclusively associated with wet or swampy sites. Some of its more common associates are northern red oak, chestnut oak, post oak, southern red oak, scarlet oak, hickories, red maple, yellow-poplar, eastern white pine, shortleaf pine, American basswood, and sweetgum. White oak can also grow in relatively pure stands, especially on medium- to low-quality sites.

Ecological Role. As with most oaks, white oak utilizes a conservative ecological strategy with heavy seed, intermediate shade tolerance, moderately slow growth, and great longevity. Another attribute of white oak that helps define its ecological role is its relative resistance to ground fires.

Seed production in white oak is very cyclical (Grisez 1975) and relates to spring weather events as well as to growing conditions in the year before flowering. Good seed crops occur on a 4- to 10-year cycle, and seed production can range from essentially none to more than 200,000 per acre. Acorns are attacked by several species of moths and weevils, which reduces the number of viable seed dramatically in some years (Oak 1993). Acorns are also consumed by white-tailed deer, wild turkey, and several species of rodents (Gribko and Hix 1993). Rodents, as well as gravity, are important in distributing seed. Acorns of white oak can germinate as soon as they drop to the ground in the fall. The radical begins growing downward over winter, and the shoot begins growing the following spring. White oak seedlings may become established as advance regeneration. As with most oaks, regeneration of white oak is most difficult on high-quality sites (Smith 1993). White oak, like other oaks, invests much of its energy during its early years into root growth, thus sprouts (both seedling and stump) are important sources of regeneration in white oak. Stump sprouting is much greater for smaller diameter stumps (Johnson 1977).

White oak is classed as intermediate in shade tolerance, being more tolerant when it is young, and is the most shade tolerant of the important oak species in the central hardwood region (Rogers 1990). Although oaks usually have a decurrent growth form with a spreading canopy, when grown under stand conditions white oaks can maintain a single, well-pruned stem up to 75 percent of its total height. In contrast, open-grown trees tend to form very spreading crowns with large and heavy limbs. White oak seedlings that have become flat-topped due to lengthy suppression will not respond well to release (Carvell and Tryon 1961). However, clipping or fire, which induces seedling sprout forma-tion, can actually benefit such suppressed seedlings, causing them to develop a new apical growing point. Sapling-sized white oaks respond well to release and thinning of sprout clumps to a single stem can be beneficial to the remaining sprout (Minkler 1967). White oak is prone to production of epicormic branches when heavily thinned or released. Therefore, releases should be gradual to avoid loss of stem quality that results from epicormic branching (Trimble 1975).

White oak seedlings and saplings are somewhat more slow growing than many of their competitors. For example, white oak will be over-topped when growing in mixed even-age stands by yellow-poplar, black cherry, northern red oak, scarlet oak, and eastern white pine. However, white oak maintains its position in competition with chestnut oak, black oak, post oak, and most hickories. Because of its intermediate shade toler-ance and great longevity, white oak can exist as a midstory species under light shade, such as that produced by shortleaf pine and black cherry, and ultimately prevail as these species die or cease growth. White oak trees can live to great age (greater than 500 years) and can respond to release up to 150 years of age.

Injurious Agents. Several defoliators (loopers and caterpillars) attack white oak. It is among the most preferred hosts of gypsy moth (Liebhold et al. 1995), and defoliation often leads to a loss of vigor and subsequent attack by secondary organisms, such as two-lined chestnut borer and shoestring fungus. Wood borers, such as the oak timberworm, often attack through wounds caused by logging, frost crack, lightning, or wind and can cause serious damage to the wood. A vascular disease, oak wilt, is a potentially damaging disease of white oak (Hepting 1971). Deer browsing is relatively heavy on white oak, and high deer populations can naturally eliminate seedlings from the understory.

26. *Quercus coccinea (Scarlet Oak)*

General. Scarlet oak is a medium- to large-sized tree, often of poor quality, that is usually found on the lower-quality sites. It is a true "red" oak (eryth-robalanus), with open wood pores, bristle-tipped leaves, and acorns that mature in 2 years. Wood of scarlet oak, although frequently of lower quality, is useful for any application that requires red oak, and it provides food and habitat for wildlife.

Distribution and Associates. Scarlet oak occurs throughout the portion of the central hardwood region east of the Mississippi River and west of the Mississippi in southeastern Missouri. Using Bailey's (1994) provinces, scarlet oak occurs in 221, M221, and 222 in the central hardwood region (Fig. 45). It is generally found in mixtures with other oaks and hickories on the poorer sites, although nearly pure stands occur in the Ozark Plateau of southeastern Missouri (Johnson 1990). Where scarlet oak does occur on better sites, it is capable of faster growth than most other oaks and hickories. The primary soil orders that support scarlet oak are Alfisols, Inceptisols, and Ultisols. Scarlet oak grows on a variety of topographic situations, from rolling hills to mountain slopes up to 5000 ft. elevation in the southern Appalachians (Johnson 1990). The primary associates are chestnut oak, black oak, white oak, northern red oak, hickories, Virginia pine, pitch pine, and shortleaf pine.

Ecological Role. Scarlet oak employs elements of both conservative and exploitive ecological strategies. For example, a conservative trait is its large seed, while exploitive traits include relative shade intolerance, fast growth, and a shorter life span than most other oaks.

Scarlet oak begins seed production at an early age (about 20). Scarlet oak acorn crops are generally lighter than other oaks (Beck 1977), and seed are dispersed by a variety of rodents as well as by gravity. Scarlet oak is a more prolific stump sprouter than most other oaks and capable of producing viable sprouts on larger and older stumps.

Scarlet oak is ranked as the most shade intolerant of the upland oaks. Johnson (1990) describes it as "very shade intolerant." Thus, scarlet oak almost always occurs in even-age stands following disturbance and cannot regenerate in the understory of closed stands. In spite of its shade intolerance, scarlet oak is a poor natural pruner. Lower limbs die but decay slowly, persisting for many years as stubs. Scarlet oak is a rapid-growing species, which enables it to stay ahead of its competitors. However, it tends to peak in growth earlier than other oaks and is the first oak species to be eliminated in mixed stands.

Injurious Agents. Scarlet oak is prone to heart rot and decay, which usually enters through basal wounds or branch stubs. Decay is especially prevalent in stands of sprout origin where sprouts arise from a high position on the stump. Scarlet oak is susceptible to a number of defoliating caterpillars, including gypsy moth. Liebhold et al. (1995) rate scarlet oak as susceptible to gypsy moth. Combined stress resulting from defoliation and drought can lead to attack by secondary organisms, such as two-lined chestnut borer and shoestring fungus, ultimately leading to "decline" and death. Scarlet oak is also susceptible to oak wilt. Although scarlet oak often invades after fire, it is also one of the most sensitive to fire damage as a mature tree. Scarlet oak, like other oaks, is a preferred deer browse species.

27. *Quercus falcata (Southern Red Oak)*

General. Southern red oak is a medium-sized upland red oak that is used extensively for lumber and as an ornamental. It is also an important source of mast for wildlife.

Distribution and Associates. Southern red oak is distributed throughout the southeastern states from southern New Jersey to northern Florida and from East Texas to the Carolina coast. In the southern portion of its range, generally coinciding with the Coastal Plains, an important variety (var. *pagodaefolia*) known as cherrybark oak occurs in bottomland situations. In the central hardwood region, southern red oak occurs through much of the Ozark/ Ouachita Region, most of the western mesophytic association of Braun, and into the mixed mesophytic association in eastern Kentucky and East Tennessee. In Bailey's (1994) provinces in the central hardwood region it is found in 221, 222, M222, 231, and M231 (Fig. 45). It is generally absent in the northern and higher Appalachian portions of the central hardwood region.

Southern red oak usually occurs in mixed stands associated with a number of upland species, including white oak, post oak, black oak, scarlet oak, chestnut oak, and hickories. It is also frequently associated with red maple, Virginia pine, shortleaf pine, eastern redcedar, and sweetgum.

Southern red oak grows most often on soils in the Ultisol and Alfisol soil orders. It is usually found on drier slopes, although it is occasionally found on alluvial lower slopes, where it makes its best growth.

Ecological Role. Southern red oak employs a conservative ecological strategy with intermediate shade tolerance and heavy seed. But its ability to adapt to dry and poor sites gives it a competitive advantage on these sites. Like other oaks, southern red oak invests much of its energy in developing a large root system.

Southern red oak trees begin acorn production at about 25 years of age and achieve maximum production between the age of 50 and 75 years (Belanger 1990). Some seed are produced almost every year. Dispersion of acorns is primarily by gravity and animals (bluejays and rodents). Southern red oak is capable of vigorous stump sprouting, especially if stumps are less than 10 in. in diameter.

Southern red oak is classed as intermediate to intolerant in shade tolerance, and advance regeneration can develop in stands, providing adequate light is available. Natural pruning is usually good in stand-grown trees, but rapid release of crop trees will result in profuse development of epicormic shoots. Southern red oak is a moderately fast-growing species. On below-average sites, it can outgrow most of its competitors. On poor sites, it may be outpaced by species such as Virginia pine or scarlet oak, and on good sites it cannot keep pace with yellow-poplar or sweetgum. Southern red oak is capable of long life (greater than 200 years) and can reach large size on good sites. But due to its frequent occurrence on poorer sites, Belanger (1990) reports that the species usually produces a medium-sized tree (80 ft. in height and 24–36 in. in dbh).

Injurious Agents. Southern red oak is susceptible to fire damage due to its relatively thin bark, and wounding from fire or logging can lead to heart rot. Southern red oak is susceptible to several defoliators (orangestriped oakworm, loopers, etc.), and Liebhold et al. (1995) rate it as susceptible to gypsy moth. Defoliation and drought can cause stress in southern red oak trees that ultimately leads to decline. Secondary organisms, such as two-lined chestnut borer and root diseases, become more important as trees decline, and these organisms are often direct mortality agents.

28. *Quercus prinus (Chestnut Oak)*

General. Chestnut oak is a medium-sized tree in the white oak group that is frequently found on poor quality upland sites. Its wood is darker in color than most other white oaks, but it is used for flooring, lumber, and specialty products and often sold as white oak (McQuilkin 1990). Chestnut oak is a valuable wildlife species, both for its acorns and as a den tree.

Distribution and Associates. Chestnut oak occurs throughout the eastern Piedmont, Appalachians, and their plateaus. In the central hardwood region, it occurs through the western mesophytic and oak-chestnut associations, as defined by Braun (1950). Chestnut oak does not occur west of the Mississippi River. Occurrence of chestnut oak in Bailey's (1994) provinces within the central hardwood region is in 212, M212, 221, M221, and 222 (Fig. 45).

Chestnut oak is capable of growing on a variety of sites, but it is most often found on poor uplands, although it makes its best growth on the alluvial soils of benches and coves (Ike and Huppuch 1968). Chestnut oak grows most commonly on soils of the orders Ultisols and Inceptisols.

Associates of chestnut oak are most frequently those species that inhabit the dry uplands. For example, chestnut oak often occurs with upland hickories, scarlet oak, black oak, white oak, northern red oak, red maple, blackgum, Virginia pine, and pitch pine. On bench areas and transitional sites, such as small headwater drainages, chestnut oak associates with species more typical of good sites, for example, yellow-poplar, sweet birch, sweetgum, and black walnut.

Ecological Role. Chestnut oak is primarily a species that utilizes a conservative ecological strategy with heavy seed and slow growth. However, it is somewhat of a specialist, being tolerant of harsh site conditions, thereby avoiding competition with many species.

As a seed producer, chestnut oak produces relatively heavy crops every 4–5 years; good years are usually followed by years of low production. As with other species in the white oak group, chestnut oak acorns mature in a single season and drop to the ground earlier than those of other oaks. Acorns begin to germinate immediately, and the radical grows into the soil over the winter. The epicotyl emerges from the acorn the following spring. Because chestnut oak trees usually have a relatively small live crown, acorn production per tree is generally lower than for white oak or northern red oak of the same age. But chestnut oak acorns are preferred by many species of wildlife due to their relatively large size, early maturation, and sweet kernel. As a result, animals, as well as gravity, play an important role in dispersal. In addition to seed, chestnut oak is a vigorous sprouter. Both stump and seedling sprouts develop. Stump sprouting is good, even from stumps of 60-year-old trees (Wendel 1975). Because chestnut oak trees have a well-developed root system, stump sprouts and seedling sprouts are often very competitive in second-growth stands.

Chestnut oak is classed as intermediate in shade tolerance (Trimble 1975) and is similar in tolerance to white oak (McQuilkin 1990). Natural pruning in chestnut oak produces a relatively clean bole and a small live crown when trees are grown in closed stands. Open-grown trees retain lower branches but are less likely to produce a spreading crown form than white oak grown in open conditions. In competition with other species, chestnut oak can maintain its canopy position with most hickories, white oak, and black oak, especially on poorer sites. On these sites, scarlet oak and northern red oak can overtop it. On good sites, chestnut oak will quickly be overtopped by yellow-poplar, northern red oak, red maple, and black cherry, which usually leads to suppression and death of chestnut oak on such sites.

Chestnut oak is somewhat slow growing but relatively long-lived as long as it can maintain a dominant position in the canopy. It is capable of living to about 200 years of age, but its growth rate peaks at an age of less than 200 years.

Injurious Agents. Chestnut oak is susceptible to a variety of defoliators, including cankerworm and loopers. It is listed as susceptible to gypsy moth by Liebhold et al. (1995). Chestnut oak, when subjected to defoliation stress, is among the most vulnerable of its associated species to mortality (Crow and Hicks 1990). The combined stresses of defoliation, drought, and suppression of chestnut oak can lead to a decline syndrome involving two-lined chestnut borer and shoestring fungus. Chestnut oak is susceptible to fire damage and subsequent decay although its thick bark can insulate large trees from damage, and chestnut oak regeneration, like that of several other oaks, seems to be selectively favored by fire. Chestnut oak seedlings are heavily browsed by white-tailed deer, particularly in areas with high deer density.

29. *Quercus rubra* (Northern Red Oak)

General. Northern red oak is a wide-ranging member of the true red oak group with a broad site tolerance. It is one of the fastest growing oaks native to North America. Northern red oak wood is used widely for lumber and veneer, and it is a valuable wildlife species as well. It is widely planted and managed for timber in France.

Distribution and Associates. Northern red oak is found from southern Canada to southern Alabama and from the Atlantic coast to the edge of the Great Plains. It occurs throughout the central hardwood region in Bailey's (1994) provinces 212, M212, 221, M221, 222, M222, 231, and M231 (Fig. 45), but its importance as a timber species is greatest in the Appalachian region from central Pennsylvania to North Carolina. Northern red oak can be found in a variety of topographic and elevational situations. It is found above 5000 ft. elevation in the southern Appalachian mountains to near sea level along the Atlantic seaboard (Sander 1990a). It grows on a variety of soils including Alfisols, Inceptisols, Mollisols, Ultisols, and Entisols.

Although northern red oak is capable of growing on a wide variety of sites, it grows best on the deep, well-drained soils of concave slopes and on north or northeastern aspects. On the ridges, upper slopes, or southern and western exposures, northern red oak usually grows in mixtures with other oaks and hickories, including chestnut oak, white oak, scarlet oak, and black oak. Red maple, blackgum, Virginia pine, and pitch pine are also among the associates of northern red oak on these lower-quality sites. On high-quality lower slopes, benches, and northeastern exposures, northern red oak becomes a component of the mesophytic hardwood mixture, including yellow-poplar, maples, and basswood. Eastern white pine and hemlock can be associated with northern red oak on a variety of sites.

Ecological Role. Northern red oak utilizes both conservative and exploitive strategies in its ecological role. It produces large, heavy seed and is somewhat shade tolerant, but is capable of stump sprouting and has a moderately fast rate of growth.

Northern red oak acorns require 2 years for flowering to maturity. Trees begin to produce acorns at age 25, although good production does not occur until about age 50 or when trees are at least 14 in. dbh (Auchmoody et al. 1993). Good acorn crops occur at intervals of two or more years with bumper crops (greater than 250,000 acorns per acre) being produced about every 5–7 years. Acorns are heavily preyed upon by insects and vertebrates. Up to 35 percent may be lost to insects in certain years (Beck 1977). Animals, including small rodents and white-tailed deer, may consume up to almost 40 percent of northern red oak acorns (Gribko and Hix 1993; Steiner 1995). In spite of their roles as acorn predators, animals (especially rodents) are important vectors of distribution for red oak acorns. As is the case with all oaks, acorn production and even seed germination are not the most limiting problems to successful regeneration. Oaks rely heavily on advance regeneration, but in closed canopy stands, seedlings seldom maintain adequate vigor to enable them to compete satisfactorily when the overstory is removed (Sander and Clark 1971). Thus, to ensure an adequate stocking of vigorous seedlings, it appears that removal of the overstory must be done in stages. Northern red oak sprouts readily from cut stumps, with smaller stumps being more vigorous in sprout production. Seedling sprouts are an important source of regeneration, and seedlings, when damaged by fire or breakage, sprout readily and grow fast, producing a well-formed stem. The fact that oaks invest heavily in their root systems gives such sprouts an advantage over competing species.

Northern red oak is considered intermediate in shade tolerance. It is less tolerant than maples, beech, hemlock, or white pine, but more tolerant than yellow-poplar, black cherry, or ashes. Among the oaks, it is less tolerant than white or chestnut oak and about the same as the other red oaks, with the possible exception of scarlet oak, which is frequently listed as the least tolerant oak (Trimble 1975). Northern red oak responds well to release if the trees are at or above the high intermediate crown class (Graney 1987). Northern red oak trees grown in fully stocked stands usually develop straight, well-pruned boles but develop into limby trees with spreading crowns if open-grown. Northern red oak is listed by Trimble (1975) as prone to develop epicormic branches when released, but trees that have few side branches before release are less likely to develop epicormics after release.

Northern red oak is capable of excellent growth on good sites. It cannot compete in early height growth with white pine, yellow-poplar, or black cherry. However, with black cherry, which produces a light shade, red oak may overtake it since the cherry is a fast starter but red oak tends to sustain its growth rate longer. Red oak generally cannot overtake white pine, but it can compete favorably with yellow-poplar on the poorer sites. Red oak outgrows all other oaks in height with the possible exception of scarlet oak. Dominant and codominant northern red oak are capable of diameter growth rates of up to 0.4 in. per year (Trimble 1969).

Northern red oak is a long-lived species and can be managed on relatively long rotations. It sustains a good rate of growth up to 100 years of age and is capable of living longer than 200 years. Red oak is longer-lived than black cherry or yellow-poplar but shorter-lived than white oak.

Injurious Agents. Northern red oak, like other oaks, is susceptible to a number of insects and diseases. Defoliators, such as cankerworm, loopers, and gypsy moth, are capable of completely defoliating red oak trees. When this occurs repeatedly or in conjunction with a drought, a decline syndrome that involves secondary organisms, such as the two-lined chestnut borer and shoestring fungus, is capable of causing heavy mortality in stands containing northern red oak. Liebhold et al. (1995) list northern red oak as susceptible to gypsy moth, although Campbell and Sloan (1977) classify northern red oak as less susceptible than white and chestnut oaks.

Northern red oak is susceptible to damage by fire, which can kill young seedlings and initiate decay in older trees. The time of year of burning and the temperature of the burn are critical in determining how severe the damage will be. McGee et al. (1995) recommended spring fires for prescribed burning in oak stands.

Northern red oak is very susceptible to oak wilt, which is a vascular wilt disease that spreads through root grafts, producing groups of affected trees. Northern red oak is also a preferred deer browse species, and in areas with large deer herds, such as the Allegheny Plateau of Pennsylvania, deer sometimes completely eliminate regeneration of red oak from the understory.

30. *Quercus stellata (Post Oak)*

General. Post oak is a member of the white oak group. It is a relatively slow-growing oak typical of dry woodlands in the southeast. Its wood has the same uses as white oak and is reputed to be more resistant to decay than other oaks (Stransky 1990). Post oak, because of its poorer form and slower growth, is less important for timber production than white oak, but it is a very valuable species for wildlife habitat.

Distribution and Associates. Post oak is generally regarded as a southern oak, but it is found through most of the central hardwood region, with the exception of the northeastern portion. Using Bailey's (1994) provinces, post oak occurs in the following parts of the central hardwood region: 221, M221, 222, M222, 231, and M231 (Fig. 45). Post oak usually grows on poor sites. It can exist on rocky and exposed ridges or on sandy and excessively drained soils. It is most common on soils belonging to the orders Alfisols and Ultisols. Post oak is associated with other species, such as black oak, chestnut oak, and scarlet oak, that can tolerate harsh sites. It is also associated with blackjack oak, pitch pine, shortleaf pine, white oak, northern red oak, blackgum, sourwood, and occasionally with white pine and hemlock at higher elevations.

Ecological Role. Post oak utilizes a mixture of conservative and exploitive ecological strategies, and due to its adaptation to poor sites it competes in a niche where the site has excluded many would-be competitors.

Post oak begins seed production at about age 25 and produces good crops of seed about every 2–3 years. Post oak acorns, like other white oaks, germinate immediately after falling to the ground, and the radical grows into the soil over winter. Primary vectors of dissemination are gravity and animals. Post oak trees up to 10 in. dbh are capable of sprouting but produce fewer sprouts per stump than black oak, chestnut oak, white oak, or scarlet oak (Stransky 1990).

Post oak is intolerant of shade, but due to its ability to survive on harsh sites, it competes well with its associates. Post oak generally grows slower than most of its associated trees. It can persist in the understory of shortleaf pine stands, especially in stands with low stocking and after trees have grown tall and produce a high shade. But generally post oak is forced to compete on the very dry sites where its deep taproot enables it to outcompete its associates.

Injurious Agents. Post oak is susceptible to injury from insects and diseases that affect other oaks. It is classed as susceptible to gypsy moth by Liebhold et al. (1995). It is susceptible to oak wilt but less so than red oaks. Deer will browse post oak and can be a serious factor in reducing post oak abundance, particularly in view of the species' inherently slow rate of growth. Post oak is also sensitive to damage from air pollutants, such as sulfur dioxide, fluoride, and ammonia (Stransky 1990).

31. *Quercus velutina (Black Oak)*

General. Black oak is a true red oak, often occurring on lower-quality uplands. It produces wood of similar quality to northern red oak, and therefore it is useful for timber production. It is also a valuable wildlife species.

Distribution and Associates. Black oak is widely distributed throughout the eastern United States, except for New England. It occurs throughout the central hardwood region with the exception of the Allegheny Plateau in northcentral Pennsylvania. In Bailey's (1994) provinces within the central hardwood region, black oak can be found in 212, M212, 221, M221, 222, M222, 231, and M231 (Fig. 45). Black oak grows in steep mountainous terrain, as well as in level and rolling land forms, but is more abundant on lower-quality sites (Sander 1990b). However, black oak makes its best development on well-drained loamy soils of coves and benches (Hannah 1968). Black oak is found on the soil orders Alfisols, Ultisols, Entisols, and Inceptisols.

Black oak is typically found in mixture with other oaks, such as white oak, southern red oak, scarlet oak, chestnut oak, and northern red oak. It is also associated with hickories, shortleaf pine, and Virginia pine, and less frequently with yellow-poplar.

Ecological Role. Black oak employs primarily a conservative ecological strategy. It produces heavy seed, grows somewhat slowly, and lives a long time.

As a seed producer, black oak begins acorn production at about age 20, but optimum production begins at about age 40 and continues to age 75 (Sander 1990b). Good crops of acorns occur about every 2–3 years but are highly variable. Acorns require two full years from development of flower primordia to ripening, and they drop to the ground in the fall and sprout the following spring. Animals consume large numbers of acorns as do insects, but rodents are an important vector of spread for black oak.

Black oak is very dependent on advance regeneration, from either seedlings, seedling sprouts, or stump sprouts. Thus, sources of advance regeneration must be established during the rotation if regeneration is to be successful. Black oak is a relatively prolific stump sprout producer, but less so than red, scarlet, or chestnut oaks (Sander 1977).

Black oak is listed as intermediate in shade tolerance; thus, seedlings require some canopy opening in order to develop into viable advance regeneration (Sander 1972). Black oak produces a clean, well-pruned bole when grown in fully stocked stands, especially on higher-quality sites. On poorer sites or in more open-grown situations, black oak is often more limby and of poorer form than northern red oak. Black oak trees respond well to release if they are codominant or above in crown class but are prone to epicormic shoots.

Black oak grows at about the same rate as white oak and chestnut oak but slower than northern red or scarlet oak. It is somewhat shorter-lived than most other oaks, seldom reaching more than 200 years of age. It is physiologically mature at about 100 years of age (Sander 1990b).

Injurious Agents. Black oak is susceptible to the same agents as northern red oak. It is classed as susceptible to gypsy moth by Liebhold et al. (1995). It is also susceptible to damage from fire and deer browsing.

32. *Robinia pseudoacacia (Black Locust)*

General. Black locust is a relatively fast-growing and short-lived species that is capable of colonizing harsh sites. It is a legume with nodulated roots, which enables it to fix atmospheric nitrogen and incorporate it into the soil. Therefore, black locust has been used extensively in reclamation of disturbed sites. The wood of black locust is very hard, and its heartwood is resistant to decay, making it useful for making fence posts. Susceptibility to insects and disease are limiting factors to the culture of black locust.

Distribution and Associates. Black locust is naturally distributed through the Appalachians (from central Pennsylvania to northern Alabama) and the Ozark/Ouachita Regions, but it has been widely planted throughout the eastern United States and has escaped cultivation and become established throughout the eastern states (Huntley 1990). Thus, black locust is widely distributed throughout most of the central hardwood region. It occurs within the region in the following provinces (Bailey 1994): 221, M221, M222, 231, and M231 (Fig. 45).

Black locust grows best on rich, moist limestone soils but will grow well on a wide variety of soils. It is most common on soils of the orders Inceptisols, Ultisols, and Alfisols (Huntley 1990). It can tolerate a wide pH range (4.6–8.2) but is intolerant of poor drainage or poor aeration, such as occurs in heavy clay soils with a high water table.

Black locust is associated with a broad array of species because of its wide geographic distribution and site tolerance. It is particularly abundant in pioneer communities but occurs as scattered individuals in mature mixed mesophytic forest communities (Braun 1950).

Ecological Role. Black locust employs an exploitive strategy. It is a prolific seed producer, beginning about age 6, with good crops at 1 to 2-year intervals. Seed are enclosed in legumes that ripen in the fall with seed being dispersed from September to April. Gravity and animals are important vectors of seed dispersion for black locust. Black locust is a prolific stump sprout producer, and sprouts grow rapidly from the root collar of cut stumps.

Black locust is very shade intolerant, and seedlings require full sun to maintain good growth. In mixed stands on good sites, black locust seedlings and sprouts can outgrow most of their competitors (with the exception of yellow-poplar) up through the sapling stage. During the pole to small sawtimber stage, black locust is frequently overtaken by competing species (oaks, red maple, sweetgum, ash, black cherry, etc.), and as soon as crown closure occurs they begin to lose vigor, ultimately becoming suppressed and dropping out of the stand. Black locust is a relatively short-lived species, seldom exceeding 50 years of age in natural stands. Ornamental black locust trees can live as long as 100 years when maintained free of competition.

Injurious Agents. Black locust is susceptible to several insect and disease organisms that severely limit its management. The locust borer is a wood-boring insect that damages the wood and provides entry ports for diseases, such as heart rot fungi (*Phellinus* or *Polyporus*). The locust leaf miner is a beetle that excavates the soft leaf tissue, leaving the leaf veins. Leaf miner outbreaks occur almost annually and weaken trees, which may ultimately lead to mortality due to secondary organisms. Twig borers often cause heavy damage to black locust, particularly young seedlings. Black locust is ranked as resistant to gypsy moth by Liebhold et al. (1995). Black locust is highly susceptible to fire and is also a preferred species for deer browse.

33. *Tilia americana (American Basswood)*

General. American basswood is a large, relatively long-lived tree that occurs in mixed stands. Like white basswood, American basswood produces lightweight wood that is dimensionally stable and useful for a variety of purposes. Basswoods make excellent ornamentals and also produce wood that is popular for carving.

Distribution and Associates. American basswood occurs from southern Canada across most of the northeastern quarter of the United States. In the East, it extends from the Appalachians into Tennessee and North Carolina but is not generally found in the Piedmont or Coastal Plains. In the western portion of its range, it borders the Great Plains and extends down into northern Arkansas. In the central hardwood region, American basswood occurs in the following provinces of Bailey (1994): 212, M212, 221, M221, and 222 (Fig. 45).

American basswood generally occurs as a component of mixed stands. In the northern part of its range, it may be part of the northern hardwood community, in mixture with American beech, maples, eastern hemlock, and yellow birch. In the Appalachians, American basswood is more often a component of the mixed mesophytic hardwood community, occurring with yellow-poplar, maples, and northern red oak. In the southern portion of its range, American basswood may be replaced by white basswood (*Tilia heterophylla*), a species that is very similar in appearance, use, and ecological role.

Ecological Role. American basswood is a species that utilizes a conservative ecological strategy. It is a prolific seed producer with good seed crops being produced over 50 percent of the time (Godman and Mattson 1976). The nutlike drupes are borne on a leafy bract that aids in wind dissemination. Seed are also spread by gravity and animals. Although many seed are produced each year, the proportion of sound seed is quite low in basswood (Ashby 1962). Basswood is a prolific sprouter, and a high proportion of basswood in second-growth central hardwood forests is apparently of sprout origin, by virtue of the abundance of multiple-stemmed trees.

American basswood is shade tolerant, ranking as more tolerant than oaks, birches, or yellow-poplar and less tolerant than sugar maple or eastern hemlock. It is relatively fast-growing and can maintain a dominant canopy position over most of its competitors, the exception being yellow-poplar. Basswoods of sprout origin are particularly fast-growing. Basswood is relatively long-lived and is capable of growing to large size (dbh to 48 in., height to 130 ft.). The maximum longevity is about the same as yellow-poplar or about 200 years (Crow 1990), however, growth slows dramatically after 100 years of age.

Injurious Agents. Several insects attack basswoods, including stem borers and defoliators, but few cause severe problems. The linden looper and gypsy moth can cause serious damage to basswood, but due to the fact that basswoods often grow in diverse mixed stands, they are seldom exposed to outbreak-level defoliation episodes. Liebhold et al. (1995) rate American basswood as susceptible to gypsy moth. Basswood is very susceptible to fire, which causes tree death due to cambial damage. It is also a preferred species for deer browsing.

34. *Tsuga canadensis (Eastern Hemlock)*

General. Eastern hemlock is a slow-growing, long-lived conifer that is a very shade tolerant climax species in many situations. It ranges throughout the northern and higher elevation portions of the central hardwood region east to the Mississippi River. Hemlock wood is usable for pulpwood, but trees are prone to ring shake (or separation of the grain), which limits its usefulness as lumber. Hemlock is also a good landscaping ornamental species.

Distribution and Associates. Eastern hemlock grows from Maine to the Lake States and along the Appalachians into northern Georgia. It does not occur west of the Mississippi River. In the central hardwood region, it occurs in the following provinces of Bailey (1994): 212, M212, 221, and M221 (Fig. 45). Although eastern hemlock is generally found on moist well-drained soils, it can grow on soils from a wide array of parent materials and chemical composition. It grows from sea level in Maine to 5000 ft. in North Carolina and Tennessee. Hemlocks can be found on soils in the orders Spodosols and Alfisols. Leaf litter under hemlock tends to build up due to the slow rate of decomposition of hemlock needles (Godman and Lancaster 1990). Hemlock usually grows in mixed stands and is associated with members of the northern hardwoods and mixed mesophytic associations. Typical associates include sugar maple, red maple, yellow birch, American beech, northern red oak, yellow-poplar, black cherry, and eastern white pine.

Ecological Role. Eastern hemlock employs a conservative ecological strategy. It is long-lived, shade tolerant, and slow-growing. However, reproductive characteristics are not completely consistent with the conservative strategy. For example, hemlock trees begin seed production by about age 35. Trees in full sun are more prolific than those in shade. Good seed crops are produced more than 50 percent of the time and small winged seeds are widely disseminated in the wind. Trees 450 years in age are capable of producing good crops of seed. Seed are shed in the fall and germinate the following spring after stratification. Eastern hemlock does not reproduce by sprouting or root suckers; thus, advance seedling regeneration is the predominant strategy.

Eastern hemlock is very shade tolerant and capable of surviving in a closed canopy for many years. Almost all old-growth hemlock trees show evidence in their growth rings of episodes of suppression. The growth rate of eastern hemlock is generally slow; therefore, most of the competing tree species are capable of overtopping hemlocks. But the shade tolerance and longevity of hemlock enable it to outlive its competitors. Hemlock trees of over 300 years of age are not uncommon and the species is reputed to be capable of living 800 years or more (Godman and Lancaster 1990).

Injurious Agents. Eastern hemlock is attacked by several species of insects, including borers and defoliators. An introduced pest, hemlock wooly adelgid, is capable of causing mortality and severe damage in hemlock. Hemlock is classed as intermediate in susceptibility to gypsy moth (Liebhold et al. 1995), and defoliations often result in tree death (Stephens 1988). Large hemlock trees are relatively resistant to light fires, but seedlings are readily killed. Hotter fires may cause cambial damage, and resulting wounds are entry ports for heart rotting fungi. Deer readily browse hemlock seedlings, but it is not their preferred food (Anderson and Loucks 1979).

SILVICAL CHARACTERISTICS OF CENTRAL HARDWOODS, SUMMARY

Literally interpreted, the term silvical characteristics means **tree characteristics**, but silvics is defined by Ford-Robertson (1983) as "the study of the life history and general characteristics of forest trees and stands, with particular reference to locality factors, as a basis for the practice of silviculture." Thus, silvical characteristics are the inherent characteristics of trees that determine their ecological role. In other words, silvical characteristics define the ecological requirements of a species, determine if individuals of a species can tolerate excesses or deficiencies in resources of a given site, and determine how efficient members of a species might be in utilizing the resources of the site. Furthermore, silvical characteristics determine how trees function in stands where they compete with other trees in varying mixtures of species on different sites, at a variety of stocking densities, and in varying age structures. These functional relationships form the basis for what is called forest stand dynamics (Oliver and Larson 1996).

In this chapter, silvical characteristics of 34 species, which are important in the central hardwood region, are provided. Table 10 is a summary of some of these characteristics for the 34 species. This summary may be helpful when looking for potential interactions and incompatibilities that may occur among the various species.

Chapters 3, 4, and 5 of this book provide the biological rationale for the management of central hardwoods. Chapter 3 on ecology is the foundation, which discusses the principles governing functional relationships between organisms and their environment and among different organisms. Chapter 4 discusses the characteristics of the important species that determine how they function (site specificity, regeneration strategy, shade tolerance, growth rate, longevity, and injurious agents). As indicated by Ford-Robertson's definition, knowledge of ecology, silvics, and silvical characteristics is essential to the application of silviculture. Thus, Chapter 5 is an application of these principles and facts to the manipulation of central hardwood stands in order to produce the desired results.

5

Silviculture of Central Hardwoods

INTRODUCTION

The greatest challenge to foresters is to blend biological and socioeconomic principles in a way that facilitates the production of useful products from the forest while ensuring the sustainability of the system. Hawley and Smith (1954) describe forestry as "an applied science which may be likened to an arch resting on a foundation of fundamental sciences." They state that "**silviculture** is the keystone of the arch." Smith (1986) further states: "The immediate foundation of silviculture in the natural sciences is **silvics**, which deals with the principles underlying the growth and development of single trees and of the forest as a biological unit." Nyland (1996) defines silviculture as "the science and art of growing and tending forest crops." He also states: "Silviculturists draw upon the principles and theory amassed through scientific inquiry in botany, zoology, soil science, physical sciences, ecology, silvics, managerial science, economics and quantitative methods."

There are certain parallels between silviculture and agriculture, but there are also distinct differences. Silviculture is to forestry what agriculture is to farming. However, agriculture, as practiced in Western cultures, involves a very intensive approach. In crop production, sites are manipulated to achieve maximum yield per unit area, often involving use of monocultures, genetic manipulation, competition control, and pest control. Only four species of plant crops comprise approximately 67 percent of the world's agricultural crop production (Evans 1980). Silviculture, unlike agriculture, is a discipline that is less intensive and involves manipulating natural processes by directing them toward the desired end. This is particularly true in the central hardwood region due to the constraints of topography, landownership, and the prevailing economy. In addition, forests in the central hardwood region are diverse mixtures of species that regenerate naturally, although not always predictably.

In the central hardwood region agricultural-style monocultures are problematic for numerous reasons. First, the topography is frequently steep, making it difficult to operate equipment for site preparation. Soil stoniness is also a problem for equipment operation, impeding site preparation and mechanized planting. Deer browsing is also a serious problem for planted seedlings, and due to the vigor and fecundity of natural vegetative growth (woody and herbaceous) it almost always outgrows planted seedlings, since it requires at least a year for planted trees to recover from planting shock and begin rapid height extension. Plantations established in old agricultural fields may avoid some of these problems, but such sites are becoming rare, since most abandoned agricultural fields have already regrown to native tree species. Finally, with the possible exception of black walnut on high-quality sites, the economic incentives are scant for establishing plantation-style monocultures in the central hardwood region. Most such monocultures use fast-growing conifer species, which are more plantable and generally more available through forest nurseries. But due to the small conifer content of natural forests in the central hardwood region, there are few markets for these species. Many of the conifer plantings that have been established in the central hardwood region were done as a result of government subsidies or mandates. Examples are the CCC plantations of the late 1930s and early 1940s; the plantings established on cropland in the 1960s through the USDA, Conservation Reserve Program; more recent plantings on farms subsidized through the Forest Incentives and Stewardship Incentives Programs; and plantings mandated for reclamation of disturbed sites (surface mining, etc.).

Thus, the silvicultural approach that is appropriate to most central hardwood forests is one that emulates natural processes, including natural regeneration and intermediate management that involves the manipulation of growing space to favor the desired trees. In addition, several factors, both historic and current, have a bearing on selecting the appropriate silvicultural strategy. Most of the current stands in the central hardwood region are between 50 and 90 years of age (Seymour et al. 1986) and resulted from either commercial logging or regrowth from agricultural abandonment. These stands are even-aged (commercially logged stands may contain residuals) with a mixture of species, and fire has affected many of them. A significant number of central hardwood stands have been selectively logged or high-graded at least once. Others have been damaged by pests, wind storms, or ice and snow. Many of the present-day forests are owned by private individuals in small tracts and, until recently, have had little financial value and therefore have received virtually no management.

The combination of a maturing resource, strong demand for hardwood products, and an emerging industry to utilize hardwoods for alternative products has created an environment with many opportunities to apply silviculture and management to central hardwood stands. In previous chapters, the description of the central hardwood region, discussion of historical development of stands, and discussions of ecological relationships and silvical characteristics of species have provided a background for prescribing the culture and management of existing stands. The present chapter, dealing with silviculture, begins with intermediate management (since many central hardwood stands are in need of such treatments) and progresses to a discussion of appropriate regeneration systems.

SILVICULTURAL TREATMENTS

Silviculture is broadly subdivided into treatments that are used to "tend" existing stands (**intermediate management**) and treatments that are designed to regenerate new stands (**regeneration systems**). The **stand** is the basic unit of the forest to which a silvicultural treatment is applied. A stand is a group of trees with similar age structure, species composition, site quality, and condition. Although the casual observer sees a green mantle of trees over much of the central hardwood region, due to factors such as aspect, slope position, geology, past land use practices, patchy forest fires, and past cutting practices, the forests of the central hardwood region actually consist of a complex mosaic of stands with a high degree of internal diversity. A typical 80-year-old hardwood stand in the Appalachian Plateau contains 100–200 trees per acre, over 3 in. dbh, consisting of 15–20 different species. Although these stands are generally even-aged (single cohort), they have a rotated-S-shaped diameter distribution (number of trees vs. dbh) more reminiscent of the theoretical distribution for all-age stands (Fig. 93).

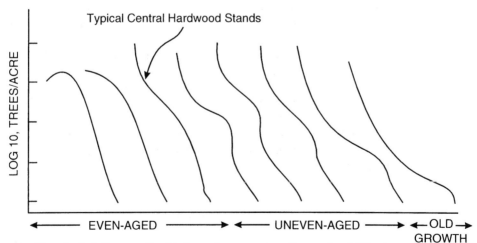

FIGURE 93   Hypothetical diameter distributions of trees per acre for stands of different age structures (from Goff and West 1975).

This is a function of the mixture of faster-growing species, for example, yellow-poplar, with slower-growing species, such as white and chestnut oaks, and the presence of shade tolerant species, such as the maples, that can form a midstory in combination with more intolerant species, like black cherry, in the overstory. Because of the life histories (growth rate, longevity, etc.) of the various species in these complex mixtures and other factors, such as timber market conditions and forest pest conditions, the appropriate cultural treatments are themselves complex. Many, if not most, central hardwood stands are not ready for a final harvest at 60–80 years of age. Thus, some intermediate treatment (or treatments) is appropriate. But it is also true that when these treatments involve canopy manipulation, they will have an effect on the understory and, therefore, will have an impact on future regeneration. It is realistic to assume that small private forest landowners will not likely apply intermediate management to their stands unless it produces a commercial product. But it is also important that these landowners involve a forester in conducting such management and avoid a strict "timber sale" mentality, where the cutting only involves harvesting the existing stand without regard to planning for future regeneration.

In the absence of severe problems, such as overpopulation by deer, some kind of natural regeneration is almost inevitable in any given stand. The challenge is to build toward the final harvest or regeneration cut by looking at regeneration as a *process* rather than an *event*. Many poorly managed stands in the central hardwood region, either through repeated high-grading, fires, animal damage, or pest attacks, have become **impoverished stands**. Carvell (1987) pointed out that selective removal has reduced the abundance of preferred species, such as black walnut, white oak, and black cherry, in many stands while increasing the frequency of less desirable species, such as scarlet oak, blackgum, and sweet birch. He further states that repeated high-grading and wildfire have produced low-quality sprout-origin stands with little potential for future management.

Although the practice of high-grading (cutting good and leaving undesirable trees) is the most blatant form of mismanagement, straight diameter-limit cutting, with no silvicultural considerations, can also lead to stand impoverishment. For example, in stands of mixed-species composition, valuable species such as oaks and yellow-poplar are often mixed with less valuable species such as red maple and hickories. Depending on the site, the more valuable species usually grow faster in diameter than the less desirable species. Thus straight diameter-limit cutting removes proportionately more of the faster-growing species, such as northern red oak and yellow-poplar, leaving proportionately more red maples and other lower-quality species.

Intermediate Management

Intermediate management is often described as "tending" the forest. Although tending can involve practices that improve the site (fertilization, etc.) or pest management (pesticides, etc.), it often involves cutting some trees to reallocate resources to the residual stand. In a typical developmental scenario for an even-age, single-cohort central hardwood stand, a large number of seedlings become established initially. Plants begin to interact competitively as soon as their roots or crowns come into contact. Competition occurs for resources (light, water, oxygen, carbon dioxide, minerals) that are limiting. The type of competition that occurs among plants is generally indirect in nature. That is, when one plant takes in or utilizes a resource, it is no longer available to be used by another plant. Competition that occurs among different species often reflects the adaptive strategies of the species. For example, two species may be growing on the same site and competing for a limited water resource. One species that possesses a shallow, fibrous root system will intercept water from summer showers before it percolates into the soil, while another species with a deep, unbranched root system can extract water from a deeper water table that would be unavailable to species with shallow root systems. Although less common, plants also engage in direct competition such as occurs when one plant produces a toxin that kills or retards the growth of another. This process is termed **allelopathy**, and chemicals that function in this way are called **allelochemicals**.

The resources that are subject to competition can be obtained through the root system (soil resources) or the crown (above ground). Soil resources generally include minerals, water, and oxygen, while above-ground resources include light, carbon dioxide, and oxygen.

Trees have various morphological and physiological adaptations that enable them to compete more or less efficiently in a given environment. For example, some species, such as yellow-poplar and sweetgum, which have an excurrent growth form when they are young, are capable of rapid upward extension to get above their competitors to obtain direct sunlight. Others, such as maples and American beech, have a low light compensation point and, therefore, are more shade tolerant and can persist in the shade of faster-growing species, awaiting their opportunity to dominate the site. In mixed, unmanaged stands, competition leads to a natural thinning process that occurs as a result of the various ecological strategies of the species in the stand and their interactions with each other and the particular conditions of the site. Trees with poor commercial potential consume site resources in the same way that valuable ones do. Thus, as foresters and managers, we have the opportunity to direct this process toward particular goals. For example, trees with crooked or knotty stems may not suffer a competitive disadvantage as a result of these conditions, but they can be removed in a silvicultural operation to favor trees with greater commercial potential.

Intermediate cuttings can be grouped into those aimed at controlling stand density (**thinnings**), those designed to improve stand quality (**release operations, improvement cuttings**), and those designed to cope with pest problems (**sanitation** and **salvage cuttings**) (Smith 1986). In addition to the above treatments, which tend to focus on the condition of the stand, **crop-tree management** is a type of intermediate management that focuses on individual trees of potentially high value to the landowner. In most cases, these operations direct the allocation of above- and below-ground resources by making more resources available to the residual trees. Sometimes silviculturists refer to this in the context of "growing space." For example, Gingrich (1967) has developed **stocking guides** for upland hardwoods in the central states. These guides are based on the concept that there is a theoretical "full stocking" condition that exists for a given stand, considering its age, species composition, and site quality. Stocking can be expressed in terms of number of stems per unit area, basal area per unit area, or percent of full stocking. The natural tendency of forest stands is to maintain full stocking. Thus, cuttings conducted in the stand that reduce it below full stocking will result in unused resources that become available to the residual trees, enabling them to expand their crowns and grow faster. Sometimes when cuttings are heavy enough or when shade tolerant species are present, some of these resources will be redirected to understory trees, trees that already exist, or new regeneration (Kirkham and Carvell 1980). The resources can also be utilized by the residual trees in the main canopy, since these trees already possess a competitive advantage due to their crown position. It is important when deciding which trees to leave in an intermediate cutting to leave those that are *capable of responding* to **release**. This may be a function of the tree's species, age, condition, site, and so on and generally relates to its silvical characteristics.

Response of a tree to release is usually a process that begins gradually and gains momentum until full stocking is achieved, at which time growth slows again. The mechanism for this process begins with increased availability of resources (light, water, minerals) to the residual trees. But before they can take full advantage of these resources the residual trees must expand their root systems and crowns to occupy the vacated space. Thus, the initial phase involves the expansion of surfaces (leaves, roots). Once expanded, these new tissues can fully exploit the additional resources to enhance the physiological functions of the tree, therefore initiating more rapid growth. The slowing of growth that follows is a result of reestablishment of competition as crowns and root systems of adjoining trees fill the unoccupied space.

Response of any given tree to release depends on several factors. First, younger trees are generally more responsive to release than older ones. This relates, in part, to the process of senescence where gradual changes take place at the cellular level, making trees less vigorous. Another factor in the aging process for large perennial woody plants relates to the fact that as they age the amount of living tissue (cambium, apical meristems, ray cells, etc.) increases faster than the area of leaf tissue is capable of increasing. Thus, the ratio of living tissue that must be maintained to the amount of photosynthetic tissue tends to increase with age. At some point, the requirement for maintenance of respiration approaches or exceeds the capability of the tree to produce food. When this happens, any environmental stress (drought, defoliation, fire, etc.) can lead to declining vigor and ultimately death (Manion 1981). The aging process occurs at different rates for different species, and, therefore, the chronological age for a given tree does not necessarily equate to its physiological age.

Response to release is also a function of the general health and vigor of a tree, which, in turn, is related to factors such as site quality, stocking density, crown position, fire history, climatic stresses (drought, frost, ice), and insect and disease conditions. Thus, when choosing trees to carry in the residual stand, trees that are immature, thrifty, vigorous, and healthy should be selected. In forest trees, there are few quantitative measures for overall health and vigor of trees. At the stand level, there are quantitative measures of relative stocking and site quality, but when choosing individual trees to leave in stands, much subjectivity is involved. Criteria, such as species or crown position, are straightforward, but there is still a large degree of subjectivity in evaluating tree vigor, and there is no substitute for experience in assessing it. Characteristics of low-vigor trees in central hardwood stands include upper crown dieback, presence of basal wounds or scars, deeply furrowed or thick bark, narrow and/or shallow crowns, presence of fungal fruiting bodies and presence of sucker growth on the bole or larger branches. Characteristics of vigorous trees are generally those with well-developed crowns, dark green foliage, thin bark, and absence of wounds, insects, diseases, and dieback. Again, many of these characteristics are variable, depending on the particular silvical characteristics of the species, and the ability of the forester to assess tree vigor improves with experience.

Thinning. Thinnings are cuttings made in immature stands to control stand density and redirect the resources of the site to the residual trees (Beck 1986). Thinning is based on the concept that the trees in the residual stand will respond to the additional resources by putting on more growth. But because there is a period of time immediately after thinning during which the residual trees have not fully utilized the site, gross initial biomass production in thinned stands is less than gross production of unthinned stands (Fig. 94). But as Lamson (1985) demonstrated, thinning can dramatically increase diameter growth of residual trees and can result in greater production of merchantable products and lower stand mortality rates. The primary objectives of thinning in central hardwoods are therefore to:

1. Increase the rate of diameter growth of the residuals so that the stand can be harvested at an earlier age.
2. Concentrate growth on more desirable trees in order to produce a greater volume of higher quality and value products.
3. Obtain income during the rotation.
4. Salvage losses that would occur as a result of impending suppression and normal mortality.

To better understand these objectives, a few basic concepts need to be

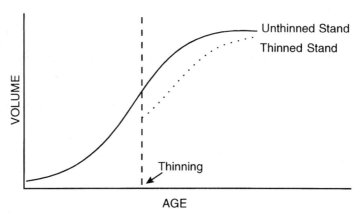

FIGURE 94   Graph of gross production of wood over time for thinned and unthinned stands.

discussed. Although trees compete with each other above and below ground, the condition and position of the crown are often useful indicators of competition and vigor. To facilitate monitoring of forest and tree health, several researchers have developed and applied **visual crown rating** (VCR) systems (Belanger et al. 1991; Kelley et al. 1992). These ratings are based on assessments of crown density, depth, foliage color, and amount of dieback. Like bark characteristics, these ratings are useful as subjective measures when applied by an experienced evaluator. Another crown characteristic that foresters have traditionally used to assess the severity of competition among trees is **live crown ratio** or the percent of the tree's bole occupied by live branches. Lower **live crown ratios** reflect a greater degree of competition. It should also be noted that shade tolerant species have inherently deeper crowns than intolerant species; thus, live crown ratio is only useful as a relative measure and then only for a given species. Foresters also use the relative position of tree crowns in a stand to indicate the degree of competition experienced by individual trees. This is often called **crown class**. The following four crown classes are generally recognized (Fig. 95):

1. *Dominant*: Tree crowns receive direct light from above and the sides; trees are clearly emergent above the general canopy.
2. *Codominant*: Trees receive direct light from above and some direct light from the sides; the crowns of codominant trees generally form the main canopy level.
3. *Intermediate*: Intermediate trees receive direct light from the top only; their crowns are usually narrow and slightly below the main canopy level.

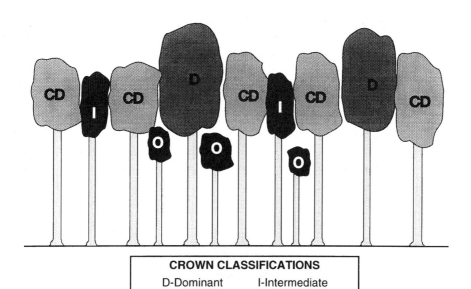

CROWN CLASSIFICATIONS	
D-Dominant	I-Intermediate
CD-Codominant	O-Overtopped

FIGURE 95   Diagram of crown classes in a mixed hardwood stand.

4. *Overtopped*: These trees receive only indirect sunlight and their crowns are clearly below the main canopy.

Canopy positions of trees develop in a predictable way over time in single-cohort stands, which describes most central hardwood stands. The stands usually begin with thousands of seedlings and/or sprouts per acre, and through a process of **natural thinning**, the result, after 7–10 years, is a stand with one to a few thousand stems per acre. Once a given tree establishes a competitive advantage over its neighbors (either due to vigor or position) it begins to deprive the competitors of resources. The disadvantaged trees lose canopy position and, given time, may die. Mixed-species stands, such as those that generally occur in the central hardwood region, undergo a more complex scenario due to the different silvical characteristics of the species involved. For example, shade intolerant species, such as sassafras, black locust, and sweet birch, may grow faster than their competitors in the early years but will likely be overtopped and die as the stand progresses. Species such as red maple may not be able to keep pace with faster-growing species (yellow-poplar, northern red oak, etc.) but, due to their shade tolerance, persist in a midstory position, often with intermediate or overtopped crowns. Crown position is a critical consideration in management of mixed stands, where cuttings, such as diameter-limit cuts, can inadvertently result in alterations of species diversity and regeneration potential (McGill et al. 1995). Much of the rationale for thinning is based on crown characteristics as indicators of competition and growth or quality potential of residual trees.

Another concept that is important relative to thinning is **stocking**. Stocking refers to the number of trees or basal area present relative to the optimum a site can carry. Roach (1977) and Gingrich (1967) have developed stocking guides for upland hardwoods in the central hardwood region, and these can graphically be displayed as overstocked, fully stocked, or understocked (Fig. 96). The lines separating these categories, shown in Figure 96, are referred to as the A, B, and C lines, respectively. Thinning would be recommended for stands above the A

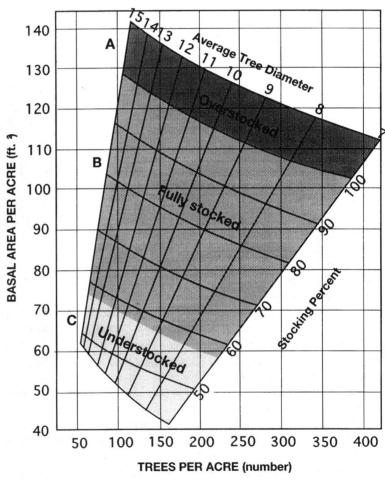

FIGURE 96   Hardwood stocking guide (from Roach and Gingrich 1968).

line, and Sander (1977) recommends thinning to 60 percent of the original basal area.

The relationship of tree characteristics such as diameter and natural pruning to competition is also important relative to thinning. Although it has been demonstrated many times that, over a broad range of stand densities, the height growth of trees is independent of stocking density and more a function of site quality, diameter growth is another matter. When trees are grown in closely competitive stands, their diameters are much smaller than when grown in less dense stands, although total biomass productivity of the stands may be similar for both situations, providing the stands have fully closed canopies. In addition, trees grown in dense stands tend to lose their lower branches, which die and fall off at a point in the canopy below which the light compensation point is reached. This response is called **natural pruning**, and it is more dramatic in shade intolerant species than in tolerant species. Thus, thinning can be used to modify the morphology of trees such that trees in heavily thinned stands will grow larger in diameter but will retain their branches lower in the canopy. In fact, very heavy thinning can result in the production of epicormic branches developing on the bole from dormant buds (Miller 1996).

Thinning is generally more effective when done at a young age (Hilt 1979). But trees of 60 years of age and older are capable of responding to thinning. Lamson (1985) documented a positive response to thinning for 60-year-old black cherry and maples. Hilt (1979) found that upland oaks responded to thinning regardless of age, although younger stands were more responsive. Yellow-poplar responds to thinning, up to age 70 (Beck and Della-Bianca 1981).

There are three thinning methods that are appropriate for use in central hardwoods. These are low thinnings, crown thinnings, and selection thinnings. Smith (1986) describes these methods in detail. These techniques were developed in Europe and seem most applicable to stands of pure species. When working in mixed-species stands of central hardwoods, such techniques are useful to illustrate concepts, but, in reality, intermediate cuttings applied to mixed-species stands are often a blend of more than one method.

Low thinning or thinning from below is a method that involves removal of trees from lower canopy positions (overtopped, intermediate, and sometimes codominant). The primary objective of a low thinning is to salvage anticipated mortality losses. This technique is most applicable to pure, even-aged stands of relatively shade intolerant species and would remove trees that have become, or are becoming, overtopped. Such trees are presumably destined to eventually succumb to shading, and their removal simply salvages the anticipated loss. This technique seems most applicable to fully stocked or overstocked stands of species, such as sweetgum and yellow-poplar (Beck and Della-Bianca 1972). Black cherry may also be thinned using this method, but it has proved to be less responsive to release than some other species, especially as it matures (Philips and Ward 1971). A disincentive to low thinning is the lack of markets for small-diameter products. An exception to this is the decorative fence-rail market that has developed in recent years for yellow-poplar. Otherwise, such low thinnings, where hardwood pulpwood or firewood is the only product produced, would have little commercial potential. Since low thinnings result in little release of the overstory, the benefit to the residuals is limited; therefore, if non-commercial, such thinnings would be difficult to justify.

Marquis et al. (1984) and Smith and Lamson (1986a) discuss a variation of this technique applied to sapling stands called "thinning with basal area control," where most trees are removed from the intermediate or weak codominant positions to favor good quality codominant or dominant trees. This type of low thinning equates to a "Grade C" as defined by Nyland (1996) but without the removal of overtopped trees (Fig. 97). Such thinnings are recommended for fully stocked stands, and a residual basal area of one-half to

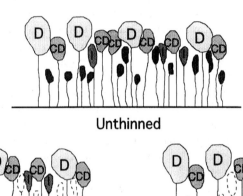

Unthinned

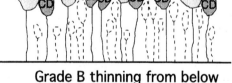

Grade A thinning from below
(Only cut overtopped trees)

Grade B thinning from below
(Cut overtopped and intermediates)

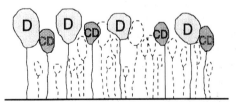

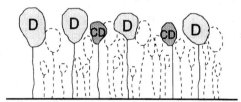

Grade C thinning from below
(Cut overtopped intermediates,
and some codominantes)

Grade D thinning from below
(Cut overtopped, intermediates,
and most codominantes)

FIGURE 97   Illustration of different levels of low thinning (from Nyland 1996).

two-thirds of the original basal area is suggested. This type of thinning may be a marginal commercial operation, especially when applied to younger stands with dbh values averaging less than 12 in. But even when applied to a 60-year-old cherry-maple stand in West Virginia, this thinning method resulted in increased diameter and volume growth of residual dominant and codominant trees (Lamson 1985).

A second type of thinning applicable to central hardwood stands is **crown thinning**. Crown thinning involves removal of trees from the middle and upper strata of the canopy to favor desirable trees in the same canopy range (Smith 1986). Figure 98 illustrates a stand before and after crown thinning. This type of thinning allows for selectivity in favoring trees that may show potential for production of higher quality or grade. The removal of trees in the codominant

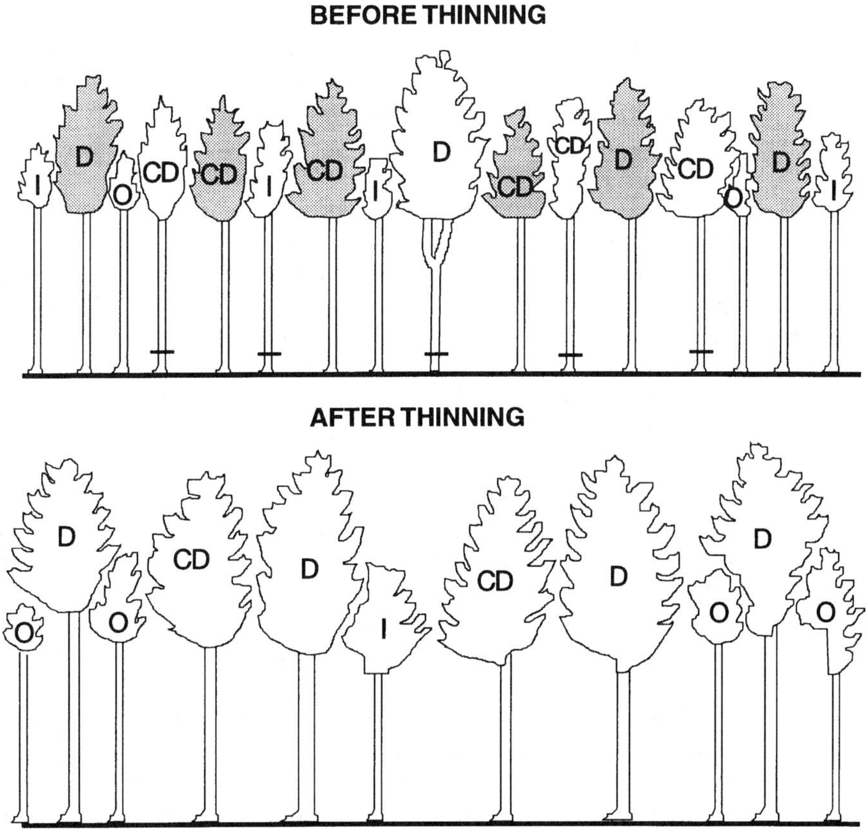

FIGURE 98 Illustration of a stand before and after crown thinning. Potential crop trees are shaded and trees removed are marked with a horizontal bar (from D. M. Smith, *The Practice of Silviculture*, 1986; reprinted by permission of John Wiley & Sons, Inc.).

and dominant crown classes creates growing space and additional resources for residual trees with greater potential. The primary difference between low and crown thinning is that crown thinning focuses on the upper strata of the canopy, and following such a thinning, almost all the intermediate and overtopped trees will be retained. These residual trees are presumably not in direct competition with the final crop trees.

Crown thinning is most appropriate in even-aged mixed oak or mixed mesophytic hardwood stands. In these stands, the dominant and codominant layer is often composed of oaks (northern red oak, white oak, etc.) or yellow-poplar, black cherry, ashes, cucumber magnolia, or basswood. All these species are capable of responding to release up to age 50 and beyond, when growing on average to above-average sites. On below-average sites, release will not result in as dramatic a response, and this is especially true for older trees. As Nyland (1996) indicates, crown thinning allows the manager to concentrate almost all the growth potential of the site on the crop trees. Crown thinning appears to be especially appropriate in fully stocked stands on good sites with a good number of potential crop trees that are well distributed throughout the stand. Crown thinning is more likely to produce a commercial product than low thinning, especially when applied to small sawtimber stands.

There are several potential hazards of crown thinning. First, if applied to younger stands (e.g., pole-sized), especially in oaks or other moderately shade tolerant species, thinning results in retention of lower limbs on residual trees, which can lower the future quality of crop trees. Second, a heavy crown thinning may allow enough light into the canopy to encourage the development of epicormic branches. Based on thinning studies in mixed oak stands in southeastern Ohio, Sonderman and Rast (1988) recommended moderate to light thinnings in such stands due to the higher incidence of branch-related defects in heavier thinnings. Another hazard of crown thinning in many mixed stands is the existence of a midstory consisting of slower-growing species or less vigorous individuals. For example, red maple, sweet birch, blackgum, and hickories often occupy the midstory position, and trees of these lower-value species would benefit from crown thinning in the present stand and could increase in abundance in future stands. Nyland (1996) recommends crown thinning to follow an earlier low thinning, which should help alleviate several of the problems alluded to previously. For most landowners, such a practice would only be feasible if the thinnings produced a commercial product.

Selection thinning is described by Smith (1986) as a technique that removes poorly formed dominants to favor crop trees in the upper canopy stratum (Fig. 99). Such poor quality dominants, in addition to occupying a large area of growing space, have no potential for improvement; thus, they should be removed as soon as possible. Smith (1986) recommends that if selection thinning is appropriate, it should be done as early as possible in the life of the stand, as soon as dominance is clearly expressed.

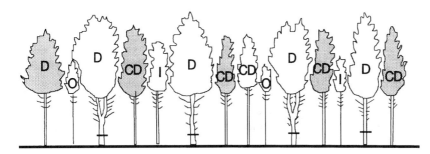

FIGURE 99   Diagram of selection thinning showing trees to remove (horizontal bar) and potential crop trees (darkened crowns) (from D. M. Smith, *The Practice of Silviculture*, 1986; reprinted by permission of John Wiley & Sons, Inc.).

A situation where selection thinning seems appropriate in central hardwood stands is old-field stands of species, such as yellow-poplar, sweetgum, or black cherry. Such stands often contain scattered trees that, for various reasons, obtained an early advantage over their competitors. These trees become dominant early, due to their advantage in canopy position, and develop spreading and limby crowns. In many instances such open-grown trees, due to their large crown area, have also received glaze or snow damage and have poor stem form in addition to limbiness. Another situation where selection thinning might be appropriate is in even-aged mixed oak stands with a component of scarlet oak. Scarlet oak is capable of faster initial height growth than black, chestnut, or white oaks, and when mixed with these species, scarlet oaks often establish early dominance. Scarlet oak, however, is prone to limbiness, occurrence of branch canker, and several other defects, making it a low-value species in comparison to most other oaks.

There are financial incentives to selection thinning in addition to the silvicultural rationale previously discussed. For example, cutting the largest trees provides economic advantages and makes it likely that the thinning will yield a commercial product. But as Nyland (1996) states, selection thinning "essentially amounts to a diameter-limit cutting." Cutting the largest trees, if repeated over and over, can revert to **high-grading** and will leave stands of trees with low growth potential and can ultimately lead to an impoverished condition where few good management alternatives remain. The problem discussed in relation to crown thinning, where a low-grade or poor-species understory develops in the stand, can occur with repeated high-grading, disguised as selection thinning (Figs. 100 and 101).

Release Operations. Smith (1986) defines release operations as those designed to "free young stands, not past the sapling stage, from undesirable trees that threaten to suppress them." The term "crop-tree release" has been applied

FIGURE 100   An impoverished central hardwood stand, western North Carolina.

recently to canopy release of selected crop trees (at any age), and use of this term has led to some confusion. In this chapter, such operations will be covered under "crop-tree management," and Smith's (1986) definition (above) will be adhered to in referring to release operations.

Nyland (1996) cautions foresters to look for the ecological *reason* that undesirable trees are dominating the desirable individuals in the first place. In other words, a release operation conducted without such analysis may treat only the symptom of a problem and could lead to future problems as the stand develops.

There are three basic types of release operations: **weeding**, **cleaning**, and **liberation**. Weedings are performed in seedling stands to remove herbaceous plants and shrubs that overtop desirable tree seedlings. Cleanings remove overtopping trees of similar age to favor trees of better species or quality. Cleanings are done during the *sapling stage*. Liberation operations are done to remove *older trees* that are overtopping sapling-size or younger trees.

All release operations can be accomplished either by **cutting/felling** of the competing vegetation or by **killing** the competing plants and leaving them standing (Carvell 1983). The latter is usually accomplished by stem girdling,

FIGURE 101   An impoverished hardwood stand resulting from repeated high-grading, Preston County, West Virginia.

herbicide treatment, or both. There are two primary problems with cutting unwanted trees. First, in young stands, there may be thousands of undesirable seedlings/saplings/weeds per acre, and finding and removing them can be expensive and time consuming. The second major problem is the fact that many species, which might be considered undesirable, have the capacity to resprout. Sprouts are capable of rapid height extension, which allows them to overtake the desirable stems quickly. This problem can be avoided by waiting until the stand averages 25 ft. in height before performing release operations. Some species, considered to be weeds, such as ferns, are capable of allelopathic inhibition of their competitors, and unless the roots are killed this inhibition can persist even after cutting the tops, which allows new above-ground growth to reestablish dominance over desirable tree seedlings. **Herbicides** can be an effective means of accomplishing release operations (Carvell 1983). Broadcast application of herbicides is appropriate when all the vegetation is undesirable or when using a herbicide that selectively kills only the undesirable plants. For species such as black walnut that remain dormant later than many of their competitors, selectivity may be achieved by applying a herbicide like glyphosate that is absorbed through leaves of actively growing plants early in the season before the desired species has leafed out.

There are several practical problems associated with the use of herbicides in release operations. Due to a variety of factors (timing of application, species treated, formulation, etc.), herbicides may not effectively control the targeted vegetation. In addition, it is more difficult to visualize the result of herbicide release compared to the actual cutting of undesirable vegetation. Miller (1991) and Hamel (1983) have reviewed the various chemicals that are available and their effect on different plant species. Before a herbicide or a method of application is selected, great care should be taken to be certain that the proper effect will be achieved without causing environmental damage.

In central hardwoods, there are several common situations calling for release operations. One such situation where a **weeding** is appropriate occurs when a fern, bramble, or herbaceous understory develops during the regeneration phase following harvest of the overstory (Fig. 102). Plants, seed, or other propagules of competing vegetation are usually already present and are capable of rapid response to the additional resources. The problem of unwanted competition can be exacerbated in the presence of an overpopulation of white-tailed deer, which often selectively browse the desirable tree seedlings (Marquis 1981a). Fern competition can suppress tree growth for a number of

FIGURE 102   Dense bramble and weedy undergrowth that developed after clearcutting on an above-average site.

years and is difficult and often impractical to control. Use of foliar-active herbicides and controlling selectivity either by timing of application or spot application is an effective treatment for ferns. In the case of brambles and annual weeds, these plants are usually temporary and tree seedlings will ultimately prevail. However, spot application to release individual tree seedlings can speed this process and affords a degree of selectivity in controlling composition of the stand. Any weeding operation is costly and the investment must be carried for a long time since the weeding is done very early in the life of the stand.

Rhododendron and mountain laurel pose an especially difficult problem in central hardwood stands (Clinton and Vose 1996). These species are shade tolerant enough to become established in hardwood understories. When the final harvest is conducted, rhododendron or laurel can virtually dominate the site, eliminating regeneration of trees (Fig. 103). Again a herbicide cleaning or possibly prescribed fire is recommended to control rhododendron or mountain laurel.

Several situations occur in central hardwood forests where **cleanings** are appropriate. Often in the development of old-field stands, lower-value species,

FIGURE 103   Dense undergrowth of rhododendron under an oak stand in western North Carolina.

such as black locust, sassafras, eastern redcedar, sourwood, blackgum, aspens, and Virginia pine, become established at the same time as do more valuable oaks, yellow-poplar, ash, and black cherry. Old-field stands may also contain limby or poorly formed dominants of the desired species as well. Cleanings to remove undesirable trees can be effective either using cutting, girdling, or herbicide injection. If cutting is the selected technique, it is better to wait until the stand is well into the sapling stage and has a closed canopy in order to make better decisions regarding future crop trees and to assure that stump sprouts will not overtake the desirable trees.

Grapevine management is a special type of release operation that can be considered a type of cleaning. Grapevines generally regenerate with the trees, especially in even-aged stands, where they grow from stored or animal-deposited seed. Smith and Lamson (1986b) recommended that to control grapevines it is best to wait until the canopy of the sapling stand closes and then cut the grapevines off at ground level. The shade intolerant grapevines will not be able to reoccupy the canopy and will die. Grapevine can be managed in concentrated "arbors" (Fig. 104) for wildlife food production, while minimizing their impact on the forest as a whole (Smith and Lamson 1986b).

Although cleanings might be silviculturally advisable in many cases, they

FIGURE 104   Grapevine arbor engulfing trees in Fayette County, Pennsylvania.

seldom produce a merchantable product. Because, like weedings, cleanings are conducted early in the life of the stand, the cost of such operations must be carried for a long period and may be difficult to justify economically.

In central hardwoods, **liberation** operations are often appropriate in stands that have been high-graded and cut over, but leaving scattered poor-quality and unmerchantable residual trees (Fig. 105). Typically, commercial logging on private land is conducted with little or no professional input regarding silviculture or forest management. Therefore, the cutting involves removal of anything that can be economically removed by the buyer. Depending on the type of stand and the prevailing market conditions, such cuts can range from one in which a few high-value trees are selectively cut to one that is essentially a clearcut, but leaving cull or unmerchantable residuals. Under the latter scenario, the influx of light to the understory is adequate to stimulate good regeneration. It is this type of situation that results in the need for a liberation operation since the regeneration competes with the residual overstory.

In general, release operations in central hardwood forests will have costs associated with them and will be noncommercial operations. For most non-industrial private forest landowners, such operations may be difficult to justify. If these owners are encouraged to engage in planned forest management, they

FIGURE 105   A poorly done high-grade logging job in Monongalia County, West Virginia.

will be able to benefit from government cost sharing for silvicultural operations and can write off part of their cost of these operations from income taxation when utilizing the capitol gains method for calculating tax on income from future timber sales. But to do this, landowners must be aware that the opportunity exists and foresters who work with private landowners (consultants, service foresters, industrial foresters) need to be aware of these opportunities and should encourage landowners to take advantage of them.

Improvement Cutting. Smith (1986) defines improvement cutting as cuttings "made in stands *past the sapling stage* for the purpose of improving composition and quality by removing trees of undesirable species, form or condition from the main canopy." Owing to the inherent nature of mixed species stands, and to the past history of many central hardwood stands, improvement cutting, as defined above, seems to be a frequently needed operation. Improvement cutting seems particularly applicable to mixed oak, oak-hickory, and mixed mesophytic hardwood stands (Carvell 1973). It is less applicable to pure stands, especially in the small sawtimber or sapling-size range.

The focus of an improvement cutting is directly opposite that of crop-tree management, a method that will be discussed later. In improvement cutting, the forester identifies the *undesirable trees* for removal whereas in crop-tree management, the *crop trees* are identified for release. Improvement cutting seems most appropriate where there are enough acceptable trees to form a stand once the unacceptable trees are removed. In other words, removal of the unacceptable trees will not result in a stand that is appreciably below the B line of Roach and Gingrich (1968). In planning improvement cuttings, it is necessary to conduct an inventory to determine the basal area of acceptable and unacceptable growing stock in the stand and to prescribe a target basal area for the residual stand. Improvement cutting in central hardwood stands can be applied anytime after the sapling stage and before the desirable trees lose their vigor. Sander (1977) estimates that mixed oaks are capable of responding to release up to an age equaling three-fourths the rotation age.

In mixed oak stands, on above-average sites, improvement cutting can be conducted well past 80 years of age, but it is important to consider the silvical characteristics of the various species when deciding which trees to leave. In mixed oak stands, with mixtures of northern red oak, white oak, chestnut oak, and scarlet oak, as a general rule, the latter two species will be selectively removed, leaving the former two. But decisions will still be made to leave better individuals of the least-favored species while removing poor individuals of the favored species. For black oak and southern red oak, which usually occur as scattered trees in mixed stands, the quality and potential of individual trees can vary widely and decisions as to cutting or leaving trees should be made on a case-by-case basis. Most of the oaks are relatively long-lived and capable of release well past 100 years of age, with the possible exception of scarlet oak. However, individual oaks that have been growing in a suppressed canopy position, were subjected to heavy stress (defoliation, drought, etc.), or have small narrow crowns due to long-term overcrowding will not respond to release and should not be considered as acceptable growing stock, irrespective of species. Site quality is also important when contemplating improvement cuts in oak stands. In general, oaks growing on poor sites lose their ability to respond to release sooner than on good sites. Therefore, on poor quality sites (50-year oak site index <65), improvement cutting will not be as likely to result in the desired effect, especially in stands older than 80 years. The situation is complicated by the fact that on such sites oaks grow slowly and do not develop into sawtimber-sized trees before about age 80, which affects the commercial feasibility of improvement cutting in oaks on poorer sites. Carvell (1973) indicated that precommercial improvement cuttings in oak and cove hardwood stands cannot be justified economically.

Improvement cutting in oak-hickory stands should employ an approach similar to that of mixed oak stands where the oaks are concerned. But in stands managed for sawtimber, hickories, as a group, would be considered undesirable since they are generally of lower potential future value than the oaks. Occasional high-quality hickories may be left on a case-by-case basis where they achieve spacing objectives or where hickories have the best vigor and quality in a group of trees. Oak-hickory stands often contain a red maple midstory. Due to the smaller average diameter, therefore lower commercial value of red maples, there is a tendency to remove fewer of them in improvement operations. The consequences of this could be an overall gain in importance of red maple in future stands where oaks should be the favored species. Thus, although improvement cutting does not theoretically deal with regeneration, it may have an inadvertent impact on the future stand composition. Foresters should take this into consideration when implementing intermediate management. For example, in the case of red maple as discussed above, it may be necessary to remove or deaden noncommercial trees of undesirable species during an improvement cutting where commercial products are being produced from other species. With the development of markets for wood composite products, smaller diameter trees of species, such as red maple, blackgum, sweet birch, sycamore, aspens, and American beech, may have a better market in the future, and foresters planning improvement cuttings will have more flexibility in designing these operations.

In mixed stands on more mesic sites, improvement cuttings will generally remove poor quality trees of a variety of species, leaving high quality oaks, maples, yellow-poplar, sweetgum, ashes, or black cherry. In such mixed stands, shorter-lived species, such as sweet birch, black locust, and aspens, will have a high priority for removal, providing they are merchantable. Some black cherry, yellow-poplar, or sweetgum trees growing in these stands should also be targeted for removal, especially if they occur in the intermediate or weak codominant crown class, since such trees are unlikely to respond to release. Vigorous high-quality trees in codominant and dominant canopy positions of the desirable species will form the residual stand. These stands generally occur on above-average sites; therefore, improvement cuttings in them should produce merchantable products.

There are several aspects to improvement cutting that differentiate it from thinnings, release operations, and crop-tree management in central hardwoods. Improvement cutting, like thinning, is designed to redirect resources to better trees, but it is *always* applied to stands past the sapling stage and is most applicable to mixed-species stands, whereas thinning is best conceptualized for pure-species situations. Improvement cutting tends to focus on the poor quality trees for removal, whereas thinning and crop-tree management tend to focus on the residual stand or crop trees. Improvement cutting generally produces a merchantable product since it is done after the sapling stage. Carvell (1983) suggests cutting cycles of 15 and 20 years in mesophytic hardwood and oak stands, respectively. In the life of a stand, different types of intermediate cuttings may be applied at different stages of development, depending on the species composition, tree quality, and site, and at times more than one cutting or operation will be applied to the stand at the same entry. Obviously the **landowner objectives** are the real determining factor in designing a silvicultural prescription for intermediate management. A separate discussion on decision-making and silvicultural prescriptions is presented at the end of this chapter and in Chapter 6.

Sanitation and Salvage Operations. Smith (1986) defines sanitation cuttings as those made to "reduce the spread of damaging organisms to the residual stand" and salvage cuttings as those "made for the purpose of removing trees that have been or are in imminent danger of being killed or damaged by injurious agencies other than competition between trees."

The philosophy behind sanitation cutting is more *proactive*, whereas salvage cutting is more *reactive*. In order to apply **sanitation cutting**, the forester must be aware of the potential hazards from injurious agents that a particular stand is likely to experience. This knowledge can be gained from using "hazard rating" (Hedden 1981; Hicks et al. 1987). For gypsy moth management, Gottschalk (1982) proposed a treatment for reducing gypsy moth susceptibility of hard-wood stands that he termed a "sanitation thinning." This example of sanitation cutting would be applied to fully stocked stands where more than 50 percent of the basal area is in species that are not preferred by gypsy moth. Almost all species in the central hardwood region are susceptible to some injurious agent or agents. Many of these are referenced in the preceding chapter on silvical characteristics. But although trees may be *susceptible* (capable of being attacked) they can be more or less *vulnerable* (capable of being killed or damaged), depending on certain other predisposing conditions. For example, overstocking, droughts, air pollution, site conditions, and other insects and diseases can predispose trees to greater vulnerability when they are exposed to injurious agents. It is important to take into account the "stress loading" of a given stand

and the particular individuals within the stand when making decisions about which trees to favor in a sanitation cutting.

In central hardwood stands, sanitation cutting is most applicable to mixed stands containing species or species groups that vary in vulnerability to a pest buildup. It is also most likely to be applied where it can be accomplished in conjunction with another needed silvicultural operation (thinning, improvement cutting, etc.), and a merchantable product can be obtained. Gottschalk's (1982) sanitation thinning for gypsy moth is a good example. In fully stocked mixed mesophytic stands with a substantial oak composition (but less than 50 percent), the prescription would be to thin the stand by selectively removing oaks (the preferred host), especially individuals of lower vigor, thereby reducing the susceptibility and vulnerability of the residual stand. Some oaks, as well as better quality individuals of other less vulnerable species, would be left. The net result would be to lower susceptibility by reducing the proportion of preferred host species, to reduce vulnerability by leaving more vigorous trees, to improve the overall quality of the stand, to redirect resources to the residual trees for faster growth, and to generate income from the sale of the thinned trees.

Salvage cutting is often done after damage has occurred, and the primary objective, as the name implies, would be to recover value from trees that have died or are expected to die. For some "uses" of the forest, the value of a tree is completely lost when it dies; therefore, salvage is impossible. For example, the aesthetic value of trees is generally greatly reduced when they are no longer alive. In oak stands where gypsy moth mortality exceeded 30 percent, Brock et al. (1990) found that scenic preference ratings of residents dropped sharply. In other situations, a stand's value may actually be enhanced by the presence of dead trees. A good example of this would be the value of dead "snags" for denning, nesting, and feeding habitat for several species of wildlife.

Where timber production is concerned, damage to, or death of, trees usually constitutes a loss in value of the tree. Labosky (1987) found that lumber and pulp recovery from oak trees that had been standing dead for 5 years was about half that of a control sample of live trees. Not only did the recovery volume decrease, but the quality was also reduced. These losses translate into lower stumpage value for standing dead timber. Thus, landowners should try to avoid salvage sales wherever possible. One way to avoid salvage of dead material is to engage in pest suppression activities. Spraying for gypsy moth is an example of an activity that can dramatically reduce losses and, in strict economic terms, has a very high benefit/cost ratio, particularly when done later in the rotation (50+ years) and no repeat application is needed (Hicks et al. 1989a).

A type of salvage cutting that removes trees that are highly vulnerable to mortality *before* they die is called **presalvage**. The advantages of presalvage are to obtain live-timber value for material sold and to avoid depressed markets that often accompany large-scale pest outbreaks. Regarding presalvage, Smith (1986) states: "Usually more complete command of the situation can ultimately be taken by shifting to species or age classes less vulnerable to damage." Presalvage thinning is described by Gottschalk (1987), in reference to gypsy moth management, as being appropriate in situations where the forest is immature and more than 50 percent of the stand basal area is in preferred host species. He also listed a "presalvage harvest" treatment as being appropriate in mature stands that had adequate advance regeneration and a "presalvage shelterwood" cut for mature stands lacking adequate advance regeneration. As with sanitation cuttings, these presalvage operations are recommended in conjunction with other silvicultural operations. In mixed-species stands, salvage or presalvage operations may focus on a particular age class, species, or species group. Because salvage cuttings involve recovery of economic value, these operations would, by definition, produce a merchantable product.

Crop-Tree Management. Crop-tree management is a type of silvicultural operation that focuses on *individual trees* that have the potential of developing into high-value crop trees. Perkey et al. (1993) emphasize that crop-tree value should be defined by the landowner's objectives, and a tree that enhances wildlife habitat may have more value than one that is valuable for timber production. Some of the benefits of crop-tree management include:

- Crop-tree management permits the designation to fit the landowner's objectives.
- It is simple to apply and fits well with the management of private nonindustrial land holdings (Perkey 1991).
- It provides for an even flow of forest products from the land.
- It provides for continuous forest cover.
- Management efforts are concentrated on the trees with greatest potential for future gain in value.

Potential disadvantages of crop-tree management include the following:

- It is not a "silvicultural system" in that regeneration is not considered; thus, there is no plan for the future stand.
- In stands where commercial crop trees are retained, the competing trees may be non-commercial; therefore, removing them will incur a cost.

Houston et al. (1995) indicate that there are two phases in crop-tree management: crop-tree **assessment** and crop-tree **enhancement**. In regard to assessment, Perkey et al. (1993) provide guidelines for crop-tree selection for timber management, wildlife management, and aesthetic and water management goals. *Timber* crop trees should meet the following criteria:

1. Dominant or codominant crown class and at least 25 ft. tall
   - A large healthy crown
   - No dead branches in the upper crown
   - Of low sprout or seedling origin (if multiple trunks of sprout origin, avoid **V**-shaped connections)

2. High-quality stems
   - Butt log potential grade 1 or 2 sawlogs
   - No evidence of epicormic branches on butt log
   - No indication of high risk (lean, splitting, forks, etc.)
3. High-value commercial species
4. Expected longevity greater than 20 years
5. Species adapted to the site

For *wildlife* crop trees, the selection criteria are as follows (Perkey et al. 1993):

A. Mast producing species
   1. Dominant/codominant trees with:
      - Large healthy crowns
      - None or few dead branches in upper canopy
      - Sprout or seedling origin
   2. Hard mast preferred over soft mast; maintain diversity
   3. Expected longevity greater than 20 years
   4. Dead branches and cavities acceptable
B. Cavity trees*
   1. No restriction on size, species, or canopy position
   2. Dead branches in upper crown acceptable
   3. Expected longevity unimportant

For crop trees to meet *aesthetic* objectives, Perkey et al. (1993) set forth the following criteria:

1. Species producing attractive flowers or foliage
   - Healthy large crowns
   - Few dead branches in the upper crown
   - Sprout or seedling origin
   - Understory trees are acceptable if they are capable of responding to release
2. Trees visible from roads, trails, and streams
3. Expected longevity greater than 20 years
4. Unique or unusual trees (e.g., large, old, unusually shaped, spreading)

Some silviculturists have proposed simple spacing methods for determining whether or not an adequate number of crop trees exist to produce a well-stocked stand. For example, Lamson et al. (1988) recommended leaving 50–70 residual crop trees per acre, but not more than 100. They recommend trying to space crop trees an average of 25 ft. apart. Houston et al. (1995) recommend dividing the stand into 35-ft. square "cells," with the goal being to release one crop tree per cell. Perkey (1992) indicates that a more common scenario is the occurrence of fewer trees that meet crop-tree criteria than would be needed to form a

*Cavity trees need not be released unless they are also mast producers

well-stocked stand. He recommends that as few as 5 high-quality red oak crop trees per acre are worth releasing by cutting lower-quality trees that are competing with them. In such a case, leaving the non-crop trees in the stand may help make future harvest operations more economically viable because of the additional (low-quality) volume available for harvest. This strategy fits in well with the previously stated observation that in mixed-species stands, typical of the central hardwood region, more than one silvicultural operation may apply to a given stand at any point in time. For example, where potential high-value crop trees exist, but are sparsely dispersed in the stand, the forester may focus on releasing them, while treating the remaining trees (non-crop trees) as a separate stand, within a stand. The non-crop-tree stand may qualify for an improvement cutting or, in the case of mixed oak stands, may be in need of a presalvage harvest where gypsy moth or other problems are imminent. In situations where the non-crop-tree portion of the stand is noncommercial, the best strategy may be to ignore this part of the stand, except for trees interfering with crop trees.

Enhancement of crop trees is usually accomplished by *releasing* them from competition, although Houston et al. (1995) include **fertilization** as a possible enhancement activity and **pruning** of butt logs could also be added to the list.

The recommended method for releasing crop trees is the so-called **crown-touching** method (Lamson et al. 1988). To employ this method, the crop-tree crown is divided into quadrants and these "sides" are evaluated as to whether or not the tree is free-to-grow (Fig. 106). The recommendation is to release

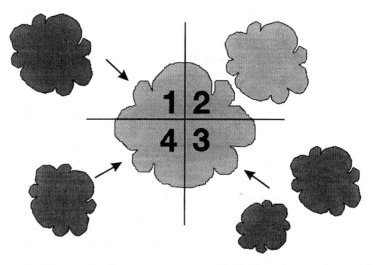

FIGURE 106 Diagram showing a crop-tree crown divided into four quadrants. After release, this tree would be free-to-grow on three sides (from Wilkins 1994).

potential crop trees on at least three sides (Lamson et al. 1990; Wilkins 1994). Figure 107 provides a diagrammatic view of a crop-tree management operation where crop trees are released on at least three sides. Crop trees can be released by mechanical means (cutting) or by deading competing trees (either by girdling or herbicides) and leaving them standing. Miller (1984) found that in precommercial release of 12-year-old black cherry crop trees, the cost of herbicide treatments exceeded that of cutting with a chain saw. Chain-saw girdling with

BEFORE TREATMENT

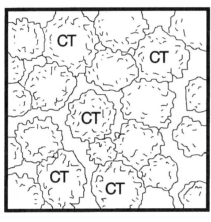 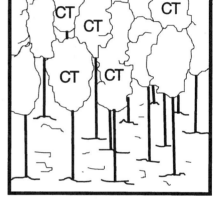

VIEW FROM ABOVE VIEW FROM SIDE

AFTER TREATMENT

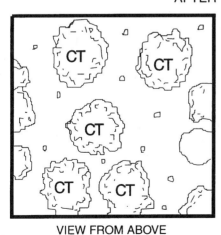 

VIEW FROM ABOVE VIEW FROM SIDE

FIGURE 107   Overhead and side views of stand before and after a crown-touching release of crop trees (from Wilkins 1994).

a double cut through the cambium and into the wood is an alternative to herbicides, but with diffuse-porous species, such as maples, birches, and yellow-poplar, girdling is not as effective as it is with ring-porous species, such as oaks, elms, ashes, and American beech.

Release of crop trees appears feasible over a wide age range from sapling to sawtimber size (Stringer et al. 1988; Lamson and Smith 1989; Smith et al. 1994a; Johnson et al. 1997). The important criteria in deciding whether or not to release crop trees have been listed previously, but they generally relate to the quality, vigor, and potential of the crop-tree candidate. Release of crop trees at too early an age can lead to misclassification, where effort and money are expended on trees that ultimately will not be of good quality. The use of a 25-ft. minimum height for selecting crop trees is designed to help avert such misclassification (Perkey et al. 1993). An advantage of early crop-tree identification and release is that the forester can influence performance over a greater portion of the rotation. In addition to the misclassification problem alluded to above, another disadvantage of early crop-tree enhancement is that such operations seldom result in merchantable products, and the cost must be carried for a relatively long time before harvest. Conversely, waiting to release crop trees until they are more mature (at least small sawtimber size) has the advantage of producing a merchantable harvest when the competing trees are removed. Perkey (1992) estimated that the rate-of-return on released red oak crop trees was greater for a commercial cut conducted at the small sawtimber stage as compared either to a precommercial release or a commercial release performed at the medium sawtimber stage. Thus, release at the small sawtimber stage takes advantage of the greater physiological response of younger trees while providing adequate time for trees to begin to express their potential as well as producing merchantable products.

One appealing strategy in crop-tree management would be to select trees with high-quality butt logs for enhancement, with the intent of producing high-value veneer-grade logs. In addition to being straight, straight-grained, and free of defects (knots, frost cracks, etc.) a veneer log must have a relatively large diameter. The temptation might be to choose high-quality stems on good sites at the small sawtimber stage (10–14 inches, dbh) and, using a four-sided crown-touching release, to stimulate them to produce veneer-sized trees as quickly as possible. The problem with such an approach is that when growth rings of trees change abruptly from narrow to wider, this change results in "multitextured veneer," which tends to break along the growth ring (Kesner 1986). Another problem that could greatly reduce the veneer potential of crop trees is the development of epicormic branches as a result of a drastic opening up of the stand. For species that are particularly prone to epicormic branching (e.g., white oak), the problem can be minimized by carefully choosing crop trees and avoiding those with existing epicormic branches or indications of such branching in the past. Both the uneven growth and epicormic branching problems can be minimized when veneer production is the desired goal by making more frequent but lighter releases (two-sided, but no more than three-sided) using a cutting cycle of 10–15 years. This will result in a less dramatic response than heavier releases and may not provide the economic incentive for commercial release cuts, especially when applied to the smaller tracts typical of nonindustrial private landowners.

Another concern of crop-tree release, and of all partial cuttings, is the occurrence of logging damage to residual trees. Where valuable crop trees are being retained, the wounding of such trees and potential decay that results can negate the purpose of crop-tree management. Smith et al. (1994b) found that logging wounds of 1–50 square inches on yellow-poplar, northern red oak, white oak, and black cherry closed within 10 years. But wounds of 50–200 square inches were not expected to close in less than 15–20 years. They recommend extreme care be taken to avoid wounding crop trees.

In summary, crop-tree management is a system that seems particularly suited for use in immature stands of mixed species that are managed by small, nonindustrial private owners. Crop trees can be selected to accommodate a variety of goals and management efforts can be concentrated on the trees in the stand with highest potential value. Crown-touching release of crop trees is easy to apply, and when applied to small sawtimber stands the release often produces a merchantable yield, which enables the landowner to accomplish it without cash outlay. Finally, crop-tree management can easily be integrated with other silvicultural treatments that may be appropriate for the non-crop-tree component of the stand.

Regeneration Systems

General Concepts and Decision-making. This section deals with activities designed to initiate regeneration as distinct from operations designed to tend existing stands. Smith (1986) defines a reproduction method as "a procedure by which a stand is established or renewed; the process is accomplished during the regeneration period by artificial or natural reproduction." He separates reproduction from the so-called **silvicultural system**. The latter is "more comprehensive and designates a planned program of silvicultural treatment during the whole life of the stand; it not only includes reproduction cuttings but also any tending operations or intermediate cuttings." In reference to the central hardwoods, I have maintained a separation between the discussion of regeneration and those of tending the forest, even though in reality they have a great effect on each other (Kirkham and Carvell 1980). There are two reasons for keeping them separate. First, most central hardwood stands are currently in a stage of development where tending operations are still appropriate, providing such operations fit the landowner's objectives. Planning for regeneration is appropriate for these stands, but such plans must be adaptable to biological and economic events that occur. Second, owing to the complexity of many central hardwood stands, the volatility of the timber market, and the varied goals of the forest landowners, it is unrealistic to formulate comprehensive and long-term plans that include regeneration and tending operations for an entire rotation. Nyland (1996) states that "foresters develop a unique silvicultural system for each forest stand." He lists the three main functions of a silvicultural system as (1) regeneration, (2) tending, and (3) harvesting. Figure 108 is a diagram showing the relationship of these functions according to Nyland et al. (1983). The harvesting method, therefore, becomes part of the mechanism through which regeneration is obtained. Considering landowner objectives, characteristics of the site, and existing stand conditions, the forester can use the harvest cut to create an environment that favors regeneration of certain species over others. Additional treatments may be necessary as part of the regeneration scheme. These include fire, mechanical site preparation, and weed control.

Many current central hardwood stands are even-aged in the 60- to 90-year age range. For some species, such as black cherry, sweetgum, yellow-poplar, yellow pines, and birches, this age is at or approaching maturity, and planning for regeneration should be in progress. Other species, such as hickories, scarlet oak, and red maple, are in stages of maturity where regeneration is still not necessary, but activities in preparation for regeneration should be planned or already taking place. For the longer-lived species, such as eastern hemlock, sugar maple, and most oaks, the decision to regenerate is not imminent in most

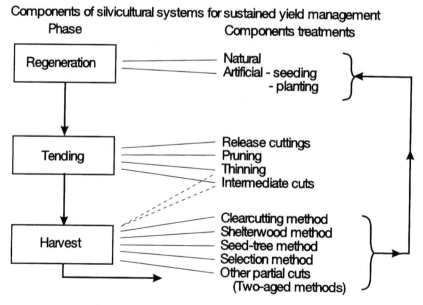

FIGURE 108   Interrelationships of activities constituting silvicultural systems (from Nyland 1996).

central hardwood stands, but, depending on landowner objectives, harvest and regeneration activities can be ongoing or in the planning stages.

A number of authors have discussed the process of selecting a regeneration system. Marquis et al. (1992) have developed a computerized system (SILVAH) to aid in this decision-making process for Allegheny hardwoods. Various authors have pointed out that development of **advance regeneration** is critical to success in central hardwood stands. This may not be problematic where shade tolerant species, such as sugar maple, are being regenerated. But for oaks, development of advance regeneration depends on having adequate sunlight on the forest floor, as well as having a source of seed or sprouts. Marquis and Twery (1992) diagrammed the decision process leading up to making the removal cut (Fig. 109). Regenerating oaks on high-quality sites is more difficult than on poor-quality sites and is even more dependent on advance regeneration (Loftis 1988a; Sims and Loftis 1990). As Beck (1988) points out, the silvicultural criteria for selecting a regeneration system depend on site quality, silvical characteristics of the desired species, presence or absence of advance regeneration, and ability to regenerate from seed or sprouts. Species in the cove (mesophytic) hardwood group, such as yellow-poplar, sweet birch, and black cherry, can regenerate adequately from seed without the need for advance seedlings, and success of these species is dependent on having a high-quality growing site.

Loftis (1989) sums up the ecological rationale for successful regeneration of hardwoods as being best viewed using the concepts of "initial floristic composition" (Egler 1954) and "vital ecological attributes." Unlike the classical successional concept of "relay floristics," which assumes that sites undergo a series of vegetational changes, the initial floristics model is based on the presence of sources of regeneration of the *currently existing species* (seed, advance seedlings, sprouting potential), and the vital ecological attributes are the conditions that promote growth and survival of the propagules (site conditions, sunlight, etc.). Thus, decisions regarding the appropriate silvicultural system are based on a combination of landowner goals/constraints, site characteristics, current stand conditions, and potential or existing sources of regeneration.

Most central hardwood stands are even-aged, single-cohort stands, but they are usually mixtures of species that are growing and maturing at different rates. Decisions regarding harvest/regeneration are further complicated by differing landowner objectives, pest management constraints, and potential for future economic value of different species (Marquis and Twery 1992).

Silviculturists generally classify silvicultural systems into two broad groups—**high-forest** methods that rely on reproduction from seed and **coppice** methods that rely on sprout reproduction (Smith 1986). In central hardwood stands, regeneration often results from both seed and sprouts, although there are few instances when a true coppice system, where "dependence is placed mainly on vegetative reproduction" (Smith 1986), would be practiced. Thus,

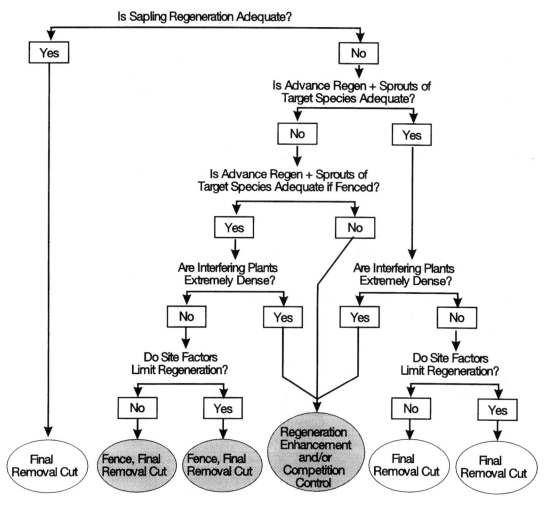

FIGURE 109   Example of a decision diagram for a mature stand to be regenerated in an area with high deer density (from Marquis and Twery 1992).

even though sprouting may form a component of the regeneration, most central hardwood stands would be managed as if by a high-forest method.

Within high-forest methods, there are two subdivisions—**even-age** systems and **uneven age** systems. Both types are applicable to central hardwood stands, although uneven-age systems are best applied where shade tolerant species exist in the stand. A system that provides for regeneration of shade intolerant species while carrying a sparse overstory of mature trees to offset the negative aesthetic impact of clearcutting is the so-called **two-aged** silvicultural system (Miller and Schuler 1995).

In even-age silvicultural systems, there are three that are described by Smith (1986). These are **clearcutting**, **seed-tree**, and **shelterwood** systems.

Even-Age Management. Even-age management is designed to totally remove an existing stand and create a new single-cohort stand. With regard to central hardwood stands, there are two primary regeneration systems (**clearcutting** and **shelterwood**) that are appropriate for even-age management. In the vast majority of cases, both of these systems rely on natural regeneration, but in some cases artificial regeneration (planting or direct seeding) are used as a primary or supplemental source of regeneration. A situation where artificial regeneration is especially applicable in the central hardwood region is planting of abandoned fields with pines (eastern white pine, southern or hybrid pines). Another example is the planting of black walnut in previously agricultural alluvial meadows, which are ideal sites for this valuable species (Schlesinger and Funk 1977). Finally, supplemental underplanting with oaks has been recommended in conjunction with tree shelters in areas with high deer impact.

Clearcutting Method. Clearcutting involves the felling of *all* trees in a stand in one operation and is perhaps the most maligned and misunderstood silvicultural technique in existence (Roach 1968). The basic ecological premise behind clearcutting is to redistribute the resources of the site to a new crop by removing the existing stand. This type of cutting is designed to mimic natural disturbances, such as fire, windstorms, and catastrophic insect and disease outbreaks, which promote regeneration of species that have evolved to exploit these conditions. Such disturbances are relatively common in the central hardwood region and occurred historically at intervals of one to a few hundred years (Abrams 1992). Patric and Schell (1990) indicated that much of the controversy over clearcutting is based on the misconception that clearcutting is simply a harvest-only technique and is frequently confused with the practice of high-grading, or cutting only good and valuable trees and leaving the poor ones. The concern expressed that clearcutting leads to erosion is also generally unfounded, and most of the soil movement that occurs during forest harvesting is preventable since it results from poorly constructed roads (Kochenderfer 1970). But clearcutting, when properly applied, is designed to *regenerate* the new stand as much as to harvest the old one (Roach and Gingrich 1968).

The USDA, Forest Service has conducted numerous tests of clearcutting throughout the central hardwood region. In addition, even-age management was, until recently, almost universally practiced on National Forests in the central hardwood region, and much of this was in the form of clearcutting. Thus, considerable research and experience have accumulated to document the success or failure of regeneration following clearcutting. Several generalities seem to apply. First, an opening of at least 1–2 acres in size is required in order to create the openness needed to produce the characteristics of a clearcut (Sander 1992; Dale et al. 1994). Second, clearcutting tends to promote regeneration of fast-growing, shade intolerant exploitive species, such as yellow-poplar, black cherry, sweet birch, sweetgum, sassafras, pines, and aspens (Beck and Della-Bianca 1981; Parker and Swank 1982; Beck and Hooper 1986; George and Fischer 1989). In regions that are dominated by oaks (Ozark Plateau, Ridge and Valley Province) and on poorer sites, oaks appear to regenerate well after clearcutting (Dawson et al. 1989; Smith 1992). And finally, if advance seedlings are the primary source of regeneration after clearcutting, older and larger sized seedlings have a much better chance of surviving in the new stand than do younger, smaller ones (Marquis 1982).

In spite of the above generalities, Loftis (1988b) points out that the type and amount of regeneration following clearcutting can be quite variable. Loftis (1990a) and Sander et al. (1984) have developed models that incorporate site, ecological, and historical variables with the number and type of advance regeneration in order to better predict the outcome of clearcutting. A computerized model for predicting oak regeneration success in the central hardwood region is reported by Dey (1992).

Although clearcutting is a viable regeneration system and has proved to be successful in regenerating central hardwoods, there are instances where clearcutting has not achieved the desired objectives. Elliott and Swank (1994) reported on a southern Appalachian hardwood stand that was clearcut in 1939 and again in 1962. Their results show that species diversity has declined after the second clearcut mostly due to the increase in yellow-poplar at the expense of oaks and other species. They concluded that this decrease in frequency of oaks was due in part to the regeneration strategies that are initiated when clearcutting a very young stand (notably sprouting). Most managers would agree that maintaining a degree of diversity is desirable in hardwood stands to ensure market flexibility, reduce vulnerability to pests, and provide habitat for wildlife.

Another hazard of clearcutting is the possibility that it will, at least temporarily, redirect site resources to unwanted competing or allelopathic vegetation, rather than the desired species (Boring et al. 1981; Leopold and Parker 1985). Horsley (1988) lists a number of woody and herbaceous species as undesirable competing vegetation in central hardwoods. These include ferns, grasses, brambles, rhododendron, mountain laurel, grapevine, striped maple, sourwood, dogwood, pin cherry, sassafras, and blackgum (Fig. 110). It is important, *before* clearcutting, to assess the potential for such competitive interactions. The options for control vary, according to the type of competing vegetation being managed, but may range from selecting an alternative regeneration system to use of prescribed fire or herbicide treatment before or after cutting. Since most selective herbicides kill broadleaf species, it is not practical to use broadcast spraying after hardwood regeneration has already become established. In such cases, spot spraying, injection, or basal spraying may be required. Cutting competing vegetation without the use of herbicides is not effective due to the rapid regrowth of sprouts from the cut stumps. All the above treatments are expensive and labor intensive; thus, where such problems are expected, it may not be economically feasible

FIGURE 110   Dense growth of competing vegetation resulting after clearcutting on an above-average site.

to do them, particularly for the small private landowners, typical of the central hardwood region.

Another growing problem throughout the central hardwood region that is contributing to the failure of regeneration is deer browsing (Marquis and Grisez 1978). Smaller isolated clearcuts are particularly vulnerable since they serve as an attractant for deer. In addition to using larger clearcuts, leaving slash piles scattered through the clearcut helps in discouraging deer and promotes regeneration success. A last resort for obtaining regeneration in areas with very high deer population levels is fencing (Marquis and Grisez 1978). But these authors indicate that in failed clearcuts where browsing has promoted a fern/grass ground cover, fencing alone may not be enough to ensure regeneration. Ferns and grasses also seem to have an advantage over tree seedlings in highly compacted soils so it is imperative that soil compaction be minimized during logging by confining equipment to skid trails and landings.

In a few situations in the central hardwood region natural regeneration may need to be supplemented by planting or direct seeding (Pope 1993). Plantings can be successful but are expensive and will probably require weed control or protection from deer during the first few years, particularly on good sites (Stout 1986; Walters 1993). Although planting may be desirable under certain circumstances (Davidson 1988), cost and uncertainty of success will likely limit its practice in the central hardwood region.

Shelterwood Method. The shelterwood method is an even-age management system where the objective is to develop a standing crop of advance regeneration through a series of partial removal cuttings of the overstory (Smith 1986). Removal cuttings accomplish several goals. First, good phenotypes are selected as leave trees with the expectation that they will contribute good genotypes to the next generation. Second, these high-quality leave trees have the capacity to increase in value after they are released. Third, the opening of the canopy is expected to stimulate seed production of the residuals and to provide additional light to the forest floor that should promote growth of the new regeneration (Marquis 1979a). Finally, shelterwood cuttings can be tailored to favor particular species by creating enough light to encourage some species but not enough for others. Swank and Vose (1988) discuss how the silviculturist is able to modify the microenvironment of the site and how this can be done to favor certain species over others (Fig. 86). Once an adequate crop of desirable regeneration is established, the final removal cut can be conducted.

The shelterwood method involves two or three cutting treatments extended over a 15 to 30 year period (Nyland 1996):

1. **Preparatory cutting:** a cutting designed to remove poor quality trees and to increase vigor and seed production among the residuals.
2. **Seed cutting:** a cutting performed to open the stand sufficiently to encourage the development of regeneration.
3. **Removal cutting:** a cutting that is done after regeneration is established to remove the overstory and to allow the new stand to grow.

The shelterwood method appears to be especially well suited to regenerating species that are intermediate in shade tolerance and have slower initial growth. In the central hardwood region, this fits the description of the oaks. Thus, where an adequate quantity of vigorous oak seedlings are absent and oak regeneration is the goal, the shelterwood method seems to be a good choice (Sander and Graney 1993). It also has been demonstrated to work well for regenerating Allegheny hardwoods (Marquis 1979b). The shelterwood method also appears to be useful for regenerating one of the most valuable species in the central hardwood region—northern red oak (Loftis 1990b) (Figs. 111 and 112). Carvell and Tryon (1961) state that oak regeneration is easier to establish on the poorer upland sites, since on the better sites, competition from other relatively shade tolerant species will impede oak regeneration. Many proponents of the shelterwood method agree that a precutting treatment (usually with herbicides) may be necessary in cases where an understory of undesirable tolerant species has become established under the canopy (Horsley 1981; Loftis 1985). Red maple also commonly develops in the understory of oak stands (Lorimer 1984), and, if these understory plants are not controlled, they will inhibit the new seedling stand that is established as a result of the shelterwood seed cut.

FIGURE 111   A shelterwood cut in an oak stand at the Sinkin Experimental Forest in the Missouri Ozarks.

In a comprehensive discussion of oak regeneration, Sander (1992) recommends the shelterwood method "when the regeneration potential of the existing oak advance reproduction is not adequate to replace the stand." He further states that "oak advance reproduction is most likely to be inadequate on the middle and lower north- and east-facing slopes." Sander (1992) proposes the sequence of cuttings for a shelterwood method to regenerate oaks as follows:

1. Determine the size of shelterwood cuts based on the overall management goals for the property and the size of the property. In order to create an even-age characteristic, they should be at least 2 acres in size. Shape and arrangement of even-age stands should blend with the character of the landscape.
2. Before cutting, control any competing non-oak understory, preferably by deadening.
3. Reduce the overstory to 40–60 percent stocking. On good sites and where yellow-poplar is a likely competitor, 70–80 percent overstory stocking may be necessary to inhibit competition. Leave the best dominant and codominant oaks as evenly spaced as possible. Cut or deaden all unwanted species.

FIGURE 112   An oak shelterwood cut on the Bent Creek Experimental Forest in western North Carolina.

4. Fall is a good time to treat the understory before a good acorn crop is anticipated.
5. Monitor regeneration establishment and growth. If competing understory redevelops to an unacceptable level, treat with spot spray or injection. This may be necessary 5–10 years after the establishment of regeneration, especially on good sites.
6. When quantity and quality of oak regeneration are adequate, remove the overstory in one cut.

Sander (1992) indicates that it may take 10–20 years to complete this process.

Fosbroke and Carvell (1989) place the time frame for the shelterwood method in the 10- to 30-year range. They recommend a three-cut shelterwood method in areas where no advance regeneration currently exists.

The shelterwood method seems to provide the conditions that fit the silvical characteristics of several species in the central hardwood region. Coder et al. (1987), utilizing multivariate statistical methods to factor out the complexity of species/site interactions, recommended a "group-shelterwood" method to re-generate oak in eastern Iowa. But their analysis also concluded that there was a large portion of the variation that is unmanipulatable. Experience has demon-strated that the shelterwood method can be used successfully in central hardwoods, but as Fosbroke and Carvell (1989) point out, there have been many inconsistencies in the results of its application. In areas like the Allegheny Plateau where high levels of deer browsing are a factor, methods, such as fencing and the use of plastic tree shelters, are being tested to protect regeneration (Fig. 113). Loftis (1990b) points out the results of several authors (McGee 1975; Beck 1977; Auchmoody et al. 1993) who found that opening up oak stands did not necessarily result in increased acorn production. Thus, Loftis proposed that the preparatory and seed cuts may not be part of an oak shelterwood. Rather, the

FIGURE 113   Fencing and tree shelters used to protect planted oaks from deer browsing in the Allegheny National Forest of northwestern Pennsylvania.

first cut will be a release cut to enhance *already existing* oak regeneration and the final cut will remove the overstory from the new stand (Tryon and Carvell 1958). Loftis also recommended a lighter first cut on above-average sites, leaving 70 percent stocking, since this level of shading discourages aggressive competitors that are shade intolerant such as yellow-poplar. These conclusions seem to be verified by Schuler and Miller (1995) in central West Virginia, where they reported that 10 years after a shelterwood seed cutting, desired regeneration of northern red oak was still inadequate. Prescribed fire may serve a valuable role in oak regeneration. When burned by light surface fires, flat-topped oak seedlings resprout vigorously from their well-developed root systems and these seedling sprouts are often more competitive than the plants from which they originated.

A type of cutting that has been described as a shelterwood method is **deferment**, cutting which is similar to the "reserve-shelterwood" method (Nyland 1996). A variation that is being tested in central hardwoods is called **two-age silviculture** (Sims 1992). It appears that the differences between these methods have more to do with their objectives than with the way they are applied or the stand development that results (Figs. 114 and 115). In deferment cutting the goal is to obtain regeneration by reducing the overstory basal area

FIGURE 114   A deferment cut in an oak stand on the West Virginia University Forest.

FIGURE 115   A two-aged stand in the Bent Creek Experimental Forest, western North Carolina.

to approximately 20 ft^2 per acre and to defer harvest of the overwood through a complete rotation of the regeneration (Miller and Schuler 1995). In the reserve-shelterwood system, the overwood is maintained for more than 20 percent of the length of the rotation. In the two-age silvicultural system, the residual stand density is reduced to 20–25 ft^2 of basal area per acre, and a second age class develops as regeneration. Both cohorts are then tended as a two-aged stand.

There are several significant characteristics of these techniques. They carry a sparse stand of large residual trees for an extended period, thus mitigating some of the adverse aesthetic effects of clearcutting while maintaining some of the qualities of mature forests, such as hard mast production. Because of the low density of the residual stand, regeneration under such cutting is usually composed of shade intolerant species, similar to those that regenerate under clearcutting. Careful selection of vigorous reserve trees can result in good growth and production of large-diameter and high-quality products from these trees. Retention of the overwood ensures a seed source for a long period in case there is difficulty in obtaining regeneration immediately after cutting. Because a high percentage of the stand basal area is removed in the initial cut, such harvests can be commercially attractive.

There are several cautions that need to be addressed relative to two-age silviculture systems. First, if the objective is to regenerate oaks or some other species that are intermediate or tolerant of shading, these methods will probably not be effective, especially on above-average sites where shade intolerant species, such as yellow-poplar, are more likely to be competitive. Second, the reserve trees may be at greater risk of damage from lightning strike, windthrow, or ice damage. In soils that are prone to uprooting, these techniques are not advisable. Extreme care must be exercised to avoid logging wounds to the reserve trees since such wounds can result in decay that will greatly reduce their future value. The heavy cutting and opening of the stand can result in the formation of epicormic branches, which will reduce the value of reserve trees. Species that are prone to formation of epicormic branching, such as white oak, and individual trees with indications of a tendency toward branching on the lower bole should be avoided for two-age silvicultural methods (Miller 1996). Finally, there might be a temptation to perform a commercial high-grading and call it two-age silviculture. Smith (1995) lists the following qualities for reserve trees:

1. Expected to live at least 50–80 years after seed cutting.
2. Single, dominant stem, no major forks.
3. Dominant or codominant crown position.
4. No more than 10° lean from vertical.
5. No more than 15 percent deduction for sweep, crook, or decay.
6. No dead or dying major branches in the upper crown.
7. No signs of developing epicormic branches in the butt log.
8. A species not prone to dieback or decline following heavy cutting.

In summary, the shelterwood method is one that appears biologically well suited to regenerating a variety of central hardwood species, including oaks. But research and operational examples have not always confirmed its effectiveness. In some cases the causes for failures are apparent (deer browsing, weed competition, allelopathy, etc.) and can be corrected. In other situations the cause for poor regeneration success is unknown and is either due to uncontrollable (random) factors, such as weather, or those that are not as yet understood. Several silviculturists have noted that oak regeneration on above-average sites is more practically viewed from the standpoint of encouraging existing regeneration rather than stimulating new seedlings to develop. In the central hardwood region, the intensity of management required to apply the shelterwood method is a disincentive, especially where small non-industrial landowners are concerned. Landowners may also succumb to economic incentives to harvest the trees that should be left as shelterwood seed trees or reserve trees (where two-age silviculture is proposed). The shelterwood method also requires a long time commitment on the part of the landowner. Thus, even though the shelterwood method appears to offer a compelling biological rationale in the central hardwood region, its application on a large scale will not likely occur unless some of these disincentives are nullified.

Uneven-Age Management. Smith (1986) points out that, in the truest sense, there is no such thing as an unevenage stand. He makes this observation based on the fact that "even when a single large tree dies it is replaced not by one new tree but by many that appear nearly simultaneously." Thus, in reality, an uneven-age stand is an aggregation of small even-age patches, which, depending on the management intensity or perspective, may be seen as individual small "stands." For purposes of defining uneven-age stands, Smith uses an opening size that is smaller than twice as wide as the height of mature trees, which has at its center an area not under the "microclimate influence of adjacent mature trees." For trees with a height of 90 ft., this opening size would be slightly larger than a half acre. Thus, the range of stand sizes considered as minimum for even-age and maximum for uneven-age management appears to lie between 0.5 and 2.0 acres (see previous discussion under Even-Age Management). In naturally occurring canopy gaps in the southern Appalachians, for example, Clinton et al. (1994) noted that both density and diversity of regeneration were positively correlated with gap size. Apparently the larger gaps regenerate so as to resemble small even-age stands.

Uneven-age stands have several essential features. They contain multiple cohorts of trees that develop either as a consequence of natural mortality and canopy gap formation or as a result of cuttings made during multiple entries into the stand at relatively short intervals (10–30 years). Uneven-age stands usually possess a reverse J-shaped diameter distribution that is related to tree *age* rather than differential growth rates of various species, as sometimes occurs with mixed-species, single-cohort stands in the central hardwood region (Fig. 116). With multi-cohort stands, each cohort has its own diameter distribution, as depicted in Figure 116, and the overall stand distribution is a composite of these. Uneven-age stands managed by single-tree methods ultimately become dominated by shade tolerant species, since the size of stand openings is usually too small to permit the successful regeneration of intolerant species. However, there are variations of uneven-age systems (e.g., group selection) that are capable of regenerating more intolerant species.

There may be several reasons supporting the practice of uneven-age management. A primary one, and one often credited to small private land-owners in the central hardwood region, is aesthetics. Many landowners cannot accept the disturbed condition that results during the regeneration phase of even-age management and are unwilling to wait the relatively long interval before the newly regenerated stand takes on the characteristics of a "mature forest" with larger trees and an open understory. Another reason for practicing uneven-age management is the more continuous flow of products it produces at the stand level. For small forest landowners who do not have the luxury of spreading their sustained yield over a large number of stands, uneven-age management provides an alternative with a more even flow of income. Sometimes the desired species for management are best managed in an uneven-age system. This is especially true of stands with a substantial northern

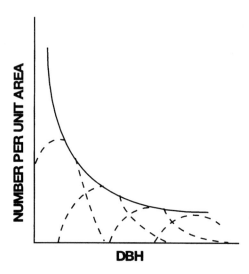

FIGURE 116   Graphic of idealized diameter distribution of a balanced uneven-age stand (solid line) and distributions of individual cohorts of regeneration established by cuttings at regularly spaced intervals (dashed lines) (from D. M. Smith, *The Practice of Silviculture*, 1986; reprinted by permission of John Wiley & Sons, Inc.).

hardwood component. On unstable sites where continuous tree cover is required, uneven-age management may be necessary to protect the site. Because uneven-age management favors mixtures of large and smaller trees, this type of habitat is desirable for maintaining certain species of wildlife (gray squirrel, scarlet tanninger, etc.).

Uneven-age management has certain disadvantages in central hardwood forestry. It is very difficult to apply on the ground and requires a commitment to intensive management that many small private landowners lack. It may not be the *best* way to manage the particular species that are desirable on given sites. For example, where regeneration of yellow-poplar, black cherry, pines, and sweetgum is concerned, even-age systems work well and are generally easier to apply than uneven-age systems. Uneven-age management requires numerous entries into the stand, therefore, a greater investment in a road system and more opportunity for logging damage to the residual stand.

The primary methods for uneven-age management applicable to central hardwoods are the **single-tree selection** system and **group selection**. These methods in addition to some other **partial cutting** (or adaptive silvicultural) methods will be addressed in the succeeding discussions.

Single-Tree Selection. As the name implies, single-tree selection is based on the removal of single mature trees. This technique simulates the natural gap dynamics that occur in mature unmanaged natural stands (Bormann and Likens

1979). This technique theoretically results in a **"balanced" uneven-age stand**, where an array of diameter classes related to tree age are represented and possess a reverse J-shaped diameter distribution (Fig. 116). This method leaves relatively small canopy gaps that can close fairly rapidly due to crown expansion of residual trees. Thus, it promotes the regeneration of shade tolerant species. Crow and Metzger (1987) indicated that by dropping the residual basal area in single-tree selection cutting to 40 ft.^{2} per acre considerable regeneration of intermediate and intolerant species can be obtained, but this may seriously compromise the ability to grow very large diameter trees. In mixed oak or mesophytic hardwood stands, single-tree selection does not regenerate shade intolerant species, such as yellow-poplar, black cherry, sweetgum, or ash, nor would oaks be favored under such a regime. Therefore, maples, buckeye, American beech, and basswoods would most likely be promoted at the expense of the less tolerant species. In general, this would be an undesirable scenario and would make single-tree selection a poor choice for a majority of central hardwood stands.

The real challenge in establishing a single-tree selection system is in creating and perpetuating the balanced uneven-age condition. Most existing stands in the central hardwood region are even-aged and resulted from some prominent disturbance event. Although many central hardwood stands currently have a diameter distribution that approximates a reverse S shape, which differs from the theoretical bell-shaped distribution of pure even-age stands, this occurs as a result of the mixed-species composition and differential growth rates of different species as opposed to different age classes. Due to the developmental history of these stands, they currently contain many species that are intermediate or intolerant of shade and have few representatives of the shade tolerant species that would be needed for single-tree selection management. In order to create the desired balanced diameter distribution, the first cuttings need to be directed toward this goal, and the idealized notion of harvesting large, mature trees may have to be delayed until the appropriate diameter distribution is achieved. Nyland (1996) lists five elements that need to be addressed:

1. Determine the maximum diameter of trees to grow.
2. Decide on the stand density and length of cutting cycle that should be used.
3. Conduct an inventory to determine how the existing stand diameter compares with the idealized distribution.
4. Construct a marking guide to remove "surplus trees" from diameter classes where they exist.
5. Mark and cut the stand to move toward the idealized distribution.

One of the first steps in accomplishing this is to conduct an inventory of the existing diameter distribution of the stand and to compare this to the ideal distribution (Fig. 117). To establish the ideal distribution, three things must be

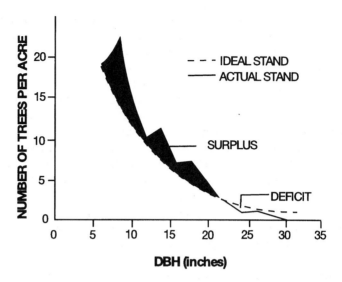

FIGURE 117   Example of actual and desired stand stocking by diameter using stand structure control: q ratio = 1.3, largest tree-to-grow = 32 in., residual basal area = 80 sq. ft./ac. (from Smith and Lamson 1982).

decided. First, what is the desired relationship of one diameter class to the next (q *ratio*)? For example, if $q = 1.5$, there should be 1.5 times more trees in each succeeding smaller 2-in. diameter class than in the previous class. Second, what is the target residual basal area after cutting? And third, what is the diameter of the largest-tree-to-grow? Hansen and Nyland (1986) conducted a simulation of uneven-aged sugar maple stands and recommended a q ratio of 1.2 and a maximum diameter of approximately 20 in. for growing large sawtimber. However, a larger q value is needed to assure adequate growing stock in lower diameter classes and perpetuate the stand. They recommended a largest diameter of 16.5 in. where total volume production is to be maximized. Smith and Lamson (1982) have developed tables for ideal diameter distributions for central hardwoods using several combinations of q ratio, residual basal area, and largest-tree-to-grow. In addition, they discuss the construction of a marking guide to determine the number of trees to cut per acre and the ratio of cut to leave trees in each diameter class (Table 11). Such a guide is essential for marking stands in order to achieve the proper stand structure for single-tree selection management. As can be seen from the foregoing, application of the single-tree selection system is complex and intensive and requires a great deal of information, skill, and effort. Because of these facts and because it is driven by the attainment of a balanced uneven-aged stand, as opposed to economic considerations, single-tree selection, although it has several advantages, will be

TABLE 11 Number of Trees to be Cut by Individual-Tree Selection Cutting for a 50-acre Stand (for *q* ratio of 1.3, residual basal area of 80 sq. ft., and largest tree-to-grow of 32 in.)

DBH CLASS (inches)	ACTUAL NUMBER OF TREES/ACRE	NUMBER OF DESIRED TREES/ACRE	NUMBER OF SURPLUS TO CUT/ACRE	NUMBER OF TREES TO CUT/DBH CLASS	PERCENT OF TREES TO CUT/ DBH CLASS	RATIO
6	32.8	19.4	13.2	660	40[a]	2:5[b]
8	17.0	14.9	2.1	105	12	3:25
10	14.0	11.5	2.5	125	18	2:11
Subtotal				890		
12	8.8	8.8	0	0	0	—
14	9.6	6.8	2.8	140	29	2:7
16	6.0	5.2	0.8	40	13	1:8
18	6.0	4.0	2.0	100	33	1:3
20	4.9	3.1	1.8	90	37	3:8
22	4.2	2.4	1.8	90	43	3:7
24	3.3	1.8	1.5	75	45	5:11
26	1.0	1.4	−0.4[c]	0	0	—
28	1.2	1.1	0.1	5	8	2:25
30	0.2	0.8	−0.6	0	0	—
32	0.4	0.6	−0.2	0	0	—
Subtotal				540		
Total				1430		

[a]Percent cut = 13.2/32.6 = 40.
[b]Two of every five 6-inch trees can be marked for cut.
[c]Deficit dbh class; usually, no trees should be cut.
Source: Smith and Lamson (1982).

difficult to promote for small private landowners. As stated previously, single-tree selection works best in regenerating shade tolerant species. Clatterbuck (1993) concluded that white oaks that had grown in subordinate canopy positions, as in a single-tree selection system, were poor candidates for future management; therefore, single-tree selection would be inappropriate for white oak. Conversely, after examining a 156,000-acre oak-hickory forest in the Missouri Ozarks that has been managed for 40 years by single-tree selection, Loewenstein et al. (1995) concluded that the selection system was *not* converting the forest to shade tolerant species, and the system was maintaining a healthy, vigorous, and productive mixed oak forest. Carvell (1967) noted that a single-tree selection cutting did release understory oak seedlings in a West Virginia study, but seedlings with a defined apical growing point responded better than flat-topped seedlings. These conflicting conclusions regarding oak regeneration may derive from the historical background of the different stands; therefore, in each situation where silvicultural options are being considered in the central hardwood region the final decision will be an informed judgment based on ownership goals, economic factors, historical factors, site, and current stand conditions.

Group Selection. Group selection is a method in which larger openings are created (than in single-tree selection), which has the advantage of permitting regeneration of intermediate and shade intolerant species (Roach 1974). Smith (1986) sets an upper limit on group selection openings as having a diameter smaller than twice the height of the tallest trees. In central hardwood stands, a half-acre circular opening is a reasonable approximation. Smith describes a "group" as a uniform inclusion of trees within a stand. The recognition of a group has more to do with the management perspective than with biological factors that characterize similar groups.

Some of the characteristics of group selection are as follows. Like single-tree selection, a goal of group selection is to produce a balanced uneven-age stand by creating cohorts of regeneration in canopy gaps. With the larger-sized openings of group selection, more resources are made available to the regeneration within the opening, which enables shade intolerant species to reproduce. Murphy et al. (1993) suggest using the silvical characteristics of the species to be regenerated in deciding opening size rather than an arbitrary size based on height of the residual stand. They also recommended using area control (cutting a set area in each cutting cycle). Group selection, like single-tree selection, is supposed to produce an even flow of products and to maintain an overall "forest" character, as opposed to even-age systems.

Experience with applying group selection in the central hardwood region has been varied. When applying the technique to northern hardwood stands, Leak and Filip (1977) found that most of the regeneration was from tolerant species, in spite of the larger openings created using the technique. Weigel and Parker (1995) found that group selection did regenerate intolerant hardwoods (yellow-poplar, black cherry, ash) when applied in southern Indiana. Aspect and time-since-cutting had the greatest effect on stem density of the various species of regeneration. Based on their experience with selection management of southern Appalachian hardwoods, Della-Bianca and Beck (1985) recommended using larger openings than were created by single-tree removal, coupled with herbicide control of unwanted competition to obtain regeneration.

Although group selection may work well under some circumstances, it also has several limitations. Nyland (1996) summarizes these as follows:

1. Initiating group selection management is complicated by the fact that trees in uneven-age stands do not occur naturally in well-distributed groups that occupy comparable areas of the stand.
2. The edge-to-center ratio of openings is very high, especially for smaller openings typical of group selection; thus, the adjacent trees have a disproportionate effect on the group of regeneration.
3. Because of the shape and size of "natural" groups, the surrounding trees may quickly close the opening as crowns spread.

4. The size and arrangement of groups within a forest make them difficult to map, locate, and mark.
5. Standard inventory techniques do not provide the needed information to manage a stand using group selection.
6. Difficulty in seeing from one mature group to the next makes it difficult to establish proper ground control.
7. The scattering of groups through the forest makes it difficult to construct and maintain an access system through the forest.

Miller and Smith (1991), however, point out several advantages to group selection and they advocate an approach whereby volume of growth is used to regulate harvested volume in a given operation and opening size is flexible based on local stand conditions.

In summary, group selection, like single-tree selection, can provide many benefits, such as aesthetics and even flow of products, while making it possible to regenerate shade intolerant species. Unfortunately, for non-industrial private owners in the central hardwood region, group selection shares many of the same limitations of single-tree selection, such as being difficult to apply and requiring long-term commitment. Also, because group selection focuses more on moving the stand toward its idealized uneven-age state than on current markets and economic considerations, small landowners may find it difficult to adopt in lieu of a lucrative timber sale when high market conditions exist.

Adaptive Silviculture (an Alternative for the NIPF)

It has been stated throughout the preceding discussions of silvicultural treatments that many practices have limited appeal to nonindustrial private forest (NIPF) owners. Although silviculture is based on state-of-the-art knowledge regarding the biological principles that affect the ecosystem, the NIPF owners will only practice "good" silviculture as long as it fits their personal goals and is economically attractive. Many NIPF owners do not want to practice even-age silviculture for aesthetic reasons, and many are poorly informed regarding the benefits of practices such as clearcutting and prescribed burning. Although education can help to inform landowners about the benefits of even-age management, clearcuts will always produce an unsightly result that remains until regeneration is well established. Furthermore, on small forest tracts it is difficult to provide for an even flow of products over time using even-age methods. Uneven-age systems tend to regenerate the more shade tolerant species, which are frequently less desirable for timber or wildlife goals than intermediate and intolerant species and uneven-age methods require great skill to apply and a long-term commitment to the methods. Smith (1988) sums it up as follows: "In many cases, neither clearcutting nor selection harvesting is acceptable to the landowner or the forester."

FIGURE 118   An impoverished hardwood stand in central Pennsylvania.

What distinguishes management from non-management is that management involves *planning, goal setting,* and *implementation.* Thus, even though partial cuttings, such as diameter-limit cutting, may occasionally produce good results, they are *not* management. Nyland et al. (1993) differentiate sustained yield (sustention) from exploitation, where sustention involves application of planned silvicultural treatments to individual stands and exploitation is simply removal of salable timber. The latter practice is often what occurs on NIPF ownerships and can lead to impoverished stands with no good options for future management (McWilliams et al. 1995) (Fig. 118). Nyland (1996) points out that every stand is unique and requires a tailored silvicultural system. But much of forestry is based on application of ecological principles (e.g., silvical characteristics and experience rather than on irrefutable experimental proof or a set of rigid principles). Thus, although a planned silvicultural regime is still the best we can offer, there remains a considerable probability that undesirable results will occur (Fig. 119). Based on this conceptual diagram, there is a substantial proportion of the variation in the natural regeneration process that is either uncontrollable (Coder et al. 1987), not understood, or not feasible to be controlled. As Loftis (1990b) points out in regard to regenerating oak, the

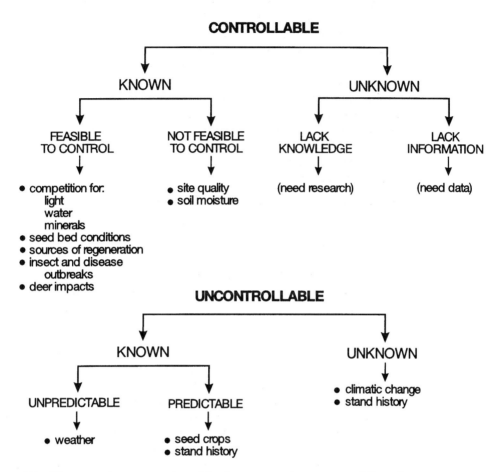

FIGURE 119 Diagrammatic representation of the factors governing success or failure of silvicultural operations, especially with regard to regeneration.

process of releasing already-existing regeneration may be more realistic than that of creating new regeneration in many cases. Thus, there is an element of opportunism in the practice of **adaptive silviculture**.

One of the dilemmas regarding NIPF owners in the central hardwood region is that, given all the choices, they will take the economic certainty of a timber sale over the uncertainty of "best" silvicultural management much of the time. Thus, as foresters and knowledgeable experts, we must seek ways to adapt to the reality of the situation and develop scenarios in which silvicultural management can be accomplished within the overall framework of ownerships, taking into account not only what is *best*, but what is *realistic*.

With adaptive silviculture, the forester can assist landowners to achieve their goals in the context of changing market conditions and occurrence of unpredictable events, such as pest outbreaks, unusual weather events, or bumper seed crops. McGee (1975) noted that in uncut southern Appalachian hardwood stands, which ranged in age between 60 and 80 years, there developed consistently large numbers of seedlings and sprouts in the understories, even though a closed canopy existed. Therefore, if partial cuts are scheduled to coincide with good crops of desirable regeneration, these seedlings can be released to serve as advance regeneration for future stands. Smith (1988) discusses several ways of rejuvenating central hardwoods that are practical and adaptable to landowner objectives. These are as follows:

Alternatives to Clearcutting (for Even-Age Management)

- Leave residual trees—such as occurs with deferment cutting or two-age management.
- Utilize more shelterwood management. Shelterwoods can be conducted where more light cuts are made or residuals are retained until the regeneration is 10–15 ft. tall. An irregular shelterwood (Smith 1986) may be applied in mixed species where one or more species has a much longer rotation age than the others.

Alternatives to Single-Tree Selection (for Uneven-Age Management)

- Use a longer cutting cycle for selection management, which allows a greater harvest volume at each entry.
- Consider pole-sized trees in silvicultural operations, even in the absence of markets for smaller products. In many instances, due to market conditions, only large sawtimber can be marketed in a sale. Often, in central hardwood stands, the pole/timber class in sawtimber stands consists of inferior trees of poor species or vigor. Thus, culling of these trees should be done when valuable sawtimber is being harvested to prevent them from dominating the future stand. With development of composite markets, it may become more economically feasible to manage these trees.
- Apply a financial maturity concept. This method of "selection" management was developed by Trimble et al. (1974) and involves harvesting trees when they reach their "financial maturity" diameter (Table 12, Chapter 6). Financial maturity is based on expected rates of return. In addition to harvesting these larger trees, silvicultural improvement cuttings are recommended throughout all sawtimber diameter classes. With emerging markets for smaller diameter materials, this may be extended down into the pole-size diameters. With this technique, Trimble et al. (1974) reported obtaining adequate regeneration over 20 years, although it

tended to favor more shade tolerant species. Smith (1988) recommends adjusting minimum cutting diameters based on experience, with undesirable or short-lived species adjusted downward and valuable and long-lived species adjusted upward.

- Use differential diameter-limit cutting (Smith 1986). Diameter-limit cutting is the practice used by most NIPF owners. With proper modifications it can be made silviculturally acceptable (Smith 1988). These include raising the diameter limit to 16–18 in. for valuable sawtimber species. At each cutting cycle, harvest no more than can be sustained by growth. Apply appropriate cultural practices (improvement cutting, thinning, presalvage) during each entry, irrespective of diameter. The major limitation of diameter-limit methods is the fact that they do not *provide* for regeneration; however, as with any cutting that opens the canopy, regeneration does occur (Fig. 120). The "adaptive silviculture" approach would enable the manager to respond by favoring desirable regeneration or alternatively discouraging or killing undesirable regeneration as it develops.

FIGURE 120 Dense regeneration of black cherry and other species that developed after a diameter-limit cut in the central Appalachians.

Alternative to Group Selection (for Uneven-Age Management)

- "Small opening" concept versus "group selection." Smith (1988) points out that there is a confusion of semantics between group selection and "patch clearcutting," the former being applied to inclusions within a stand and the latter being applied to *small stands*. He recommends dropping this terminology in favor of a "small opening" concept with the size of openings being tailored to fit the silvical characteristics of the species to be regenerated. Small openings can provide edge and habitat for wildlife and are generally aesthetically acceptable to most owners. He suggests a relatively long cutting cycle of 20–40 years, which creates a less patchy appearance of the forest. These longer cutting cycles may make it necessary to perform intermediate cuttings within the small groups, which can be done at the same time as harvests of mature trees are done at an entry period.

In summary, adaptive silviculture can be tailored to meet landowner objectives/constraints, take advantage of favorable market conditions, exploit good crops of desirable regeneration, and respond to insect or disease outbreaks. Several systematic methods for initiating an adaptive approach to managing central hardwoods are discussed, as proposed by Smith (1988). The adaptive silviculture approach becomes "management" when management plans are developed, goals are set, and work is done to implement these goals. However, to retain an *adaptive* approach, goals must remain flexible and the forest manager must be ready to take advantage of opportunities. There are three cardinal rules that must be observed in any silvicultural management, and this is especially true of the adaptive silviculture approach. First, cuttings should never be conducted in a manner that amounts to high-grading — that is, never cut only good trees and leave poor ones. Second, regeneration should always be a consideration of management, either implicit or explicit. Finally, the harvest should be designed to remove only the growth, which is a basic tenet of sustained-yield management. For the first harvest of central hardwood stands with little prior cutting, harvest volume can exceed the periodic growth for the cutting cycle since growth has accumulated over a long term. If these general rules are kept, foresters should be able to interpose good silvicultural management on most non-industrial private forests of the central hardwood region.

CULTURAL METHODS, SUMMARY

Most existing central hardwood stands are even-aged, of mixed-species composition, and between 60 and 90 years of age; therefore, they are in stages of development where intermediate management methods (thinning, release, improvement, and sanitation cutting) are appropriate. However, in mixed-species stands, different species mature at different rates, and species such as black cherry, yellow-poplar, sweetgum, and yellow pines are more mature than oaks or maples. Because of the different pest problems and silvical characteristics (longevity, growth rate, shade tolerance) of different species in mixed stands, several different intermediate operations may be needed at the same time. Crop-tree management is a straightforward and promising intermediate management technique which may be used to facilitate the production of high-value, grade sawlogs or veneer logs.

Both even-age and uneven-age systems of regeneration have been recommended for regenerating central hardwoods. The former is better for shade intolerant species and the latter for tolerant species. Clearcutting has been used successfully to regenerate shade intolerant species, such as black cherry and yellow-poplar, on better sites. On poorer sites, clearcutting has been used successfully to regenerate oaks, providing an adequate source of regeneration (advance seedlings, sprouts) is present. Where advance oak regeneration is inadequate, the shelterwood system of regeneration has been used successfully, especially on the poorer sites. Oak regeneration on good sites is problematical, and success may be improved by treating the understory with herbicides. A two-age variation of the shelterwood system, where reserve trees are retained in low-density over the regenerating stand, is a way of obtaining regeneration of intermediate or intolerant species while retaining some large trees.

Uneven-age management methods appropriate to central hardwoods include single-tree and group selection. The former is best suited to regenerating shade tolerant species while the latter is a variation that allows the manager to achieve an uneven-age stand condition while managing for more intolerant species.

All the standard silvicultural regeneration methods have been shown to work for some species on some sites and at a given time. But failures occur with some degree of regularity, which suggests that silviculturists either lack necessary information or knowledge, or some factors that affect the process of regeneration cannot be controlled (or all the above.) This uncertainty is particularly debilitating for the small non-industrial landowners, who cannot afford to invest heavily in management, who do not own land solely for timber and revenue production, and whose management goals may change over a time scale shorter than that required to complete a silvicultural program. Several methods of "adaptive silviculture" are discussed, which enable the NIPF owners to derive income, have a sustained yield and not high-grade their stands, a practice that ultimately leads to impoverished stands.

Numerous papers and reports are cited in the foregoing section to support and illustrate the discussion of various silvicultural techniques, but to avoid becoming too involved in details, the discussions presented here have omitted much. For foresters who need to know more about particular silvicultural methods, in addition to references cited in the text, I have listed below a list of silvicultural references dealing with central hardwoods.

ADDITIONAL READING

Burns, R. M. 1983. Silvicultural systems for the major forest types of the United States. *USDA, For. Serv. Agric. Handb.*, No. 445, 191 pp.

Dey, D. C., Ter-Mikaelian, M., Johnson, P. S., and Shifley, S. R. 1996. Users guide to ACORN: a comprehensive Ozark regeneration simulator. *USDA For. Serv. Gen. Tech. Rep.* NC-180, 35 pp.

Harlow, R. F., Downing, R. L., and Van Lear, D. H. 1997. Response of wildlife to clearcutting and associated treatments in the eastern United States. *Dept. For. Res. Tech. Pap.* No. 19. Clemson University, Clemson, SC, 66 pp.

Isebrands, J. C. and Dickson, R. E. (compilers). 1994. Biology and silviculture of northern red oak in the north central region: a synopsis. *USDA For. Serv. Gen. Tech. Rep.* NC-173, 68 pp.

Leak, W. B. and Filip, S. M. 1975. Uneven-aged management of northern hardwoods in New England. *USDA For. Serv. Res. Pap.* NE-322, 15 pp.

Leak, W. B., Solomon, D. S., and Debald, P. S. 1987. Silvicultural guide for northern hardwood types in the northeast (revised). *USDA For. Serv. Res. Pap.* NE-603, 36 pp.

Loftis, D. L. and McGee, C. E. (eds.). 1992. Proceedings: Oak regeneration: serious problems, practical recommendations. *USDA For. Serv. Gen. Tech. Rep.* SE-84, 319 pp.

McCauley, O. D. and Trimble, G. R. Jr. 1975. Site quality in Appalachian hardwoods: the biological and economic response under selection silviculture. *USDA For. Serv. Res. Pap.* NE-312, 22 pp.

Parker, G. R. and Merritt, C. 1995. The Central Region. In: *Regional Silviculture of the United States*, J. W. Barrett (ed.), John Wiley & Sons, New York, pp. 129–172.

Rogers, R. Johnson, P. S. and Loftis, D. L. 1993. An overview of Oak Silviculture in the United States: the past, present, and future. *Ann. Sci. For.* 50: 535–542.

Smith, H. C., Perkey, A. W., and Kidd, W. E. Jr. (eds.). 1988. Proceedings: Guidelines for regenerating Appalachian hardwood stands. Morgantown, WV, SAF Publ. 88-03, 293 pp.

Walker, L. C. 1972. Silviculture of southern upland hardwoods. *Stephen F. Austin State Univ. Bull.* 22, 137 pp.

6

Management of Central Hardwoods

INTRODUCTION

One of the philosophical bases of forestry is the concept of **management**. A desire to manage, control, or organize is innate to human behavior, and it is one of the qualities that has led to the success of our species and the development of our complex technological society. Agriculture, including cultivation, fertilization, and genetic improvement, has provided management innovations that have enabled human populations to grow dramatically beyond the apparent carrying capacity of an unmanaged earth. Production forestry, although often less intense in its application, has generally embraced similar approaches.

Forestry developed in America from its European roots in an era marked by rapid technological changes and a blossoming capitalist economy. The conception of forest management reflected this wider context. Chapman (1950) cites the following from a 1905 lecture by (European-trained) Bernard Fernow at the newly established Yale School of Forestry "Forest management includes all that goes toward making the technical art of forest production successful as a business, selecting the practical means and measures for realizing the object of crop production, namely, continuous revenue." The evolution of the concept of forest resource management is exemplified by a passage from Davis (1954) "Forested lands are managed for a multiplicity of purposes, with usually one use dominant, most often timber production." Duerr et al. (1979) later stated: "Looking at the whole sweep of forest resource management history in the United States, we see a more-or-less steady progression from exploitation in pursuit of economic development, toward the conservation that a developed nation can afford." It was during this era that the concept of multiple-use, sustained-yield management came into full flower (Behan 1967). In the 1990s some views of forest resources management (vis-à-vis **ecosystem management**) have taken a decidedly sociological turn, as can be seen from this definition of Cornett (1994): "Ecosystem management defines a paradigm that weaves biophysical and social threads into a tapestry of beauty, health and sustainability. It embraces both social and ecological dynamics in a flexible and adaptive process." Other elements that have been proposed for incorporation into ecosystem management are biodiversity, forest water quality, forest health, and so on. (Society of American Foresters 1993).

As suggested by Bernard Fernow, traditional forest management focuses on commodity yield and economic gain. Davis (1954) describes "**sustained yield**" as a basic tenet of forest management. He further stated that "the idea of maintaining forest productivity distinguishes forestry as a profession from forest liquidation, no matter how skillfully the latter may be accomplished." Ford-Robinson (1983) defines sustained yield as "the yield that a forest can produce continuously at a given intensity of management." As forestry in America evolved as a profession, the term "yield" has been broadened to include more than one commodity or use. The concept of **multiple use** was advanced to embrace this broader view. Ford-Robinson (1983) defines multiple-use forestry as "any practice of forestry fulfilling two or more objects of management." When the two concepts (sustained yield and multiple use) are married, the result is "multiple-use, sustained-yield forestry." Such a philosophy, combining the two concepts, has prevailed in American forestry for almost three decades. But by the late 1980s, the concept of sustained yield was being criticized by those who believed that it did not fully meet the needs of society and that sustained yield of commodity products did not ensure that the ecological conditions of the forest were sustained (Dunlap 1991).

Forest management has developed from a field with a narrow definition and a simple goal to one that encompasses a broad base and has a complex set of goals. The early concept of forest management was obviously too limited to account for the current full range of activities, but the newer concept of ecosystem management, in an attempt to be all-inclusive, inhibits the process of focusing on tangible goals. In the central hardwood region the nature of forest ownership, the land use history, the physical features of the landscape, and the emerging economic climate all play a role in determining how resource management will be defined and how management decisions can be made in the twenty-first century. As previously discussed in Chapter 1, the majority of forestland within the central hardwood region is owned by small nonindustrial private forest (NIPF) landowners, and most of these do not hold land with the primary purpose of growing timber. However, many such landowners, when confronted with the opportunity for what seems to be "windfall" income, can be enticed to cut and sell timber. Oftentimes such sales take the form of so-called selective cuttings such as diameter-limit cuts or high-grading. Timber buyers may point out that such practices sometimes result in desirable outcomes (high-realized income, adequate forest regeneration, etc.), but the attainment of desirable outcomes, when they are not planned for is *not* management. The current situation with NIPF owners is one in which great opportunity exists for them to practice resource *management*. Expanding markets provide economic incentives for management but a large effort to educate NIPF owners to avail themselves of forestry assistance is needed. Much of the discussion within this chapter will key on forest resource management for the NIPF owner, since it is especially applicable in the central hardwood region. Foresters and other resource professionals have an important role to play in this scenario.

The forester's role in management planning is to ascertain landowner objectives, to conduct a resource inventory and analysis, and to interpret to the landowner what activities are biologically possible and economically feasible. Therefore, the forester has an obligation and an ethical responsibility to inform landowners of management options of which they may not have been aware. Once the landowner is apprised of the full range of options, the forester can advise the landowner regarding the best and most sustainable practices available to achieve their goals.

LANDOWNER CHARACTERISTICS

There are three broad classes of property owners that control the bulk of forest property within the central hardwood forest region. These are **public, industrial private**, and **nonindustrial private**. Management goals and management planning on these may differ dramatically. For example, public land is held in the public trust and usually managed by a public agency. The USDA Forest Service, Bureau of Land Management, National Park Service, and a variety of state forestry and wildlife agencies are involved in management of forests, parks, and wildlife areas. Management goals for public lands are usually mandated by law and management is governed by regulations. Therefore, the forester is restricted in developing and implementing management plans by these mandates and regulations. In many cases management plans must also be subjected to public review prior to final adoption. Although public lands are significant in the fact that they *are* public lands and that they are often found in areas that are remote, unique, and scenic, they constitute but a small part of the forestland in the central hardwood region (less than 10 percent). It is this public land that seems to fit best into the ecosystem management model where societal needs form an integral component of the decision process.

Industrial private land is also a relatively small part (less than 20 percent) of the overall forestland in the central hardwood region. Most of this belongs to timber companies whose primary goal is to make a profit. These companies are often vertically integrated and the profit-making divisions are usually situated at the end where the final product is manufactured (paper, lumber, composite products, millwork, etc.); therefore, the woodlands divisions are not necessarily required to show a profit. However, the more efficient and profitable the woodlands are, naturally the more profit the whole company makes. The forest industry model seems to best fit the definition of forest management as proposed by the early forest managers, such as Fernow, where management focuses on a single resource (timber) and profit motivates the decision-making process.

The NIPF owner, as stated earlier, owns the largest portion of forestland of any ownership group in the central hardwood region (MacCleery 1990). These owners hold land in small tracts (less than 500 acres) for a wide variety of reasons including farming (generally not full-time), residence, investment, and recreational use, (Birch et al. 1982). Although most NIPF owners do not own land expressly for timber production, they are often willing to sell timber if it produces a substantial income to do so. In most cases, NIPF owners want to produce wildlife on their property, either for hunting or simply to observe in the woods. Most of them engage in some sort of recreation on their property and are concerned about the aesthetic effects of logging and timber harvest.

Most NIPF owners have a limited understanding of silviculture and, since a high-grading can be a partial cut, it seems to produce a satisfactory result since it perpetuates the area in a "forested" state. Owners cannot perceive the negative changes in tree quality, vigor, and regeneration that are occurring. Furthermore, they are seldom aware that options are available, which could result in improvement of the stand while still providing reasonable immediate income. Depending on the size of the property, age and species composition of the stands, site quality, and stand history, landowners may be able to derive substantial income at reasonably short intervals, or, if they desire, receive a large lump-sum immediately, which usually precludes subsequent harvests over the near term.

Forest resource management for the NIPF owner can best be described as "**adaptive**." It should be tailored to meet the owner's specific goals, must be ecologically sound, and must work within the existing product market structure. In many cases, a multiple-resource approach will be desirable (incorporating wildlife, timber, and recreational objectives), and a plan that addresses sustainability is also required. But because the land may change hands or the present owner's goals or market conditions may change, any management plan for NIPF owners must be flexible.

DECISION-MAKING IN MANAGEMENT

Development of management plans for public lands and corporate forests are covered in detail in many sources. On public land, management planning is often a required activity that is governed by volumes of regulations and subject to a public review process. Often a part of the planning process is the development of environmental impact statements, environmental assessments, and so on. Thus, although the development of such plans is difficult and time-consuming, little is left to chance and most of the decisions are prescribed in planning manuals. For industrial forests, each corporate owner has set procedures for planning and requires its foresters to adhere to them. Techniques such as linear programming appear more appropriate for use by the corporate manager, and these methods are covered in detail in several contemporary forest management texts. It is management planning for the NIPF owner that is often open-ended and subject to more flexibility on the part of the forester.

The first phase of planning for the NIPF is to determine the goals of the owner. If the owner is a single individual or a small group, this might best be accomplished in an interview process. Before this interview takes place, the forester should, at a minimum, visit and inspect the property. It is helpful at that time to acquire maps of the boundaries and to sketch a map of the property on a topographic sheet noting the approximate location of various stands and briefly describing them. With this information, the forester can add to the discussion and can start making the owner or owners aware of options that may not have been apparent. For larger group owners, a questionnaire is helpful to find out what resources and attributes of the property are most important to them. Questions regarding the tolerance of the owners for disturbance related to timber harvest can be incorporated as well.

Before the actual planning can begin, an inventory of the tract must be made. This can take the form of a cruise (Wiant 1989) or other inventory, such as habitat evaluation procedure (HEP), depending on the goals of the land-owner. The intensity and detail of such an inventory should depend on the intensity of management desired by the owner. In some cases, a qualitative or subjective evaluation may suffice, whereas in others, quantitative field data and detailed analysis are required.

With clearly defined goals and inventory data, the decision-making phase can begin. In this phase, the forester attempts to reconcile the owner's objectives with the various ecological considerations and to provide for sustainability of the resource. The USDA, Forest Service has developed a computer-based model to assist forest owners and managers in making decisions. The Northeast Decision Model (NED) is available free of charge and is referenced in the additional readings listed at the end of this chapter (Alban et al. 1995). In following discussions, the decision process is considered in the context of various planning elements.

The final phase of the management planning process is the actual preparation of a management plan. Forest resource management plans for NIPF owners should be understandable by the owner(s). A lengthy plan with many tables of data may appear impressive but will probably not be an effective communication tool between the forester and the landowner. Brevity for its own sake is not good, but if the elements of a plan can be captured in a brief document as opposed to a lengthy one, it will more likely have the desired effect.

Establishing Management Objectives

Public Lands. Management objectives of public lands has been a subject of controversy for many years. The paternalistic tradition of foresters comes into direct conflict with the altruism of environmentalists with regard to management of public land. Currently, for most public land, some mechanism to include public input must be used in deriving management objectives. The concept of ecosystem management seems to embrace the societal dimension and is most applicable to the public lands. The Society of American Foresters (1993) contrasts ecosystem management with traditional sustained-yield management (Table 12).

Although ecosystem management is "accepted" as a strategy for public land management, a problem with the method seems to be related to the development of management goals. For example, current definitions of ecosystem management include some reference to sustaining the ecosystem's ability to produce commodities and services and a component of society's valuation of services and commodities (Grumbine 1994; Maser 1994). Two troublesome concepts are generally associated with ecosystem management, especially pertaining to the setting of management goals. The first is sustainability and the

TABLE 12 Contrast Between Ecosystem Management and Traditional Sustained-Yield Management

	TRADITIONAL SUSTAINED-YIELD MANAGEMENT	ECOSYSTEM MANAGEMENT
OBJECTIVES	Sustained flow of products (commodities) — constrained to minimize adverse effects	Maintain desired "ecological condition" while producing a sustained yield of products
STRATEGY	Agriculture-like	Mimics natural disturbances
SYSTEM CHARACTER	Emphasis on production efficiency within environmental constraints	Retains complexity of ecosystem processes
MANAGEMENT UNIT	Stands and aggregates of stands	"Landscapes" and aggregates of landscapes
TIME FRAME	Multi-rotations dictated by landowner objectives	Multi-rotations dictated by natural disturbance cycles
CURRENT STATUS	New knowledge is bringing in new values — applicable to some landscape components (owners)	Enduring — accepted for large public land ownerships

Source: The Society of American Foresters (1993).

second is biodiversity. The concept of **sustainability** apparently contains elements of sociological as well as biological and economic principles and, therefore, is not conducive to goal setting in a strict production sense. Stout (1995) cites the World Commission on Environment and Development definition of sustainable forestry as forestry that "meets the needs of the present without compromising the ability of future generations to meet their own needs." The problem with **biodiversity** is that diversity exists at different levels (within and among plots, stands, communities, etc.). Thus, diversity goals are not clearly definable. How then do managers move beyond the rhetorical realm to the application of ecosystem management? If the goal is to preserve and sustain ecosystem functions, how is this accomplished? What are the measures of success? In short, how do managers define and attain the goals of ecosystem management?

Wiant (1995) describes ecosystem management as a "retreat from reality." As this concept evolves in resource management, the ability to set ecosystem management goals will be crucial to the success or failure of the method.

Private Industrial Lands. Setting management goals for industrial owners is somewhat more straightforward than for public land, if only because the "owner" speaks with one voice and the forest is usually owned in order to supply a *product* or *products* required by the company. Goals may focus on total yield or yield per unit area, as well as quality of product. Some forest industry landowners have designated "ecosystem management demonstration areas" (Brenneman 1995) to serve as research and demonstration sites. However, ecosystem management may mean different things to different owners, and the management goals of industrial owners will necessarily focus on the product or products they produce.

Nonindustrial Private Lands. NIPF owners account for over 75 percent of all forestland within the central hardwood region, and their management goals are as varied as they are. There are many elements that are unique to NIPF owners. They often own relatively small tracts; for example, typical of much of the region, the average size of ownership in Pennsylvania is about 23 acres (Birch and Stelter 1993). Very small ownerships do not permit the ready application of such concepts as sustained-yield or ecosystem management. Because of the phenomenon of urban encroachment, many NIPF owners in the central hardwood region find their property closer to the urban fringe (Shands 1991). Another characteristic of NIPF owners is the fact that, either due to being unaware of the potential, or due to prejudice or preference, many of them do not include timber production as an important management goal. For many NIPF owners, land is owned primarily for a place of residence or to provide recreational opportunities or wildlife habitat. Egan and Jones (1993) suggest that although many NIPF owners do not state timber management as a primary goal, they often do harvest timber. Becoming aware of timber value usually occurs due to an offer to buy timber, often extended by a logger or procurement forester. Because the owner frequently has not had a direct involvement in tending the forest, the income opportunity may be considered as a windfall. A typical scenario that occurs in the central hardwood region is one in which forestland, as part of a family farm, is inherited by heirs who no longer live on the land — often not even near it. Forest stands, resulting from regrowth after logging or field abandonment and ranging in age between 40 and 90 years, exist on the property. Some of the stands may have been high-graded 20–50 years ago. Timber buyers, looking for timber, see the property containing merchantable timber and check the local county tax map to identify the owner(s). The buyer makes a bid, often based on a cursory look at the timber, which is far below the market value and is based on a high-grading or diameter-limit cut. Because the property owner is not familiar with timber value or forest management, he/she accepts the bid and considers it a substantial windfall for something that here-to-fore had little or no value associated with it. Even forestry consultants often do little more than get a better *price* for the timber, being paid a percentage of the sale. Thus, their role is essentially that of a timber broker and not one of a forest manager. Many foresters and forestry extension personnel perceive a "NIPF problem." Nyland (1992) has alerted foresters to the potential for a greed-driven exploitation of NIPF forests, and this potential seems particularly applicable in the central hardwood region. Egan and Jones (1993) propose that the focus of management goals for the NIPF owner needs to be on the full resource spectrum with timber being included, but not necessarily featured.

Many times the goals of NIPF owners have not been clearly defined and are not derived from an informed point of view. The landowner's contact with a forester offers an opportunity to provide information regarding the possibilities of management versus exploitation. Jones (1995) cites the land ethic Canon of the Society of American Foresters, which states that foresters will "advocate the practice of land management consistent with ecologically sound principles." He challenges foresters to inform and educate landowners, thereby becoming part of the solution rather than part of the problem. This particularly applies to industrial and consulting foresters, who are procuring standing timber directly from landowners.

Management planning and goal setting are available to the NIPF through the Federal Forest Stewardship and Stewardship Incentives Programs (SIP). In providing these services, either state service foresters or consultants who are specifically trained and authorized to perform the planning activities are involved.

Flow of Products

Public Lands. The "products" of forests can take the form of commodities, such as timber, animal/fish, or water units, or they can be intangible noncommodities, such as scenic beauty, wilderness, or solitude. Public lands offer one of the only opportunities to provide these noncommodity products, and as America becomes more urbanized people seem to crave a "wild nature" experience —even if they experience it indirectly through magazine articles, television, or other media. This trend is part of the motivation for so-called ecosystem management on public lands. Ecosystem management is supposed to incorporate society's needs in producing products while protecting the ecosystem. A difficulty similar to the one alluded to earlier, regarding setting goals for ecosystem management, occurs when it comes to measuring the outputs that flow from forests managed in this way. In the western United States, the conflict between production forestry and noncommodity use is magnified by the fact that the majority of forestland is in public ownership. In the central hardwood region, the situation is reversed. We have a large human population in the urban centers of the East, most of whom live either within or near the central hardwood region. But public landownership is generally less than 10 percent of the forestland base. The National Forest System has developed a method of classifying their land base that is compatible with their needs. This system, "Ecogeographic Analysis" (Bailey 1988), is a hierarchical approach that may be useful for land management by private land managers as well. The hierarchy (Table 13) has three broad scales: macro, meso, and micro (Fig. 121). At the macroscale, regional climate is the driving variable. Landform is more important at the meso or "landscape" scale, and microclimate and soil are more important at the micro or "site" level. Such a classification permits the application of similar management to similar land units, even though they may not be spatially contiguous.

It appears that in the eastern national forests, and especially those in the central hardwood region, public opinion will continue to direct management away from commodity-based forestry and toward noncommodity uses. That is not to say that commodity uses will be excluded, but they will most likely be at a lower priority on public lands than on private lands.

TABLE 13 A Hierarchical Approach to Land Management: Mapping Criteria for Ecosystem Units at Different Scales with Examples

| | | | EXAMPLES OF UNITS | |
SCALE	NAME OF UNIT	CRITERIA	LOWLAND SERIES	HIGHLAND SERIES
Macro	Region or zone	Ecoclimatic zone (Köppen 1931)	Temperature semiarid (BSk)	Temperate semiarid regime highlands (H)
Meso	Landscape mosaic	Land-surface form class (Hammond 1954)	Nearly flat plains (AI)	High mountains (D6)
Micro	Site	Microclimate and soil	Normal microclimate over moist soil	Normal microclimate over moist soil

Source: Bailey (1988).

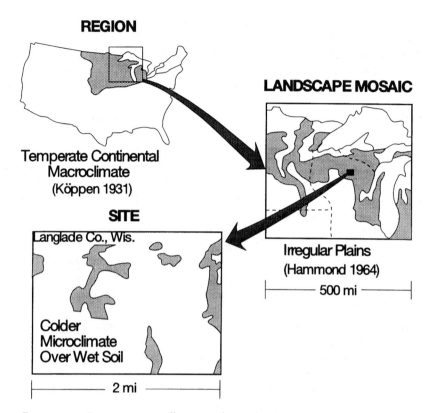

REGION

Temperate Continental
Macroclimate
(Köppen 1931)

LANDSCAPE MOSAIC

Irregular Plains
(Hammond 1964)

├───── 500 mi ─────┤

SITE

Langlade Co., Wis.

Colder
Microclimate
Over Wet Soil

├───── 2 mi ─────┤

FIGURE 121   Ecosystem maps illustrating hierarchical scales (from Bailey 1988).

Industrial Lands. In the central hardwood region, forest industry owns about 10 percent of the total forest land base (Birch 1996). But these lands are generally managed intensively and with focused management goals. Forest industries usually have good inventory data, state-of-the-art data management systems, and Geographic Information Systems. Much of the methodology taught in the current textbooks on forest management is directly applicable to monitoring and regulating the flow of products from forest industry lands. Industrial forest managers often use decision models, such as discussed by Duerr et al. (1979). They also rely on growth and yield predictions from models, such as TWIGS (Belcher 1982), OAKSIM (Hilt 1985), and STEMS (Belcher et al. 1982) to enable them to achieve a "sustained yield" of products from the forest.

For the purpose of management, it is useful to subdivide land into units. The basic unit of management is the **stand**. A stand is a group of trees that are similar in species composition, age, and condition so as to be recognizable from adjacent stands. The stand is the basic unit of siviculture and management, thus regulating the flow of products should be accomplished at the stand level. For some situations, particularly for larger ownerships, it may be convenient to subdivide a forest above the stand level (e.g., compartments, blocks).

Control of harvest is an activity that is performed to ensure a sustained yield of products and is constrained by the site productivity and the size of material needed to grow. The land manager must schedule harvest operations and adjust the cutting cycle or rotation length to provide a continuous flow of the desired type and size of products. In general, two methods are used to accomplish this goal: area control and volume control. **Area control** is a straight-forward method by which an area of the forest that represents the inverse of the rotation length (e.g., for a 100-year rotation — 1/100) can be harvested annually. A modification would be to adjust the allowable harvest for longer cutting cycles (e.g., 5 years), by summing the area allowed for harvest over the years in the cycle. For situations where a good inventory of the forest is available, site productivity is known, and the type of product desired is clearly understood, **volume control** is applicable. Volume control involves setting the allowable cut based on the annual volume growth or **mean annual increment**. This approach has some advantages over area-based systems in that it is unaffected by differential site productivity, and it is much more applicable to silvicultural systems that involve uneven-age management or partial cutting.

For private industrial land, the regulation and flow of products can be accomplished using well-documented methods of forest management. In many ways this type of management represents the prototypic methodology that has traditionally been taught in American schools of forestry throughout the twentieth century. For such lands, the ability of the owner to focus on a commodity product, the economic incentive for management, and the availability of technology to accomplish the job combine to make forest management feasible.

Nonindustrial Private Lands. Output of products and control of product flow from NIPF lands are complicated by the size and diversity of ownerships and their varied goals. Some owners may have little or no interest in timber management, although most owners are willing to sell timber under the right circumstance. Unfortunately, the NIPF owner often has difficulty connecting with the concepts of sustained yield and control of harvest. This is, in part, because they feel detached from the process of forest management and partly because of the long rotation intervals in hardwood forests. For smaller ownerships, typical of NIPF lands, control of harvests to produce a sustained flow of products may necessitate a relatively long wait between harvests, particularly if even-age silvicultural systems are used.

It is also unrealistic to fix a planning horizon that goes much beyond 30 years. NIPF owners are unlikely to relate to very long-term plans, and perhaps with good reason. Many factors, including ownership tenure, goals, and markets, can change over the longer term, and these changes will most likely result in alterations to management. Silvicultural systems that are appropriate to a given forest owner will depend on the owner's goals, the biological constraints of the forest, and the current economic climate. For example, many NIPF owners want to preserve a "forested character" to their woodland and are not willing to clearcut simply because of aesthetic considerations. Ironically, many NIPF owners will readily accept diameter-limit cutting or high-grading without regard for their potential for long-term impoverishment, since these techniques leave trees behind.

Control of product flow through management involves an interplay between silvicultural and management activities. For mixed hardwood stands in the 60- to 90-year age range (typical of many in the central hardwood region), a common suite of intermediate cuttings needed is improvement/sanitation/thinning. In cases where the forest is ready for regeneration, the management choices are more complicated, especially if regeneration of oaks or shade intolerant species is desired and a partial cutting method is favored. Some form of group selection may work to regenerate oaks, and deferment cutting or two-age silvicultural systems will permit the regeneration of intolerant species, such as yellow-poplar, while retaining a forested character to the property. If regeneration of shade tolerant species is acceptable or desirable and if the species are present (such as sugar maple or perhaps red maple) that are capable of producing a reasonable harvest, the selection system may be applied to NIPF woodlands.

The decision regarding a silvicultural prescription is only part of the complex process of management decision-making for the NIPF owner. Decisions regarding harvest control must be made amid a background of landowner objectives, size of property, economic climate, and appropriate silviculture. It is virtually impossible to optimize all of these simultaneously, especially in view of the fact that the economic climate, and possibly landowner objectives, can, and probably will, change over a time frame that is shorter than a forest rotation, or even a cutting cycle. But some principles apply to most cases. For example, for most NIPF owners, *volume control* is more appropriate than *area control*, particularly because most NIPF owners prefer partial cutting practices and uneven-age management to clearcutting and even-age management. Another factor in the decision regarding harvest control is the size of the ownership and productivity of the stands. Both these affect the *total* volume production per year, and this, in turn, affects the cutting interval over which a "viable" timber sale can be conducted. For example, if an ownership involves 100 acres with an average production rate of 200 bd. ft. per acre per year, the total annual production on the tract would be 20,000 bd. ft. If it requires at least 100,000 bd. ft. to attract buyers for a timber sale, the cutting cycle would be at least 5 years. Even though such targets are useful to help regulate and sustain production, they must also be flexible. If an insect or disease threatens a particular species or species group, it might be justified to increase the allowable cut or shorten the cutting cycle in order to perform a sanitation or presalvage cutting before the problem becomes too advanced. Likewise, when markets heat up for certain species, the owner should have the flexibility to capture the value during this "window of opportunity." Conversely, during periods when timber markets are sluggish, the owner should not feel pressured to harvest. Also, when NIPF owners need to liquidate resources to cover unforeseen expenses, they should have the flexibility to use timber sales to help cover such expenses, but doing so while being fully informed that obtaining such income today may sacrifice future income. Flexibility should never be used as an excuse to practice poor silviculture, and short-term gains never justify leaving a stand in an impoverished condition where no good management options exist.

In order to provide NIPF owners with guidelines for control of harvest, foresters need some means of forecasting growth. The literature is filled with yield studies and many mathematical models exist for predicting growth and yield of various species on different sites. These models and projections are useful, and a list of references for such methods is provided at the end of this chapter. But for the purpose of management of most NIPF lands, these techniques are unlikely to be understood or applied. What is needed is a simple, direct, and easy-to-understand set of yield tables that can provide a guideline for NIPF management. Because of the relatively short management planning horizon for NIPF owners, growth projections will probably need to be corrected fairly often. It seems logical to project yield by species, site, and stocking and to be somewhat conservative in projecting growth since it seems more in line with most NIPF objectives to undercut the growth rather than to overcut. Table 14

TABLE 14 Volume Growth (bd. ft.) for Fully Stocked Stands 50–80 Years Old on Below-Average, Average, and Above-Average Sites

SITE QUALITY[a]	UPLAND OAKS (Schnur, 1937)	CENTRAL APPALACHIAN SAWTIMBER STANDS (Miller, 1993)	YELLOW-POPLAR (Beck and Della-Bianca, 1981)
Poor site (oak site index 50)[c]	110[b]	—	—
Below average (oak site index 60)	180	200	140
Average (oak site index 70)	250	300	190
Above average (oak site index 80)	330	400	200
Excellent site (oak site index 90)	—	—	430

[a]Based on site index for upland oaks.
[b]Average annual growth in bd. ft., International scale.
[c]Oak/yellow-poplar site conversion was accomplished using the method of Doolittle (1958).

256 *Ecology and Management of Central Hardwood Forests*

indicates volume growth projections that would be applicable to oaks and yellow-poplar growing on a range of sites that approximates those typically encountered in the central hardwood region. Another way of viewing growth is in terms of mean annual growth. Schnur's (1937) graph (Fig. 122) shows that **mean annual growth** increases dramatically from age 20 to age 50, especially on good sites. It tends to level off at about age 80. Many stands in the central hardwood region range between 60 and 90 years of age and site quality is variable throughout the region, but a majority range between oak site index 55 and 75. Using Schnur's graph and an age of 70 years, mean annual growth should range between about 150 and 280 bd. ft. per acre per year. This range also is apparent in Table 14. Similar ranges for oaks can be obtained from Miller (1993) and using Gingrich's (1967) yield tables. Schlaegel et al. (1969) reported yields for yellow-poplar in West Virginia that were similar to those reported in Table 14.

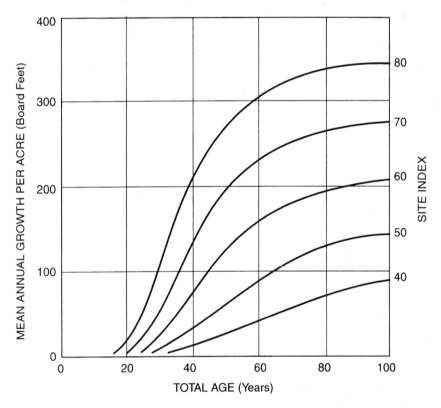

FIGURE 122   Graph showing the relationship between mean annual growth and oak site index (from Schnur 1937).

TABLE 15 Diameter Growth Rates (in inches per decade and rings per inch) for Immature Second-Growth Hardwoods in West Virginia by Three Vigor Classes

	VIGOR CLASS					
	1[a]		2		3	
SPECIES	in./decade	rings/in.	in/decade	rings/in.	in./decade	rings/in.
Yellow-poplar	3.07	6.50	2.38	8.0	1.64	12.0
Basswood	2.02	10.00	1.86	10.5	1.55	13.0
Sugar maple	2.26	8.50	1.59	12.0	1.07	18.5
Cucumber tree	1.83	11.00	1.84	11.0	1.21	17.0
Black birch	2.02	10.00	1.29	15.5	1.05	20.0
Red maple	1.13	18.00	1.55	13.0	1.07	18.5
Beech	1.65	12.00	1.26	16.0	1.25	16.0
Black walnut	1.50	13.00	1.01	20.0	0.77	26.0
White ash	2.43	8.00	1.78	11.0	1.14	17.5
Northern red oak	2.30	8.50	2.04	10.0	1.67	11.5
Shagbark hickory	1.93	10.50	1.01	20.0	0.99	20.0
Pignut hickory	3.80	5.25	1.51	13.0	1.30	15.5
Chestnut oak	—	—	1.81	11.0	0.77	26.0
Blackgum	—	—	0.98	21.0	0.60	33.5
Black cherry	3.63	5.50	1.53	13.0	1.30	15.5
Buckeye	2.55	7.50	0.60	33.5	1.22	16.5
Average (excluding yellow-poplar)	2.05	10.00	1.63	12.5	1.21	17.0

[a]Vigor class 1 trees are dominant or codominant trees with large, vigorous crowns, at least half exposed to full sun. Class 2 trees are codominant with vigorous crowns that are less than half exposed to full sun. Class 3 trees are in an intermediate to weak codominant position with some possible dead limbs and only the top exposed to full sun.
Source: Holcomb and Bickford (1952).

In Table 15, average diameter growth rates are reported in inches per decade and rings per inch (Holcomb and Bickford 1952) for immature second-growth hardwoods in the dominant canopy. These data are based on a sample of 1800 trees in West Virginia. Perky (1992) also reported 10-year diameter growth rates for yellow-poplar and northern red oak trees in West Virginia and southwestern Pennsylvania. These authors indicate a growth range for yellow-poplar, on average to above-average sites, from 2 to 4 in. per decade, with rates approaching 6.5 in. per decade for released crop trees on excellent sites. For red oak, a growth range of 2–4 in. per decade was also reported, and, consistent with this species' silvical characteristics, it outperformed yellow-poplar on average to lower-than-average sites.

TABLE 16 Growth Factor Table (bd. ft./ac./yr.)

dbh (in.)	Rings per Inch											
	4	5	6	7	8	9	10	11	12	13	14	15
4	.282	.225	.188	.161	.141	.125	.113	.102	.094	.087	.080	.075
5	.220	.177	.147	.126	.110	.098	.088	.080	.073	.068	.063	.059
6	.181	.145	.121	.103	.091	.080	.072	.065	.060	.056	.052	.048
7	.153	.123	.103	.088	.077	.068	.061	.056	.051	.047	.044	.041
8	.133	.106	.089	.076	.067	.059	.053	.048	.044	.041	.038	.035
9	.116	.093	.078	.067	.058	.052	.047	.042	.039	.036	.033	.031
10	.105	.084	.070	.060	.053	.047	.042	.038	.035	.032	.030	.028
11	.095	.076	.063	.054	.047	.042	.038	.034	.032	.029	.027	.025
12	.087	.070	.058	.050	.044	.039	.035	.032	.029	.027	.025	.023
13	.080	.064	.053	.046	.040	.035	.032	.029	.027	.025	.023	.021
14	.074	.059	.049	.042	.037	.033	.030	.027	.025	.024	.021	.020
15	.069	.055	.046	.039	.034	.031	.028	.025	.023	.021	.020	.018
16	.064	.052	.043	.037	.032	.029	.026	.023	.021	.020	.018	.017
17	.061	.048	.040	.035	.030	.027	.024	.022	.020	.019	.017	.016
18	.057	.046	.038	.033	.029	.025	.023	.021	.019	.018	.016	.015
19	.054	.043	.036	.031	.027	.024	.022	.020	.018	.017	.015	.014
20	.051	.041	.034	.029	.025	.023	.020	.019	.017	.016	.015	.014
21	.049	.039	.033	.028	.024	.022	.020	.018	.016	.015	.014	.013
22	.046	.037	.031	.027	.023	.021	.019	.017	.015	.014	.013	.012

1. Determine *growth factor* by applying average *dbh* and average rings per inch to above table.
2. Compute annual rate of growth of *basal area, cords, tons, board feet* (International), or *cubic feet* by multiplying amount of good growing stock in each category by the *growth factor.*
3. Multiply *growth factor* by 100 for growth percent.

Source: Ashley (1989).

Table 16 provides yet another simple method for estimating annual growth rates, called the **"growth factor method"** (Ashley 1989). Using this procedure, diameter growth (in rings per inch). and average dbh are the only variables needed. To illustrate the use of the growth factor method for projecting volume growth (bd. ft./ac./yr.), assume a current stand volume of 8000 bd. ft. per acre and an average stand dbh of 16 in. The volume growth would be 416 bd. ft. per acre per year if growth per decade is 4 in. (5 rings per inch) At a growth rate of 2 in. per decade (10 rings per in.), the growth per acre per year would be 208 bd. ft. These growth figures are comparable to those reported in Table 14 for central hardwood species; thus, either Table 14 or 16 or Schnur's mean annual growth curves provide reasonable growth estimates that can become the basis for projecting allowable cut.

Although an annual growth per acre of plus or minus 200 bd. ft. for central hardwood stands seems to be a reasonable average, such growth would imply that a harvest of 2000 bd. ft. of products would be possible on a 10-year cutting cycle. Practical experience of foresters in the field indicates that this is rarely possible on NIPF woodlands. Perhaps this is due to several factors. First, most NIPF forests have not received prior management and their yield is below the potential. Past high-grading may have left the stand impoverished and lacking good growing stock. Second, past overcutting may have reduced the stocking below the B-line (Fig. 123). Thus, the stand is not fully utilizing the resources of the site for productive growth. Sander (1977) discusses these problems and recommends that a prerequisite for management and control of a stand is an inventory of site quality, stand density, and growing stock quality. The total basal area of all trees is the basis for evaluating stocking with regard to the site occupancy. However, only those trees classed as **acceptable growing stock** (AGS) are contributing substantially to future growth. Thus, trees classed as **culls** or **unacceptable growing stock** (UGS) are occupying the site but are not producing a harvestable yield. In order to provide more realistic growth projections for stands on NIPF lands, an assessment of the basal area of AGS in a stand can be used in conjunction with Gingrich's (1967) stocking guide to determine the effective percent stocking. For example, if a stand has a total basal area of 110 sq. ft. per acre (55 of which is AGS), averages 12 in. dbh, and contains 150 stems per acre, the percent stocking for the whole stand is 90 percent (Fig. 123). But if only 55 sq. ft. is in AGS, the *effective* stocking is only about 48 percent. An appropriate adjustment to the annual yield projection for this stand would be to multiply the unadjusted yield projection by 0.48. For example, if an unadjusted annual yield of 200 bd. ft. per acre is obtained for a stand, the adjusted yield would be 96 bd. ft. per acre per year. Another factor reducing yield is mortality. "Normal" mortality is taken into account in the growth projections discussed earlier, but sometimes stresses, such as drought, fire, insects, and diseases, induce rates of mortality that are above the normal amount and must be accounted for.

Use of yield projections in combination with stand size can be useful in establishing the interval between successive harvests or **cutting cycle**, as well as the **allowable cut**. The allowable cut is the amount that can be harvested over time without cutting more than the net growth. The length of the cutting cycle may be, in part, dependent on the length of time required to produce a viable sale from the stand. For example, if a sawtimber volume of 25,000 bd. ft. is the minimum necessary to make a viable sale and a stand is 10 acres in size with a growth rate of 200 bd. ft. per acre, it would require 12.5 years to produce a salable volume from the stand.

A final factor that often leads to confusion regarding reported timber volumes and yields is the persistent use of the Doyle tree scale by loggers and timber buyers. The Doyle scale, because of its known inaccuracy, is seldom used in yield equations; rather, the International scale is generally used. The

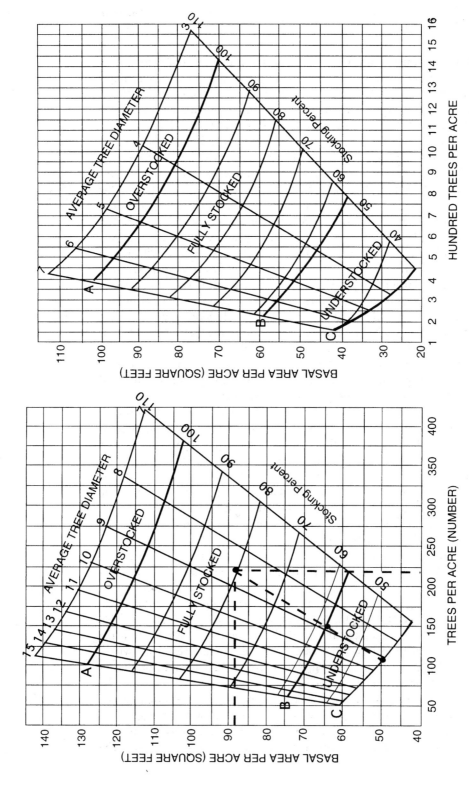

FIGURE 123 Gingrich's (1967) tree-stocking guide for upland hardwoods; relationship of basal area, number of trees, and average tree diameter (tree of average basal area). Curve C shows the lower limit of stocking necessary to reach the B level in 10 years on average sites.

International scale has proved more accurate than the Doyle scale in numerous studies where actual mill production is compared to tree scale. The Doyle scale is known to underestimate tree volumes for smaller trees and the degree of this error is inversely proportional to diameter (Fig. 124). Thus, one reason yield predictions may appear too high is the fact that most of them are given in board feet, International scale, whereas many loggers and foresters are accustomed to thinking in terms of the Doyle scale. A reasonable adjustment for conversion is to multiply International volume times a factor of 0.7 since much of the timber in the central hardwood region averages about 18 in. dbh, and at this diameter, the ratio of volume (Doyle/International) is about 0.7.

Control and scheduling of harvests for NIPF owners is a complex task, but in order for small owners to embrace forest management, rather than exploitation, it must be interpreted to them in a way they can comprehend. Such control and scheduling must also retain flexibility to enable landowners to respond to their personal needs and to capitalize on market changes. Owners must also be able to derive the nontimber benefits from the land that they desire, or they will opt for no management or mismanagement. Trimble et al. (1974) proposed an alternative system for deciding when to harvest individual trees based on their

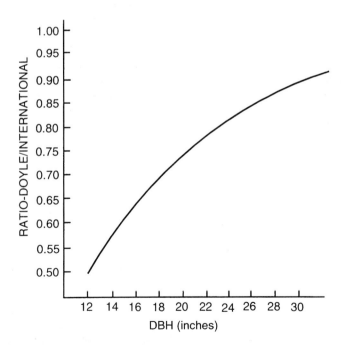

FIGURE 124 Relationship between Doyle and International 1/4-inch tree scales, by tree diameter (dbh).

financial maturity (Table 17). Using this method, one needs to know the site index and decide on an acceptable rate of desired return. When trees reach or exceed a threshold diameter, they would be cut. Perhaps the biggest shortcoming of such a system is that it lacks a true silvicultural objective.

The planning horizon for NIPF owners is generally shorter than that of industrial landowners or for public lands, and to force the issue of rotation-length plans on NIPF owners will serve only to disconnect them from the process of management. Regeneration must be an element in planning and management of stands of any ownership, but for the NIPF, where partial cutting will often be the method of choice, foresters may have to modify their thinking. First, we may need to be more receptive to managing species that are intermediate to shade tolerant, and second, we may have to learn to be more opportunistic in reacting to events like bumper seed crops, rather than laying out rigid silviculture plans that are programmed for failure without intensive management.

The methods reported in this section for predicting growth and yield are admittedly simple and, as a result, subject to error in specific situations. But taking a conservative approach to yield prediction and being careful to err on the side of safety are probably prudent steps for foresters working with NIPF owners. Yields should be adjusted downward to account for stocking, mortality, and timber scaling practices and to accommodate desired nontimber uses of the forest. Even with yields as low as 80 bd. ft. per acre per year, an owner with 50

TABLE 17 Financial Maturity Cutting Diameters for Appalachian Hardwood Species by Red Oak Site Index and Selected Rate of Return

SPECIES	2 percent			3 percent			4 percent			5 percent			6 percent		
	80	70	60	80	70	60	80	70	60	80	70	60	80	70	60
	Minimum Cutting Diameters (2-inch dbh)														
Yellow-poplar	26	26	24	24	22	22	20	18	18	20	18	18	18	18	18
Beech	24	22	22	22	20	20	20	18	18	20	18	18	18	18	18
Black cherry	32	30	30	28	26	24	22	20	18	20	20	18	18	18	18
Red maple	32	30	30	28	26	24	22	22	18	20	20	18	18	18	18
White ash	30	28	28	26	24	24	22	20	18	20	20	18	18	18	18
Sugar maple	32	32	30	28	28	24	22	22	18	20	20	18	18	18	18
Red oak	26	26	24	24	24	22	22	22	22	20	20	20	20	18	18
White oak	24	22	20	22	20	20	20	18	18	20	18	18	18	18	18
Chestnut oak	24	24	22	22	22	20	20	18	18	20	18	18	18	18	18
Other long-lived species	26	26	24	24	24	22	20	20	18	20	20	18	18	18	18

Source: Trimble et al. (1974).

acres of timberland can expect to harvest about 40,000 bd. ft. of timber on a 10-year cycle. At 1997 stumpage prices, the property owner can expect to generate about $12,000 at the end of this interval, with a very small investment. With modest management, that yield could easily double, and if management focused on improving the *quality* as well as quantity, for example, through crop-tree management, the owner could more than triple the 10-year income to over $40,000 (1997 basis).

There is a wide array of tools available for yield prediction of hardwood forests, most of which utilize computers and mathematical models to generate outputs. At the end of this chapter is a list of references for some of these tools. Certain methods will be more useful for corporate or public land management, but others were designed specifically with the NIPF owner in mind, for example, the NED SIP program. Foresters working with NIPF owners may find these tools very useful as management planning aids to improve the accuracy of yield prediction. The fact that a complex tool is used to provide an estimate does not necessitate a complex presentation of the result to the landowner.

Financial Aspects of Managing Central Hardwoods

Public Lands. Most foresters have been trained to view forestry in a capitalistic context. But the economics of forestry on public lands is often more about politics than capitalism. The U.S. National Forests exemplify this principle vividly, and ironically a national forest in the central hardwood region, one of the oldest and largest eastern national forests, was the stage for the landmark Monongahela court decision that has served as a precedent for management policy throughout the federal system. The same well-used arguments seem to resurface over and over with regard to public land management. The preservation of wildlands and noncommodity values are pitted against jobs and economic gain for local communities. And although it is true that the philosophy of Pinchot, that of "wise use" and conservation, has been a guiding principle for management of the National Forests, public lands are seldom held to the purely capitalistic dictum of making a profit. Conversely, as most foresters know, preservation is a mythical notion, borne out of naiveté and championed by biocentric thinkers, most of whom have had little direct experience with the nature they so enthusiastically espouse.

Certainly public lands are important in the financial and economic fate of many communities, and especially so in the western states. But given the political climate in the eastern states, including the central hardwood region, and the fact that public lands overall account for approximately 5 percent of the forested area, their impact on the economy of the region is often more important in terms of tourism and no-timber values than for timber production.

Private Industrial Lands. As much as public forestlands are governed by political forces, private industrial lands are controlled by capitalism. Most of the textbooks dealing with forest economics and forest finance take an approach that is inclined toward "forestry as a business," and there is very little that this book can add to these treatments.

Corporate finance is complicated by the fact that in integrated forest industries, the greatest proportion of profit is made at the downstream end of the manufacturing process. Timberlands are generally held more as a reserve and a hedge against higher stumpage prices. In reality, nearly all forest industries in the central hardwood region procure the vast majority of the wood from NIPF owners, who own over 70 percent of the timber within the region. Thus, even though forest industries are capitalistic in nature, they probably do not apply a strict profit-making requirement to their own woodlands operations. Perhaps industries' main economic impact is in how they view the NIPF lands and in the creation of demand for stumpage.

Nonindustrial Private Lands. The NIPF lands, which constitute about 75 percent of the forestland in the central hardwood region (Powell et al. 1993), are also the most vulnerable to exploitation and short-sighted management. In other words, the "quick buck" mentality is all too prevalent among NIPF owners, and forestry consultants paid on commission or industry procurement foresters sometimes exacerbate the problem. In the succeeding paragraphs, several financial and economic factors will be discussed relative to NIPF owners. These will include incentives for forest management, the impact of the tax structure, and ways of improving profitability of forestry while using silviculture and management that meets the goals of the NIPF owner.

A large array of **federal assistance programs** have been enacted to create incentives for forest management on NIPF land. Even the establishment of state forestry agencies, cooperative extension agencies, and USDA, Forest Service State and Private Forestry units are aimed at assisting NIPF owners. But in spite of this effort, only a small portion of NIPF owners have a forest management plan, and less than one-third of the timber sales surveyed by Egan (1996) in Pennsylvania utilized a forester or a timber sale contract. Many characteristics of NIPF owners affect the way they approach management. For example, in Pennsylvania, the average size of ownership is approximately 23 acres, and ownership of very small acreage is typical throughout the central hardwood region (Powell et al. 1993). A substantial portion of NIPF owners do not live on their forestland, and they have very diverse management goals often with timber production being subordinate to other goals. In a study of Arkansas owners, Williams et al. (1996) found that, in the Ozark/Ouachita region, 30 percent of the NIPF owners were retired. They also found that two-thirds of the NIPF owners had never applied any sort of silvicultural management to their forests.

There are many studies that indicate that forest management, even on small tracts, can be profitable. For example, commercial release of 5 northern red oak crop trees per acre in a 10-acre stand was projected to yield at an annual rate of return of 11.6 percent over a 10-year cutting cycle (Perkey 1992). Management for quality sawlogs can multiply income dramatically. Smith et al. (1979) reported that grade 1 sugar maple logs brought a price that was seven times greater than grade 3 logs of the same diameter.

Thus, if forestry is profitable, even on small tracts, and incentive and assistance programs are available to NIPF owners, why are they reluctant to adopt them? Egan and Jones (1993) indicate that it may be a combination of factors, ranging from the fact that the majority of NIPF landowners do not hold land for the purpose of producing timber. They also found that, although NIPF owners often articulate a stewardship ethic, the expression of this ethic was poorly correlated with application of stewardship practices, even when substantial incentives were available. They suggested that the "extension message" regarding timber management should be woven into a multiple-use fabric rather than being pushed as a separate issue. They also suggested that NIPF landowners are likely to emulate their peers, and the application of stewardship practices by one owner will likely have a snowball effect on neighboring landowners.

Gaddis (1996) reviewed the various cost sharing, incentive, and assistance programs available to NIPF owners. These include the Agriculture Conservation Program (ACP) of 1936, the Soil Bank Program of 1956, and the Forest Incentives Program (FIP) of 1973. The first two were under large agricultural programs within which tree planting and other forest practices were included. The FIP was the only one aimed specifically at forestry and where forest harvesting was an explicit objective. According to Gaddis, these programs have been successful in that most of the forest plantations established under them are being retained by their NIPF owners.

The Forest Stewardship and Stewardship Incentives Programs (SIP) were authorized under the Farm Bill of 1990. These programs are available to NIPF owners (up to 5000 acres in size) and differ from the other assistance programs in that they feature multiple-use management planning as a precursor to management practices. The program, although federally funded, is administered by the states (Lacey 1991). Cost sharing is provided both for the management planning and management practices under SIP. Some states offer property tax breaks as incentives to NIPF owners to enroll in the Stewardship Program, and response has been brisk. In Maryland, for example, after 4 years of stewardship planning, almost 70,000 acres of NIPF land have been included in the program (Van Hassent 1994), which amounts to about 3 percent of the NIPF land in the state. Although this is a significant amount, at this rate it will take 133 years to complete the job in Maryland. Stewardship planning is progressing at a similar rate in other states in the central hardwood region. Perhaps the significant question regarding planning and NIPF owners is: Are management plans being implemented? Very little information exists to confirm or refute this question.

The **tax structure** for inheritance taxes, property taxes, severance taxes, and income or capitol gains taxes on timber sale revenue is undoubtedly one of the most important factors affecting the way NIPF owners manage their land. Many such owners initially receive their land through **inheritance**. A graduated tax schedule applies to inheritance under the Federal Unified Estate and Gift Tax schedule. Haney and Siegel (1993) have prepared a detailed manual to assist forest landowners to make the best choices relative to their estate. For example, they cautioned against the practice of undervaluation of the estate, particularly the timber resource, at the time the estate is settled. Such a practice may save estate taxes but will result in higher tax if timber is sold from the estate later. They also recommended that in an estate settlement involving timberland, the value of timber should be appraised separate from the land.

The purpose of appraising timber separate from land is in order to establish a "basis" for the timber value. At the time of timber sale, tax on the income may be determined as **regular income tax** or **capitol gains tax**. For most people who have income from other sources, the latter method will prove most advantageous, especially where a high basis exists (either in the initial value or from subsequent investments made in management). The taxable income becomes the gain in value over the basis. Jenkins (1993) and Siegel et al. (1995) have developed detailed manuals for use by tax preparers in relation to timber sales and investments. These and other publications relating to taxes and forestry are referenced in the Literature Cited section. Although it is a good idea for foresters and property owners to be familiar with the tax laws, it is probably wise for the NIPF owner to use an accountant who is familiar with timber taxation, especially where relatively large amounts of income are involved.

When timber is cut, in some states, such as West Virginia, the owner of the timber at that time must pay a "**severance tax**." In West Virginia this amounts to 3.22 percent of the gross value. In cases where a lump sum sale occurs the buyer then owns the timber and is responsible to pay this tax. In so-called pay-as-you-cut sales, the property owner may be responsible to pay the severance tax.

Property tax is another obligation of the NIPF owner, and property taxes vary from state to state. Chang (1996) classified the taxation systems in all 50 states. The states within the central hardwood region utilize a variety of taxation systems. The majority use a productivity-based system, which taxes on the basis of the capitalized value of the gross or net mean annual revenue from the forest. Chang indicates that such a taxation system is probably most fair and favorable to investment in long-term forest management, whereas a straight *ad valorem* tax, or a tax on the current value of timber, tends to encourage owners to liquidate their timber earlier in order to avoid paying tax on it.

The **profitability** of forestry for the NIPF has been debated using many scenarios. As a strict economic investment, one should probably compare the income that could be generated (e.g., through interest) from the money that would be derived from the sale of land and timber to the income that can be generated from sale of forest products. But NIPF owners seldom look at land or forests in strict investment economic terms. That is, they do not hold the land and timber strictly as a business enterprise, and they have management goals that may or may not include generation of revenue as a primary concern. However, NIPF owners do incur expenses in holding land, whether or not they explicitly manage timber. For example, they must pay property taxes, and they may have to maintain improvements on the property, such as roads, fences, buildings, and fields. Thus, whether or not timber production and income generation are primary reasons for holding land, NIPF owners are often willing to sell timber to offset those expenses, especially if selling timber does not substantially diminish other uses of the land.

Foresters who work with NIPF owners should be sensitive to the owner's goals and should attempt to prioritize management so as to maximize the benefits from the forestland while providing income within the appropriate context. Miller (1993) reviews several partial cutting techniques that might be applicable to NIPF land. He indicates that some sort of uneven-age silvicultural system is often appropriate for NIPF owners. If the single-tree selection system is chosen, the owner must have suitable shade tolerant species in the stand for management. Choosing the species to favor in management operations can have a large impact on profit. For example, species such as hickory, beech, and blackgum typically bring low stumpage prices. Species such as chestnut, black and scarlet oaks, red maple, yellow-poplar, and ash bring intermediate prices, while northern red and white oaks, black cherry, and sugar maple bring higher prices. Quality and grade are also important issues. Perkey (1992) reports that rates of return in the range of 6–17 percent can be obtained on a 10-year cutting cycle for northern red oak crop trees. Northern red oak combines several attributes for crop-tree management. It is faster growing than most of its competitors, it is adaptable to a wide array of sites, and it has the potential to produce high-value veneer or grade lumber products.

Several economic considerations of uneven-age systems are discussed by Smith and DeBald (1975). They indicated several factors that could lead to greater profit. These include avoidance of logging-injury, which can destroy grade potential in residual trees, removal of poor-quality trees in the poletimber size class (improving markets for smaller diameter materials is providing commercial potential for this), raising the diameter limit to produce larger higher-quality trees, and reducing the cutting cycle as low as possible to get quicker returns on investment. Trimble et al. (1974) proposed a partial cutting system that is applied like a **flexible diameter limit**, where trees are cut when they drop below an acceptable rate of return, based on growth and quality. Table 17 is an example of how this method might be applied to some common species in the central hardwood region. In addition to cutting trees above the threshold diameter, improvement cutting can be applied at each entry to low-potential trees of any diameter.

In the final analysis, most NIPF owners are not in the "business" of forestry, and they will probably continue to hold their land, whether or not they could take the money for sale of the land and earn a higher return on it by investing it elsewhere. But, in times of personal need or when timber market conditions provide enough temptation, NIPF owners, who otherwise espouse good intentions, may exploit their timber. In many cases, they are ignorant of the consequences of practices, such as diameter-limit cutting and high-grading. The economic opportunity exists, through planned resource management, to enable many NIPF owners to increase the income from their timberland while improving it for the future and providing for an even flow and sustained yield of products. With proper management this can be accomplished while still providing the nontimber products from the land. It is one of the challenges to foresters to facilitate this process.

Forest Health Management

General Concepts. **Forest health** is "a dynamic condition that relates to the vigor of trees and associated organisms and is reflected in their physiological performance (survival, growth, reproduction, etc.) as compared to accepted norms for similar environments (Hicks and Mudrick 1994). One of the most important goals of forest management is the maintenance of a healthy forest condition. External stresses (drought, competition, fire, injury, defoliation, etc.) often weaken trees, which predispose them to secondary insects and diseases, further reducing their vigor (Manion 1981; Waring 1987). Mortality results when the tree can no longer defend against these organisms. Although tree death is a normal ecological process (Franklin et al. 1987), forest management offers an opportunity to make silvicultural and management decisions that minimize the adverse impacts of damage resulting from external stresses. Forest health management should be an integral part of the overall forest management strategy. The forest health management decision process is illustrated by the gypsy moth "action model" presented by Gansner et al. (1987) (Fig. 125). Hicks (1993) demonstrated how the forest health considerations can be integrated into the general forest management decision process (Fig. 126).

Forest health can be addressed in a preventative mode or a treatment mode. In the former, the approach may be silvicultural, where the manager attempts to control species composition, vigor, or other conditions of the stand so as to make it less **susceptible** to attack or less **vulnerable** to damage. The latter (treatment mode) attempts to treat the problem after it arises. A good example of this sort of approach is a suppression program (fire, insects, diseases, etc.). Both approaches are appropriate for use in forest health management programs under certain conditions. Forest health management, like other aspects of forest management, requires information and appropriate methods for processing and organizing the information. As an example, a decision support system (GypsES) is an so-called "expert system" developed for aiding in decision-making regarding the gypsy moth (Twery et al. 1994). **Geographic Information Systems** (GIS) are a powerful computer-mediated tool to assist in organizing information to facilitate forest management, including forest health management

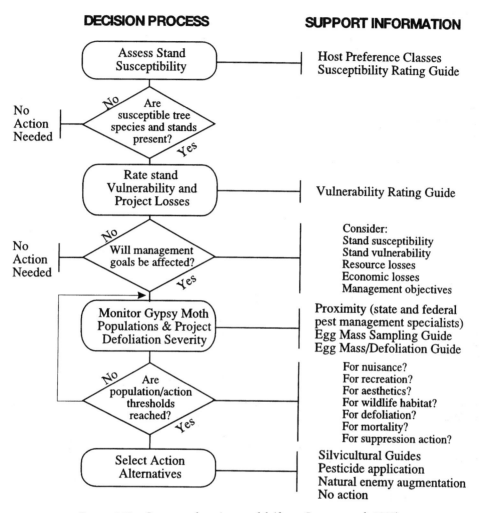

FIGURE 125 Gypsy moth action model (from Gansner et al. 1987).

(Fleischer et al. 1992b). Using GIS, databases for forest hazard conditions, insect and disease populations, resource management zones, topography, roads, and so on can be overlain in various combinations (Hutchinson and DeLost 1993) (Fig. 127).

Forests in the central hardwood region have sustained episodes of abuse, introduction of pests, and mismanagement; and as a result, some serious forest health issues exist in the region. In spite of this, the forests of the central hardwood region are generally diverse and healthy. Currently, programs exist to monitor forest health conditions in the northeast (Twardus et al. 1995), and a

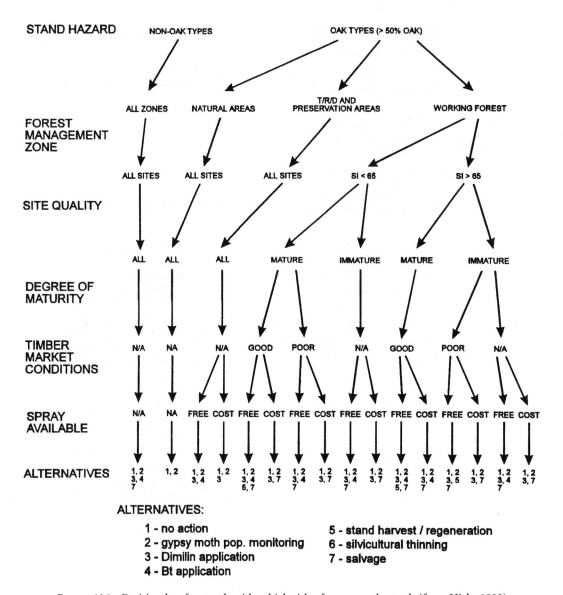

FIGURE 126   Decision key for stands with a high risk of gypsy moth attack (from Hicks 1993).

strategic plan exists to protect the health of American forests (USDA, Forest Service 1993), including those of the central hardwood region. Hardwood forests offer some unique challenges and opportunities for forest health management. The fact that species diversity is high is a positive factor in reducing the potential for pest outbreaks and devastating losses when they

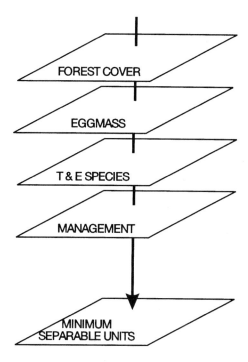

FIGURE 127   Layers of information in a typical GIS (from Hutchinson and DeLost 1993).

occur. But the ownership pattern, with NIPF owners predominating, makes it difficult to apply management uniformly over the region. Although there is a great diversity within the region, the preponderance of oaks and their local abundance on certain sites have led to some major problems with defoliating insects, such as loopers (Wagner et al. 1995) and gypsy moth (Herrick and Gansner 1987), that prefer oaks. In management discussions, several major forest health problems of the central hardwood region will be discussed. These will focus on stressing agents that have a relatively widespread and significant impact and those that have potential to be managed (Anderson 1994). They include defoliating insects, logging damage, forest fires, white-tailed deer, and air pollution.

Insect Defoliation. Defoliating insects pose a serious threat to central hardwood forests. Native defoliators, such as the eastern tent caterpillar, locust leaf miner, yellow-poplar weevil, orange-striped oakworm, fall cankerworm, and the looper complex, sometimes reach outbreak levels (Butler and Kondo 1994; Wagner et al. 1995). When these occur, widespread forest damage is often the result. But by far, the greatest threat to the central hardwoods is from the introduced **gypsy**

TABLE 18 Tree Species Fed on by Gypsy Moth by Preference Classes

---

Susceptible: Species Readily Eaten by Gypsy Moth Larvae During All Larval Stages

Overstory	Apple, basswood, bigtooth and quaking aspen, birch (gray, paper, and river), larch mountain-ash, most oak species, sweetgum, willow
Understory	Hawthorn, hazelnut, hophornbeam, serviceberry, witch-hazel

Resistant: Species Fed Upon When Preferred Foliage Is Not Available and/or Only by Some Larval Stages

Overstory	Beech, black (sweet) and yellow birch, blackgum, boxelder, buckeye, butternut and black walnut, sweet and black cherry, chestnut, eastern cottonwood, cucumbertree, American and slipper elm, hackberry, eastern hemlock, most hickory species, maple (Norway, red, and sugar), pear, most pine species, sassafras, most spruce species
Understory	Blue berries, pin cherry, chokeberry, paw paw, persimmon, redbud, sourwood, sweetfern

Immune: Species That Are Rarely Fed Upon

Overstory	Most ash species, bald cypress, norther catalpa, eastern redcedar, balsam and fraser fir, American holly, horsechestnut, Kentucky coffee-tree, black and honey locust, silver maple, mulberry, sycamore, yellow-poplar
Understory	Most azalea species, dogwood, elderberry, grape, greenbrier, juniper, mountain and striped maple, most *Rubus* species, sheep and mountain laurel, spicebush, sarsaparilla, and most viburnum species

---

Source: Ghent (1994).

moth. Gypsy moth has a wide host preference but expresses a distinct preference for certain species, such as oaks and aspens. Ghent (1994) classifies overstory and understory species into three categories relative to gypsy moth (susceptible, resistant, and immune) (Table 18). Gypsy moth was introduced to the northeastern portion of the central hardwood region in 1869 and has gradually been spreading southwestward. It is currently endemic to the area north and east of a line extending from the southern tip of Lake Michigan, across Ohio, West Virginia, Virginia, and North Carolina (Fig. 128). Gypsy moth has caused heavy tree mortality in the oak-dominated portion of Pennsylvania, which has resulted in significant changes in forest composition in these areas (Herrick and Gansner 1988). In a survey of Forest Inventory Analysis (FIA) data that included the states of Pennsylvania, Maryland, Ohio, West Virginia, and Kentucky, the USDA, Forest Service (1995) indicated that species such as maples showed an

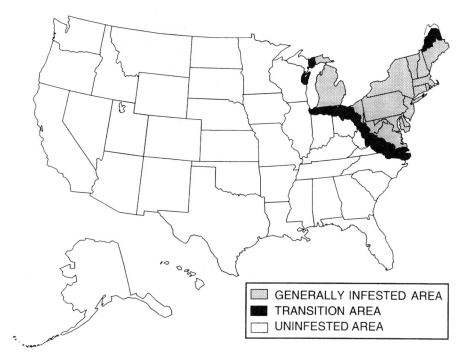

FIGURE 128   The endemic range of gypsy moth in the northeastern United States, 1992 (from USDA, Forest Service 1993).

increase in stocking, while species such as black locust and chestnut oak declined (Fig. 129). This trend was particularly prominent in the Ridge and Valley Province and in areas with a history of gypsy moth defoliation. Oak decline syndrome seems to be occurring even outside the current range of gypsy moth and may be associated with other stressors, including defoliation by native insects, drought, fire, and a variety of secondary organisms. The loss of black locust is probably a combination of the normal senescence of the species, due to aging and the recent heavy outbreaks of locust leaf miner. The increase in maples seems to be in response to the canopy gaps created when overstory trees die and the shade tolerant maples are released to grow.

As illustrated in Figures 125 and 126, **forest hazard rating** is a key component of forest health management (Hicks et al. 1987). In the context of forest health, **hazard** describes a condition of the forest under which a damaging event is likely to occur. **Risk**, on the other hand, takes into account the *probability* of occurrence. In gypsy moth management, high hazard is usually associated with stand conditions (e.g., abundance of susceptible hosts), and risk is associated with the presence of high populations of the insect in combination with high hazard (Fig. 130). Assessment of gypsy moth population status is

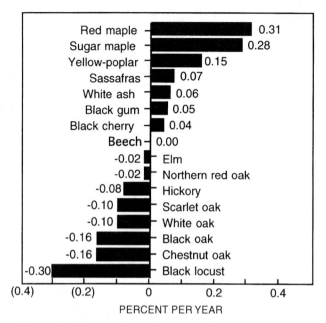

FIGURE 129   Change in relative stocking, by species (from USDA, Forest Service 1995).

usually accomplished by counting the number of current-year egg masses that are present. A sequential sampling method (Fleischer et al. 1992a) is generally used where egg masses are counted in $\frac{1}{40}$-acre plots. Sampling is continued until it is likely that the actual number of egg masses is either above or below some threshold (Fig. 131).

Establishing gypsy moth hazard involves an inventory of susceptible species and monitoring other conditions that provide habitat for gypsy moth and predispose trees to defoliation (Valentine and Houston 1979; Houston and Valentine 1985). For example, Valentine and Houston identify oaks (especially in the white oak group) and trees with numerous deep bark fissures and bark flaps (structural features for egg laying) as most susceptible. In addition to susceptibility, the concept of vulnerability is also very important in forest health management. A tree is vulnerable when the attack or exposure to a pest causes damage. Hicks and Fosbroke (1987a) indicated that susceptible oaks become more vulnerable to damage and mortality when defoliation occurs in conjunction with other stressing events, such as drought or late spring frost. Oaks growing on high-quality sites may be less susceptible to gypsy moth, but given heavy defoliation, such trees may be more vulnerable due to their relatively large crowns that must be refoliated (Fosbroke and Hicks 1989).

Choosing an appropriate management strategy for gypsy moth involves knowledge of susceptibility, vulnerability, hazard, and risk in order to predict

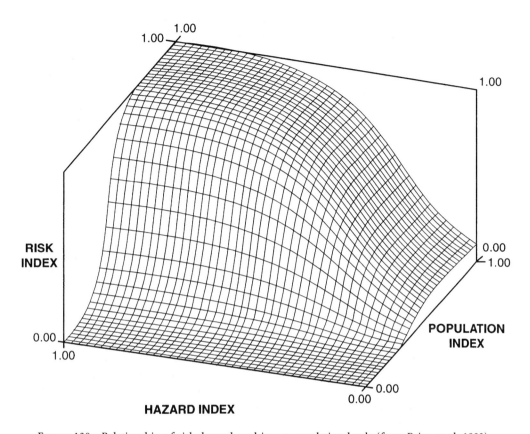

FIGURE 130 Relationship of risk, hazard, and insect population levels (from Paine et al. 1983).

the probable impact of gypsy moth on the forest. Colbert and Racin (1995) have published a stand damage model that utilizes the JABOWA forest growth model (Botkin et al. 1972) as a basis for incorporating gypsy moth impact on forest stands.

As stated earlier, a silvicultural (preventative) strategy is one approach to gypsy moth management. Such an approach is especially appropriate when adequate lead time exists and when stands contain a substantial component of manageable, immature, and resistant, or immune species. Gottschalk (1993) reported silvicultural guidelines for gypsy moth management. Figure 132 (a, b, and c) illustrates Gottschalk's approach for making silvicultural decisions. Gottschalk proposes five different silvicultural treatments for gypsy moth management. The **presalvage thinning** is a treatment recommended for immature, predominantly oak stands that are fully stocked. A **sanitation thinning** is recommended for similar stands, except with a more diverse species mixture,

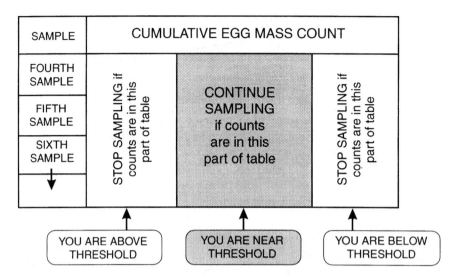

FIGURE 131   A method of sequential sampling to determine gypsy moth egg mass populations for given thresholds (from Fleischer et al. 1992a).

including non-oaks. The **presalvage harvest** is recommended for mature, oak-dominated stands. A **presalvage shelterwood** is a modification of the above where residual oaks are left to serve as a seed source to initiate a new stand. A **sanitation conversion** is a technique recommended to convert high-hazard oak stands to less susceptible species, such as pines or other mixed hardwoods. As an aid to decisions regarding which trees to remove or leave in thinnings or shelterwood operations, Gottschalk and MacFarlane (1992) have published a photographic guide to crown vigor of oaks. Perhaps the best justification for silvicultural management of gypsy moth is the fact that the recommended treatments are beneficial to the stand regardless of gypsy moth, and any reduction in hazard is a bonus.

Even when good silviculture is practiced, in stands with a substantial component of susceptible host species, gypsy moth populations may build up to defoliating levels. Because of the investment made in silvicultural treatments, it is particularly important to protect these stands from damage. Thus, it is important to maintain good information on gypsy moth population density. At such time as populations reach a threshold where damage is likely to occur (500–1,000 egg masses per acre), the manager will probably want to be ready to intervene with **direct suppression**. In a forestry context, this usually equates to some type of spray program using biological or chemical pesticides. States such as Ohio, Maryland, Virginia, and West Virginia have state spray programs that either pay fully or cost share with landowners for spraying. In strictly

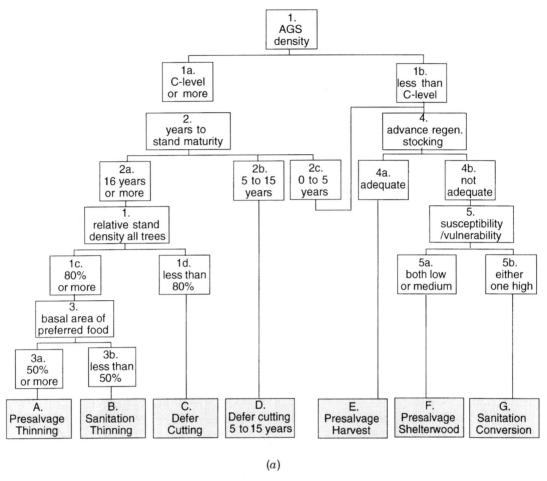

(a)

FIGURE 132a   Gypsy moth decision chart for silvicultural activities (from Gottschalk 1993). Defoliation not imminent for 1 to 3 years or longer.

economic terms, at current costs, spraying is very cost effective (Hicks et al. 1989b). Pesticides used for gypsy moth have evolved from the non-selective chlorinated hydrocarbons to biologicals, such as *Baccillus thuringiensis* and NPV virus (gypchek), and growth regulators, such as diflubenzuron. But because of their expense and the possibility that non-target organisms may be killed (Martinat et al. 1993; Butler and Kondo (1994), pesticides should be targeted and used carefully and appropriately. For example, in gypsy moth management, spraying should only be done where there is a high hazard *and* a high risk of

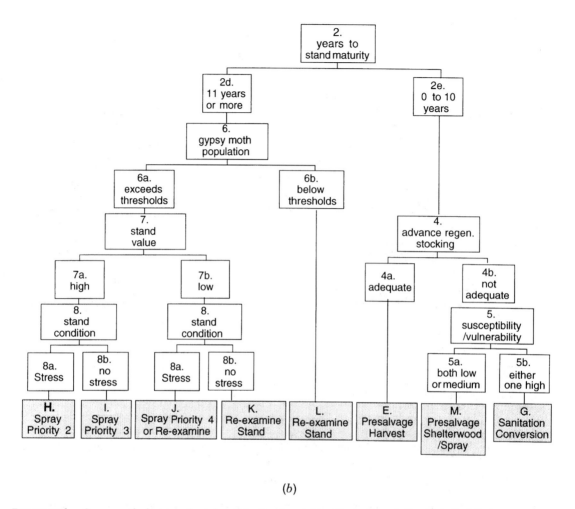

(b)

FIGURE 132b   Gypsy moth decision chart for silvicultural activities (from Gottschalk 1993). Defoliation imminent within 1 to 3 years or now occurring.

attack. And the material used should suit the need. The growth regulator, diflubenzuron, has proved very effective in controlling gypsy moth but can cause problems when used near water since it can kill non-target aquatic arthropods.

Forest Fire. Fire is generally not as common as a disturbance in the central hardwood region as in many conifer-dominated ecosystems. But, as discussed in Chapter 2 of this book, fire has had a significant role in the history of the central hardwoods. Fire was used as a tool by indigenous people of North

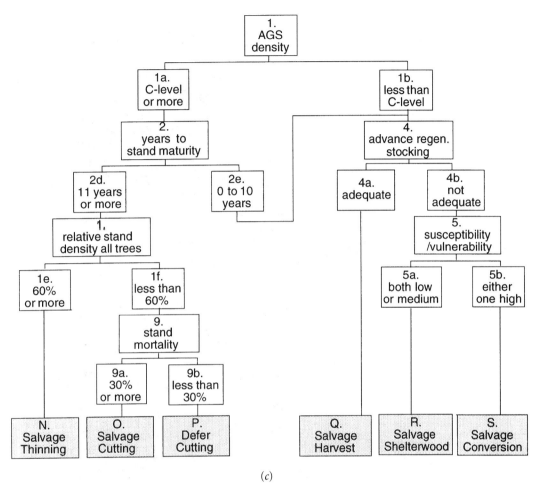

(c)

FIGURE 132c   Gypsy moth decision chart for silvicultural activities (from Gottschalk 1993). Defoliation has already occurred.

America and wildfires burned through many portions of the central hardwood region following the exploitive logging era around the turn of the century. Abrams (1992) examined the ecological role of fire in the central hardwoods and concluded that the dominance of oak in many areas can be linked to a history of fire. Stumps of American chestnut, due to their decay resistance, have preserved a record of fire since the 1930s and the presence of charcoal on chestnut stumps attests to the frequency of fire, with a high proportion of chestnut stumps having charcoal present (Fig. 133). Wildfires in the central hardwood region usually occur in the spring or fall during dry periods and are generally limited to ground fires that consume leaf litter and dead vegetation.

FIGURE 133   American chestnut stump in Doddridge County, West Virginia showing charcoal which indicates past forest fires.

The heat from these ground fires is sufficient, however, to cause basal wounding of trees, especially on thinner-barked species, such as beech, maples, and yellow-poplar, but even thicker-barked oaks can be damaged or killed by very hot fires. These wounds serve as entry courts for fungi that lead to wood decay and general stress of the trees. It may also lead to other problems, such as soil erosion, as well (Atkins and Wimer 1990). Some areas of the central hardwood region are more fire prone than others. For example, the area of southern West Virginia and eastern Kentucky is especially fire prone. In the years 1986, 1987, and 1991 Mingo County, one of the 10 fire-prone counties in southern West Virginia, sustained annual fires over 11–56 percent of its area (Fig. 70).

In spite of the fact that damaging wildfires still occur in some parts of the central hardwood region, fire control efforts have been quite successful throughout the region. Fire suppression programs are generally administered at the state level, but federal funding also provides support, dating back to the Clarke-McNary Act of 1924. Most states conduct fire surveillance via aircraft, especially during the high-risk periods and in areas of high hazard.

Management of forest fire can best be accomplished by prevention or early suppression. Most fires in central hardwoods are human-caused; therefore, prevention includes not burning during periods of high fire danger, limiting access to the property during such periods, and posting the land to discourage trespass. When fires do occur, good access is critical for suppression. The owner/manager can greatly improve the ability to control fire by providing a good road system into the property. Wildfires in central hardwood forests, although less spectacular than in conifer dominated ecosystems, often cause subtle damage that has a long residual effect. Thus, fire prevention, early detection, and suppression are critical to the management of central hardwood stands.

Logging Damage. Logging damage includes direct injury to trees resulting from falling and skidding trees, including broken tops, basal wounding and root injury. In addition, damage to the site may occur in the form of compaction or erosion along logging roads and skid trails. Several factors contribute to the occurrence of logging damage in the central hardwood region: first, many hardwood species have relatively large spreading crowns, which are likely to cause damage when felled; second, many NIPF owners prefer partial cutting methods so a residual stand is usually left; and finally, central hardwood stands often occupy steep and rocky sites. The logging method of choice in the central hardwood region is ground-based skidding because it combines high productivity with flexibility. However, ground-based skidding has a great potential for causing damage, if used improperly (Nyland 1990).

Logging injuries have the potential to cause decay and long-term decline of injured trees (Figs. 134 and 135). Nyland (1986) indicated that basal wounding of trees almost always leads to lower-grade butt logs. This type of damage is most likely to occur during skidding. Logging damage is particularly significant on potentially high-value crop trees. Breakage of limbs and crown damage are more likely when larger trees are being felled, leaving smaller ones—such as occurs with diameter-limit cuttings and crown thinnings. Lamson et al. (1984) found that as residual stand density in 60-year-old thinned stands decreased, the proportion of basal area and trees damaged increased. With a residual stand density of approximately 80 sq. ft. of basal area, 25 percent of the trees suffered damage. For a residual stand of 56 sq. ft., 60 percent of the trees were damaged.

Shovel logging, cable yarding, and truck-mounted crane systems of logging may offer alternatives that will reduce the amount of logging damage. When conventional ground-based skidding is used, care should be taken when locating

FIGURE 134   Recent logging damage (bark skinning) on a chestnut oak.

skid roads so as to avoid high-value crop trees. Skidding should never be done during periods when soils are water saturated. At these times, the potential for compaction and/or erosion is greatly increased. Skid roads should not traverse steep slopes. A subsequent discussion of forest harvesting deals with logging roads, their design, use, and reclamation in greater detail. Finally, logging done during the early spring can result in greater damage than logging done at other times. Bark tends to "slip" more easily from trees damaged at this time of year, in addition to the more likely occurrence of saturated soils at this time. When logging in the spring, extra care should be taken to avoid logging damage.

Deer Damage. Around 1900, due to loss of habitat from agricultural clearing,

FIGURE 135   A logging wound on a yellow-poplar that has begun to callus over. Prior to healing, such wounds serve as entry points for decay.

logging, and fires and uncontrolled hunting, the white-tailed deer had virtually disappeared from the central hardwood region (Smith 1966). The reestablishment of deer over the past 75 years is one of the dramatic "success" stories in wildlife management. In fact, deer management has been so successful that deer have overpopulated their habitat in portions of the central hardwood region. Their impact is primarily on regeneration (Tilghman 1989; Jones et al. 1993). But this may have a longer-term effect of reducing the future value of forest stands (Marquis 1981a; Marquis 1981b). Deer are regarded by Stout et al. (1995) as a serious threat to forest health in the Allegheny Plateau of Pennsylvania. In some areas, deer browsing has influenced species composition of forest regeneration, where preferred browse species, such as sugar maple, are all but eliminated from the understory (Redding 1995). Lorimer (1993) indicated that deer browsing was "clearly a limiting factor for oak regeneration in some places"—particularly in the Allegheny Plateau. Deer populations exceeding 30 per square mile are capable of creating savanna-like conditions (Healy and Lyons 1987).

But the deer "problem" is often one of perspective. To many deer hunters, success is viewed as sighting numerous deer or bagging a deer easily. Hunters exert political clout, and game regulation laws are often controlled by a political process. In the absence of natural predators, hunting is often the only control mechanism over deer herd size, and bucks-only seasons permit herds to increase almost unchecked.

Managing deer problems, like other forest health problems, can take a **proactive** or a **reactive** approach. The proactive approach is to manage deer numbers through hunting regulation. This is the ideal approach since it *prevents* the problem. But in view of the political nature of hunting regulation and the different perspectives on wildlife among segments of the public, it is often impossible for forest managers to achieve an ideal balance between deer and trees. Thus, foresters are frequently left with a reactive approach. Many different strategies have been attempted for improving forest regeneration where deer have become overpopulated. The signs of overpopulation are usually obvious with heavy and conspicuous browsing (Fig. 136). Deer census counts are often available from state game management agencies and these figures can give an indication as to the potential effects of deer on tree regeneration (Fig. 137).

The approaches to management of forests, given overpopulation by deer, are (1) **exclusion/protection**, (2) **inhibition**, (3) **satiation**, and (4) **repelling**. Two primary devices are utilized for exclusion/protection—fences and tree shelters (Fig. 138). Inhibition is attempted by making it difficult for deer to reach seedlings, usually by scattering and/or piling brush from logging so as to discourage browsing. Satiation is achieved by creating so much food (under-growth) that deer cannot eat it all, and, therefore, some portion is able to achieve sufficient height to get above the reach of deer. Repelling is attempted when some material (repellent) is placed on the trees or in the forest that discourages deer from browsing.

Deer exclusion and protection have been used experimentally in several studies. When fences (either electric or non-electric) are well constructed and maintained, they work well. But fences are very difficult and expensive to build. They need to be at least 8 ft. tall and must conform closely to the ground to prevent deer from going over or under. Once a fence is breached, animals will use the same point to reenter. Fences are also very hard to maintain due to falling limbs and trees. Snow can also create a problem by grounding out electric fences. Given all the above, it is seldom feasible to fence on an operational scale.

Tree shelters are another exclusion device that have been tried experimentally and operationally. Shelters must be at least 5 ft. tall in order to allow trees

FIGURE 136 Heavy deer browsing on sprouts from a red maple stump.

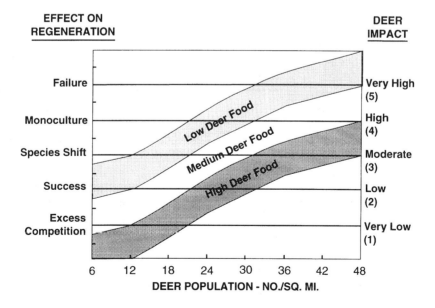

FIGURE 137 Deer impact index (from Marquis et al. 1992).

FIGURE 138   Fencing and tree shelters to protect oaks from deer browsing in Pennsylvania.

to get above the reach of deer. Shelters are also expensive and difficult to install. On an operational scale, tree shelters might be used in limited areas, where regeneration is particularly problematic, but are not likely to be used in large-scale operations because of expense and difficulty.

Inhibition by piling brush may be particularly applicable to central hardwood situations where many harvesting operations involve partial cutting, thus opening canopy gaps after trees are harvested. Indeed, regeneration systems like selection or group-selection methods are designed to emulate gap dynamics that occur in natural stands due to tree mortality. In such cases, the gaps let in light to the forest floor, which encourages regeneration. Piling or scattering brush from the trees harvested so that it is concentrated in the gap area can create an impediment to movement of deer and allow the regenerating seedlings to escape being browsed. It is often convenient to do this since the tops of the felled trees may not need to be moved very far to create appropriate brush piles.

Satiation is a technique that has application in central hardwood silviculture, particularly where even-age management is practiced. For example, clearcutting, the shelterwood method, and deferment cutting are all methods that can create an abundance of regeneration, which can overwhelm deer. When a large proportion of the deer home range area exists in a regenerating forest, it will overwhelm the resident deer herd with food (Marquis 1987). In larger cutting blocks, the ratio of interior to perimeter is greater, and the total biomass of regeneration may be large enough to satiate a local deer population. In planning and designing cuts, it should be kept in mind that size and shape of cutting blocks are important considerations. Good sites will produce more biomass per unit area; thus, a smaller area on a good site will produce as much browse as a larger area on a poor site. Shape is important because of the perimeter/interior ratio. Deer often utilize forested habitat for resting and so on, and their browsing in forest openings tends to be greater around the perimeter. Square or circular openings have a lower perimeter/interior ratio than long, narrow strips.

Repellents for deer have been developed and are commercially available. Active ingredients range from human hair, soap, and egg albumin to canine urine. In specific situations (e.g., Christmas trees), these products may be useful (Heidman 1963). The major disadvantages of such products is the difficulty and expense of deploying them and their short effective life after exposure to sun, wind, and rain. Central hardwood regeneration must achieve a height of 5 ft. or more in order to be out of reach of the deer, and this may take several years to achieve. Trees are slower to achieve a critical height on poor sites, and deer browsing may create a type of hedging effect where regeneration is almost permanently inhibited from growing out.

Air Pollution. Air pollution has been recognized as a problem in North America for the past quarter century and probably was causing damage prior to that. The USDA, Forest Service (1990) classed air pollutants into two groups, **primary pollutants** that are emitted directly into the atmosphere from their sources and **secondary pollutants** that form when primary pollutants undergo chemical changes. Included in the former are sulfur dioxide, nitrogen oxides, volatile organic compounds, and heavy metals. Photochemical oxidants and acid deposition comprise the second group. The effects of air pollutants are difficult to monitor, although the USDA, Forest Service and the National Atmospheric Deposition Program (NADP) have developed tree crown rating systems and bioindicator tests for pollutants, such as ozone (Skelly et al. 1987; Brantley et al. 1994). Central hardwood species that are sensitive to ozone — and, therefore, good bioindicators — are black cherry, green and white ash, and yellow-poplar. The physiological effect of air pollutants on trees is probably one whereby the stress "load" imposed on trees is heightened, making them more susceptible to secondary organisms, such as bark beetles or *Armillaria* root disease. Thus, the cumulative impact of all stresses is greater where air pollution is high and the resistance of trees to other primary stresses, such as defoliation, is low. A good way to view the effect of pollutants is to imagine a stress threshold that exists for trees or forests. Normal stresses are related to aging (senescence), competition, site quality, and so on. External stresses may be imposed by weather, forest fire, insects or diseases, and air pollutants. When the stress load reaches the threshold, a decline syndrome is set into motion and trees become vulnerable to secondary organisms and ultimately may die.

From 200 stations throughout the United States, the NADP program has been monitoring several air pollutants since 1977. There are approximately 40 such monitoring stations in the central hardwood region, and in 1995 the summary data from the NADP stations indicated that the central hardwood region represented a portion of the country receiving the greatest amounts of acid deposition. For example, precipitation pH in 1995 averaged between 4.3 and 4.5 for most of the central hardwood region. The Ohio Valley region into West Virginia and southeastern Pennsylvania was the region with the lowest pH, averaging about 4.3, while the Ozark/Ouachita Plateaus were the highest areas in the central hardwood region, averaging 4.6–4.8. In contrast, most of the area west of the Mississippi River averaged between pH 5.0 and 5.3.

Relative to the photochemical oxidant ozone, the USDA, Forest Service (1990) reported that "ozone is damaging the foliage and reducing the growth of some hardwood species, such as black cherry and yellow-poplar." But in the same report it was stated that "the condition of eastern hardwood forests is generally good." Hicks and Mudrick (1994) came to a similar conclusion about West Virginia's forest health.

It appears that in order to incorporate air pollution considerations into overall forest management, we need better diagnostic tests and indices of overall stress loadings on trees and forests. We need a better understanding of the physiological role of air pollutants on trees and the interactions among stresses. The USDA, Forest Service through the Forest Health Monitoring network of plots throughout the nation is beginning to assess overall forest health. Usually when trees are showing decline symptoms, such as poor or thinning crowns, it is too late to intervene and reverse the process. For example, if we could determine the stress load that trees in a given area are under at a particular time, it might provide the justification for pollution control, pest suppression (or lack of it), or preventative silvicultural actions. It is apparent that this is an area of forest health management that is in need of more basic knowledge and better strategies for management.

Forest Decline. Many of the factors discussed under the heading of forest health management directly contribute to tree mortality. But these factors in addition to weather-related phenomena (drought, flooding, ice, frost, etc.) also contribute to tree stress. Houston (1992) provides the following sequence that generally leads to so-called decline disease:

1. Healthy Trees + Stress → Altered Trees
2. Altered Trees + Stress → Severely Altered Trees
3. Severely Altered Trees + Secondary Organisms → Trees Invaded by Saprogenic Fungi
4. Invaded Trees + Irreversible Decline → Tree Death

Some declines have been linked to air pollution and acid deposition (spruce decline) and others have been linked to insects or disease outbreaks. But it is likely that multiple stress factors are responsible for many declines. From a forest management standpoint, multiple-stress declines are difficult to diagnose, therefore difficult to treat. Declines may be more acute on some sites than others and therefore management of stress on sensitive sites would be more critical. The following are recommendations for forest managers concerning overall forest health and decline.

1. Prevent stresses where possible and practical (e.g., spraying for gypsy moth in high-risk stands).
2. Stay informed on current overall forest health conditions in your region.
3. Be able to recognize the symptoms of decline.
4. Maintain a good inventory of forests being managed, including forest health conditions.
5. Prescribe silvicultural and management treatments that generally promote health and vigor of stands under management.

Harvesting and Marketing Central Hardwoods

Planning the Harvest. Harvesting products is an important outcome of the management process, but more often than not, NIPF landowners in the central hardwood region are simply selling timber rather than managing forests. Such an attitude may result from a lack of knowledge, from shortsightedness, or from a conscious pursuit of "quick money." Whatever the reason, such a 'timber sale" mentality promotes poor long-term management via high-grading and other site-damaging practices.

Even when timber is harvested as part of planned management, it is necessary to accomplish the harvesting in a way that will not significantly detract from the overall goals of management. The principal elements of a planned harvest/sale are an **inventory** and **designation of timber** to be harvested, a plan for harvest and removal including **access** and **roads** and a **sale contract**. Forest harvesting in the central hardwood region (usually on NIPF land) is often complicated by steep terrain, erodible soils, and changeable weather. The frequent practice of partial cutting adds a further complication in that hardwoods have large crowns and are difficult to fell and handle without causing damage to the residual stand.

Inventory and Designation of Timber for Harvest. If a prescribed harvest calls for removal of all trees (clearcut) or all trees in a class (all of a species or above a diameter limit), an inventory of the trees in the harvested class may be sufficient without individually designating trees to be cut. The first step in any inventory is the proper location of property boundaries. Deeds and tax maps are usually available from the county clerk or tax office in the county where the property is located. Many NIPF tracts have had farm plans developed by the

Natural Resource Conservation Service (formerly Soil Conservation Service), and some tracts have had forest management plans developed, such as Steward-ship Plans. Any of these documents, including the property owner's own knowledge, is helpful in locating boundaries. It is advisable to walk the boundaries and locate corners, line fences, and so on before inventorying the stands. If these cannot be found, a survey of the property may be necessary. A "**cruise**" of the timber is generally conducted as a sample or 100 percent cruise. A sample cruise would be conducted for larger tracts, greater than 50 acres (Wiant 1989). There are many references on methods of resource inventory. Three references that are especially useful for NIPF lands are Ashley (1991), Graves (1986), and Wiant (1989).

Actual designation of trees to be harvested is often termed "marking" and usually involves placing two paint marks on the tree (usually at dbh and below stump height). Sometimes it is more appropriate to mark the trees to *leave*, for example, with a deferment cut where all but a small number of trees are to be harvested. In such a case, timber volume to be cut is based on the volume tallied from marked timber minus the total volume from a stand cruise. In the usual case, designated trees are those that are to be harvested, and, therefore, their volume is based on the tally of marked timber. Tree volumes can be deter-mined using a tree volume table, or a variety of cruise programs available for use with personal computers. In hardwoods, most buyers of timber use the Doyle log rule, although it greatly underestimates volume of small timber (less than 16 in. dbh). The other commonly used rule is the Interna-tional $\frac{1}{4}$-in. rule, which is more accurate but may not be acceptable to buyers. Whichever log rule is used, the treatment of tree taper (form class) is import-ant in volume determination. Form class is the ratio of diameter at dbh to that at a point 16 ft. above the stump. This is especially true in central hardwoods where species like yellow-poplar often have a low amount of taper (form class greater than 80), while some oaks (e.g., chestnut oak) usually have a form class less than 75. Trees growing in tightly closed stands and on good sites generally have less taper than those growing in the open or on poor sites, but the only way to establish form class for a given stand is to measure a sample of trees. Foresters having experience in a particular region sometimes adjust form class based on their *a priori* knowledge. Assuming a single form class for all species will invariably lead to an overestimate in volumes for highly tapered species and an underestimate for species like yellow-poplar that have more cylindrical stems.

When timber marking is done in conjunction with silvicultural operations, it may be necessary to develop **marking guides**, the purpose of which is to facilitate the application of the silvicultural and management operations on the ground. Markings where particular classes of trees (species, diameter, quality, etc.) are to be harvested may not require marking guides, but other silvicultural operations, such as shelterwood cutting, group selection, or single-tree selection will require such guides. A workable approach is to determine the number and type of trees that need to be marked per unit area of land. Then the marker traverses the stand in strips of more-or-less equal width. Given the width and length of the strip, the number of trees to mark per strip can be determined. The marker must keep a tally while going through the strip and can adjust the marking intensity to meet the specified requirement, keeping in mind that within-stand variability may necessitate adjustment of marking intensity to accommodate areas of higher or lower density. When more than one silvicultural operation (crop-tree release, improvement cutting, etc.) are performed simultaneously, the objectives may best be achieved using a **marking priority** approach where classes of trees are identified and ranked regarding their priority to be removed from the stand. Establishment of a residual basal area target can be a useful supplement to this procedure.

Logging and Removal of Timber. Logging is one of the most controversial aspects of forest operations. In the central hardwood region, there are several complications to logging. First, the landscape is frequently hilly and steep, making road construction difficult. Second, hardwoods often have large crowns, making them difficult to fell and handle. Third, many operations that are silviculturally sound and acceptable to NIPF landowners require some type of partial cutting. All these factors make logging in the central hardwood region both dangerous and potentially damaging to the residual stand (Fosbroke and Meyers 1995).

   Forest roads and **skid trails** can leave areas looking unsightly (Fig. 139) or can serve as a source of siltation, which loads streams and damages water quality (Fig. 140). In fact, forest roads are the main source of pollution in the form of silt that occurs from forest harvesting (Kochenderfer 1970). But "minimum standard" forest roads may be constructed in such a way as to lessen the adverse impacts, and a well planned and constructed road network can become an asset, providing future access to the property (Kochenderfer et al. 1984). Most states within the central hardwood region have regulations that govern logging and forest road construction. Many of these are covered under guidelines that stipulate "best management practices" (BMPs) (Paff 1982; Whipkey and Bennett 1989). BMPs often include specifications for haul roads, skid trails, log landings, reclamation and reseeding, stream crossings, and filter strips and shade zones.

   There are several logging and harvesting systems, based on the type of equipment used. The typical logger in the central hardwood region uses chainsaw felling and ground-based skidding—mostly with a rubber-tired skidder. Logs are usually skidded in tree length and bucked into log length, sorted by species and product, and loaded on trucks at a landing site. This system offers great flexibility in that products of different size can be moved over a wide variety of terrains and the set-up cost is relatively small. Although the equipment for such operations is expensive, it is less expensive than some other types of specialized logging systems and can be used in almost any situation.

FIGURE 139   Aerial photograph showing logging roads in steep terrain in Pendleton County, West Virginia.

The disadvantages of ground-based skidding with rubber-tired equipment are that they require an elaborate network of skid trails, especially in steep terrain, and skidders can cause extensive site damage if used improperly (in bad weather, on steep slopes, in streams or wet areas). Ground-based skidding of tree-length material is also prone to cause damage to residual trees.

FIGURE 140   A poorly constructed skid trail running parallel to a 30 percent slope.

As an alternative, several types of cable-yarding systems are available and have special applicability to steep terrain. Le Doux et al. (1989) and Baumgras et al. (1993) have developed computer models to help in choosing the best harvesting system for particular situations. Systems such as "shovel logging" (Hemphill 1986), small tractors (Huyler and Le Doux 1991), and horse logging have been recommended for certain situations, and Baumgras et al. (1995) have reviewed the Appalachian hardwood harvesting situation and have attempted to match harvesting systems with the particular silvicultural operations in hardwood stands (Table 19).

In any timber harvest operation, it is important that the forest landowner negotiate a **timber sale contract** with the logging contractor and/or timber buyer. A sale contract should be a legal document and, at a minimum, should provide a description of the timber for sale, stipulate a sale price, stipulate a duration of the agreement, provide for adherence to BMPs, and require a performance bond to be held by the landowner to secure the proper performance of the job. Wiant (1989) included a sample contract in his publication on woodlot management. This sample agreement is reproduced below as Example 1.

TABLE 19 Estimated Net Revenue[a] for Skidder and Skyline Logging at Low, Medium, and High Prices

SILVICULTURAL PRACTICE	SITE INDEX	LOW PRICE		MEDIUM PRICE		HIGH PRICE	
		Skidder	Skyline	Skidder	Skyline	Skidder	Skyline
		----------Dollars/acre----------					
Single-tree selection	64	896	724	1853	1681	3178	3006
	70	614	518	978	882	1409	1313
Group selection	64	293	204	591	502	1006	917
	70	463	371	733	641	1165	1072
Two-age management	64	916	48	2019	1851	3598	3430
	70	1636	1437	2668	2469	4457	4258
Even-age management	64	1293	1131	2719	2557	4769	4607
	70	2095	1918	3360	3183	5507	5330
Diameter-limit	64	1184	1098	2310	2224	3829	3743
	70	1657	1539	3199	3081	5356	5238
Thinning (cherry/maple)	75	−27	−271	246	2	786	542
Thinning (yellow-poplar)/ red oak)	80	264	47	716	499	1661	1444

[a]Value of roundwood delivered to the mill minus harvesting and haul cost.

Source: Baumgras et al. (1995).

EXAMPLE 1.

Sample Timber Sales Agreement

This AGREEMENT, made and entered into this *(date) day of (month)*, 1988, by and between *(seller's name)* whose address is *(seller's complete address)* hereinafter referred to as the SELLER and *(purchaser's name)* whose address is *(purchaser's complete address)* hereinafter referred to as the PURCHASER.

WITNESSETH:

1. The SELLER hereby agrees to sell and the PURCHASER agrees to purchase, according to the conditions and requirements hereinafter mentioned, all marked timber from *(description of area from which trees are to be taken.)*
2. CONSIDERATION: The PURCHASER agrees to pay the SELLER the sum of *(price)* in full on the date of this AGREEMENT. Said payment is for all marked live timber whose volume is estimated to be *(estimated volume.)*
3. The PURCHASER will deposit a performance bond of *(amount of bond)* payable to the SELLER on the date of this AGREEMENT. In order to partially secure the satisfactory completion of this AGREEMENT by the PURCHASER,

the total amount of this performance bond so withheld shall, at the sole option of the SELLER or his field representative, be considered as liquidated damages for such breach of the terms and provisions herein contained, and shall become the property of the SELLER otherwise, this amount, so withheld, shall be returned to the PURCHASER upon the satisfactory completion of this logging chance.

4. Undesignated trees which are cut, or injured, shall be paid for at triple the average stumpage rates in #2.

5. All timber shall be cut and removed from the premises of the SELLER on or before *(two or five years)* unless an extension of this time, in writing, is granted by the SELLER.

6. The PURCHASER will not cut stumps higher than twelve (12) inches adjacent to the highest ground, and will utilize the tops to the smallest practicable diameter.

7. Temporary truck roads and skid roads necessary for the removal of said timber must be constructed in accordance with the standards as per the "Appropriate Forest Practice Standards" in regard to grade and drainage. Upon completion of logging said roads are to be graded and waterbarred to comply with the aforementioned "standards."

8. Fences and roads shall be protected from unnecessary damage during the logging operation, and if damaged by the PURCHASER, shall be repaired to their original condition by the PURCHASER.

9. Young growth and trees left standing shall be protected from unnecessary damage, and only dead or less valuable trees shall be used for construction purposes during the logging operation.

10. The PURCHASER shall do all within his and his employee's power, both independently and upon the request of the SELLER, to prevent and suppress forest fires.

11. The PURCHASER agrees to indemnity and save harmless the SELLER against all claims of loss, damage or expenses of any kind which may arise in connection with PURCHASER'S operations on the above described tract, and to take out and maintain adequate liability insurance to fully protect the SELLER from any liability whatsoever and to furnish evidence of such insurance upon request of SELLER and/or his field representative. PUR-CHASER further agrees and binds himself to comply fully with all federal and/or state laws, and amendments and supplements thereto, and any and all rules and regulations now in force, or which may hereafter be issued, for the enforcement of said laws, including the "Social Security Act," as amended, the "Fair Labor Standards Act," as amended, the applicable State "Unemployment Compensation Law," as amended, and the "Civil Rights Act," as amended.

12. In case of dispute over the terms of this agreement, final decision shall rest with arbitrators, one of whom shall be selected by each party of this contract, and in case the two selected shall disagree, they shall select a third arbitrator, and the decision of the majority shall be final with respect either to acts to be done or compensation to be paid by either party to the other.

In WITNESS WHEREOF, the parties hereto have hereunto subscribed their names on the date first written above.

EXECUTED IN DUPLICATE

WITNESS   SELLER
(signatures of *witnesses*)   By: (signature of *seller*)

WITNESS   PURCHASER
(signatures of *witnesses*)   By: (signature of *purchaser*)

Selling and marketing timber is an activity that requires general knowledge of current market trends and involves a combination of knowledge, experience, and negotiating skills. It is usually wise to obtain several bids for timber. A sealed-bidding process is often used, and the spread between high and low bids can be twofold, or even greater. Different buyers can afford to pay more or less depending on their inventory situation, their demand for specific species (Baumgras and Luppold 1993), and the haul distance to the processing facilities (Le Doux 1988).

Formulating Management Plans

Management plans can take the form of elaborate documents with environmental assessments, impact statements, detailed multiresource inventories, and several management alternatives, such as are prepared for National Forests. Or they can focus on flow of cash, products, and materials, which is more likely the case for industrial forest management plans. But the types of plans that will be emphasized in this section will be those that are appropriate to the NIPF landowners.

To provide context, it is helpful to know something about NIPF owners. Birch (1997) reported the results of a 1994 national landowner survey, which found that in the central hardwood region, 94 percent of the forest owners were private individuals, owning 76 percent of the land. Almost one-third of the owners were 65 years of age or older. The average ownership size has decreased since the last survey (1978) and is now about 31 acres. Most of the owners identified the primary reason for owning land as part of a farm or residence or for recreation and aesthetic enjoyment, while only about 1 percent stipulated timber production as the primary reason for owning land. However, even though timber production was not a primary reason for ownership, 48 percent had harvested timber. Perhaps the most significant finding of Birch's study was the fact that only about 5 percent of the owners (owning 22 percent of the land) had developed written management plans. There is obviously a need for management planning on this significant land base. Without planned management, exploitation of NIPF lands through continued high-grading will result in a depletion of the timber resource and a discontinuity of yield from forested tracts. The resulting cut-over and impoverished forests will not benefit the individual landowner, the forest industry, the local economy, or society at large.

Management planning for NIPF owners is currently being done by several entities or individuals. State service foresters, industrial foresters, consultants, and owners themselves are responsible for planning. Birch and Moulton (1997) indicate that for NIPF owners who have written plans, 18 percent wrote the plan themselves, 13 percent were written by consultants, and the remaining by state service foresters or wildlife biologists. There are also programs and agencies that assist NIPF owners in management planning. The Forest Stewardship and Stewardship Incentives Programs (SIP) were created by the 1990 Farm Bill. These programs provide cost sharing for planning and for specific forest management practices. They are administered by the states, and, as an example, Kentucky lists the following "intentions" for their Stewardship Incentives Program:

- Encourage nonindustrial private forestlands for economic, environmental, and social benefits.
- Complement and expand on existing forestry assistance programs.
- Give priority to tree planting, tree maintenance, and tree improvement practices.
- Increase the quality and quantity of the timber resource.
- Maintain and improve habitat for a diverse mixture of native wildlife species.
- Manage and improve the aesthetic value of the forest and the forest's value for diverse, quality recreation.

Stewardship plans are prepared by professional foresters or other resource management professionals. For example, in Maryland, and several other states, service foresters as well as wildlife biologists and other personnel from natural resource management agencies, who are specifically trained in Stewardship planning, develop the plans. In West Virginia, forestry consultants who are trained in Stewardship planning prepare most of the plans. Alban et al. (1995) have developed a computer program designed to guide landowners through the appropriate steps in developing their own Stewardship Plan. This tool can be used with or without the assistance of a professional forester, but without the aid of a forester, the landowner may find it difficult to interpret some of the technical aspects. Although the prescriptions generated from such programs were developed with input from forest scientists, the untrained landowner will probably lack the biological background to interpret them. An example of the type of management plan used for stewardship planning, supplied by the Kentucky Division of Forestry, is shown in Appendix 1.

Forest management plans, such as the one in Appendix 1, should provide a *usable* document to guide current management and to serve as a reference for future management activities. When a forester interacts with a landowner, the opportunity exists to inform and encourage the landowner to develop a management plan. In many instances, the contact with a forester (consultant, service forester, industrial forester) is precipitated by someone making an offer to purchase timber from the landowner. The forester is often called in to inventory the timber or to assess whether or not the owner should sell. During this contact, the forester has an opportunity, even an **ethical responsibility**, to discuss the concepts of planned management versus forest exploitation with the landowner. The forester's role should be one of providing information and guidance, and equally important in the process is the forester's willingness to *listen* to the landowner and ascertain his/her motivation. The forester can inform the landowner concerning silvicultural and management opportunities and hazards of mismanagement that may have been unknown to them. The forester can also describe the benefits of multiple-use management and can guide the landowner in deciding what management activities are practical (or impractical) to assist them in optimizing their goals.

The list below, generated by Gribko (1997), includes some basic components of a forest management plan for the NIPF owner.

I. Introduction
   A. Site Description
      1. location
      2. topography
      3. climate
      4. physiography/soils
      5. history
      6. site quality or yield class
   B. Maps
      1. topographic
      2. soils
      3. roads/access
      4. hydrologic (streams, springs, etc.)
      5. forest stands
   C. Current Forest Conditions — Stand Descriptions
      1. overstory trees (number, diameter distribution, and basal area per acre)
      2. understory stems (species, size, and quality)
      3. regeneration (type, quality)
      4. sawtimber and other product volume per acre
      5. value of timber by products
   D. Wildlife Habitat
      1. habitat suitability
      2. species present (or potentially present)
      3. habitat limitations
   E. Recreational Use
      1. potential for development
      2. types of activities
II. Management Objectives

    A. Determine Objectives with Landowner
        1. multiple resource management
        2. prioritization of resources (timber, wildlife, recreation, etc.)
        3. management time frame
    B. Special Conditions or Limitations
  III. Management Alternatives
    A. Sustainability (Method of Control)
    B. Multiple Resource Management
    C. Economic Considerations
    D. Stand Prescriptions (Itemize by Stand)
        1. silvicultural prescription
        2. method of harvest
        3. timing of operations
        4. forest health management

Leuschner (1992) lists several of these among the elements found in many forest management plans. He also includes items such as economic expectations, legal restrictions, and forest production as components of management plans. Perhaps there is no simple set of absolutes with regard to what is required for a "good" management plan for NIPF owners. In fact, a good plan is one that is effective in accomplishing the objectives, and often a **simple**, **direct**, and **flexible** approach is better than an elaborate, complex, and rigid one. **Adaptive forest management** is the management equivalent to "adaptive silviculture" proposed in Chapter 5. Adaptive management combines economic opportunities with the most practical silvicultural treatments for NIPF ownerships in the central hardwood region. Management planning must provide flexibility to accommodate adaptive management. When market opportunities develop for a particular species or product, an adaptive approach would allow the manager to capture the opportunity so long as it does not seriously compromise biological and silvicultural principles. When a good crop of desirable regeneration has become established in the understory of a stand, the adaptive management approach would permit harvesting the stand even though the harvest might exceed "allowable cuts" established under volume or area control. Just as important as having the flexibility to harvest is being able to delay harvesting when economic factors call for it.

MANAGEMENT OF CENTRAL HARDWOODS, SUMMARY

The central hardwoods, due to historical, biological, and ownership conditions, are unique in the United States and the world. The maturing hardwood forests, mostly owned by small NIPF owners, are at a management crossroads. There are potentially serious forest health problems that threaten particular components of the resource. Economic factors are aligning, which could provide incentives for better forest management (utilization of a wide array of species and sizes). But without management on the part of landowners, these same factors could lead to increased high-grading and exploitation of the resource. The taxation structure, short tenure of ownership, absentee owners, and the "windfall" attitude of many NIPF landowners regarding income from timber sales are all part of the problems facing the central hardwoods. Foresters have an opportunity and an ethical responsibility to help initiate planned management on NIPF lands. Foresters (consultants, service foresters, and industrial foresters) should focus on the management opportunity rather than the timber sale. Planning should attempt to pursue the objectives of the owner, insofar as these objectives are biologically sound. Management plans should be based on a resource inventory and should provide for sustainability, proper silvicultural management, maintenance of forest health, and adherence to best management practices and should be understandable and adaptive to suit the owner's needs. Listed below are some additional references dealing with growth and yield projection and general management planning.

ADDITIONAL READING

Dey, D. C., Johnson, P. S., and Garrett, H. E. 1996. Modeling the regeneration of oak stands in the Missouri Ozark Highlands. *Can. J. For. Res.* 26(4):573–583.

Hill, D. B. 1993. Small woodlot management in Kentucky. *Univ. Coll. Agric. Coop. Ext. Serv. For.* 15, 23 pp.

Hilt, D. E. 1983. Individual-tree diameter growth model for managed, even-aged, upland oak stands. *USDA For. Serv., Res. Pap.* NE-533, 15 pp.

Marquis, D. A. and Ernst, R. L. 1992. Users guide to SILVAH. Stand analysis, prescription and management simulator program for hardwood stands in the Alleghenies. *USDA For. Serv. Gen. Tech. Rep.* NE-162, 124 pp.

Shifley, S. R. 1987. A generalized system of models forecasting Central States tree growth. *USDA For. Serv. Res. Pap.* NC-279, 10 pp.

Simpson, B. T., Kollasch, R. P., Twery, M. J., and Schuler, T. M. 1995. NED/SIPS User's Manual. Northeast decision model stand inventory processor and simulator (version 1.00). *USDA For. Serv. Gen. Tech. Rep.* NE-205. 103 pp.

7

Synthesis and Conclusions

STATUS OF THE RESOURCE: A SUMMARY

General

The central hardwoods encompass an area of about 150 million acres. The region stretches from Cape Cod to eastern Oklahoma and generally occurs in the unglaciated mountains, valleys, and dissected plateaus of the Appalachian and Ozark/Ouachita systems. Approximately half (75 million acres) of the area within the central hardwood region is forested. This is the largest and most extensive temperate deciduous forest in the world.

Much of the topography is steep and hilly, but broad valleys and some level uplands also occur within the region. Precipitation in the central hardwood region is favorable for plant growth, varying between 35 and 60 inches (mostly between 40 and 50 inches) per year and is well distributed throughout the seasons. Most of the region is underlain by sedimentary geologic material (sandstones, limestones, and shales), and the principal soils are Udults, Udalfs, and Ochrepts (USDA, Soil Conservation Service 1981). The average freeze-free period for the region is between 180 and 200 days, being somewhat shorter in the north and at higher elevations and longer in the southern valleys. Soils and climate are conducive to good productivity and a substantial proportion of the region is currently used for agriculture. Steep slopes are the most limiting factor for agriculture, and as farming has continued to be more dependent on mechanized equipment, a trend toward agricultural abandonment has occurred throughout the region over the past several decades, with present-day forests occupying the steeper topography.

Forest Cover. Forest cover in the central hardwood region is oak dominated, although other species, such as maples, hickories, and yellow-poplar, are locally abundant. On some sites and certain aspects, these species may predominate.

The effects of site (slope position, elevation, aspect) are important in determining species composition. For example, in the northern portion of the region and at higher elevations, northern hardwood species (American beech, maples, and birches) may occur in pure stands or in mixtures with oaks, the latter predominating on south and southwest exposures and lower elevation ridges. In the southern and southwestern portion of the region, southern pines (shortleaf and Virginia pines) may be mixed with central hardwoods. The pines often occupy the southern and southwestern aspects in this portion of the region. Tree growth, as measured by oak site index, is generally rapid throughout the central hardwood region. The best sites occur in moist coves and on north- and northeast-facing slopes in the southern Appalachians where rainfall, length of growing season, and soils are all favorable for tree growth. The poorest growing sites in the central hardwood region occur in the Ridge and Valley Province and Ozark Plateau, and especially on south- and southwest-facing slopes. The foregoing summary is a brief overview of the more detailed discussion of the central hardwood region provided in Chapter 1.

Historical Development

The history of the central hardwoods is one that is marked by anthropogenic impacts, as well as other natural events. As the climate of eastern North America began to moderate after the Wisconsin Ice Age, boreal forests and tundra gave way to deciduous forests. This warmer climate enabled native people to populate the region, and these people used fire to clear agricultural land, to clean out the understory for visibility, and to create grassland habitat for game. European settlers cleared much of the forest in the central hardwood region for fields and pastures, even on steeper slopes. They also used fire to clean out forest undergrowth. With the coming of industrialization, timber was exploited for building material, and many subsistence farmers moved to cities and towns to work in mills and factories. This period of forest exploitation was ended by the Great Depression, but much of the resource had been cut over by 1930. In spite of pest introductions, such as chestnut blight, Dutch elm disease, and gypsy moth, the forests of the central hardwood region have shown amazing resilience. Regrowing forests from cut-over forestland and abandoned agricultural land have developed and are approaching commercial maturity. Chapter 2 provides a more detailed discussion of the historical development of central hardwoods, including citations of other books and references on the subject.

SUMMARY OF ECOLOGICAL AND SILVICULTURAL RELATIONSHIPS

The central hardwoods represent a diversity of ecosystems that mostly consist of second-growth forests, which developed after logging or agricultural abandonment. Bailey (1994) classifies the ecoregions of the central hardwoods as occurring in Hot Continental or Subtropical Divisions of the Humid Temperate Domain. Braun (1950) described four Forest Formations that make up the central hardwood region—oak-chestnut, mixed mesophytic, western mesophytic, and oak-hickory. Obviously, the climate of the region with warm, humid summers and cold winters is conducive to growth of deciduous species, and oaks, hickories, maples, yellow-poplar, sweetgum, black cherry, and other hardwoods make up the bulk of the forest cover in the region.

The ecological factors that govern the success of oaks and other deciduous species are those that facilitate regeneration and establishment. In addition to climate, the overriding factor affecting the composition of the current forest is disturbance (fire, logging, agriculture, pest introductions, etc.). Each species has its unique "silvical characteristics" that determine its ecological role. Oaks, for example, are somewhat shade tolerant and are able to build up propagules (seedlings, seedling-sprouts, etc.). Due to their vigorous resprouting ability and their investment in root growth early in life, oak seedling-sprouts are capable of gaining an advantage over their competitors after fire. This is especially true on average to below-average sites where oaks seem to regenerate successfully using a variety of strategies. Other exploitive species, such as yellow-poplar and sweetgum, are capable of rapid aerial growth. These species are particularly vigorous on the better quality sites where resources are abundant. In the absence of major disturbance, slower-growing shade tolerant species, such as red and sugar maples, have become major midstory components of many central hardwood stands. The ecological processes of regeneration, competition, and resource allocation are of key importance to silviculturists and managers since these are the processes that can be manipulated in order to achieve management goals. Species in the central hardwood region run the gamut from those that rely on annual crops of new seed (black cherry, yellow-poplar) to those that "build up" regeneration potential via advance seedlings (beech, sugar maple), seedling-sprouts (oaks), stored seed (pin cherry), and so on. There are fast-growing species that allocate most of their resources to upward stem growth (yellow-poplar, black cherry) as well as those that invest heavily in building their root systems (oaks and hickories). Other species rely on their shade tolerance (maples, beech) instead of rapid growth to ensure their persistence in central hardwood stands. A detailed discussion of ecological concepts and silvical characteristics appropriate to central hardwoods is given in Chapters 3 and 4, respectively.

Silvicultural methods appropriate to central hardwoods vary with management objectives, sites, and current stand conditions. Many central hardwood stands contain species that are relatively long lived, and, currently, some form of intermediate management is appropriate. Intermediate cuttings that are particularly appropriate include crown thinning, improvement cutting, and crop-tree management. For stands that are mature, some type of regeneration system is appropriate. Depending on the silvical characteristics and reproductive strategy of the desired species, methods such as clearcutting, shelterwood, and two-age systems are viable ways to accomplish even-age management of central hardwoods. Many forest owners prefer some type of partial cutting to even-age management. Thus, methods such as single-tree or group selection are useful for all-age management of central hardwoods. In the central hardwood region, due to changing owner objectives and dynamic markets, silvicultural systems must remain flexible and adaptive. These and other silvicultural methods appropriate to central hardwoods are covered in greater detail in Chapter 5.

MANAGEMENT CONTEXT

In the central hardwood region, about 75 percent of the forestland is owned by small NIPF owners. Although most owners do not state timber production as a primary goal, almost half of them reported harvesting and selling timber in the past 20 years. Only about 5 percent of the NIPF owners have any kind of written management plan, and less than 20 percent used professional foresters—even to assist in timber harvesting and marketing. Thus, there is a great need and opportunity to initiate planned management on NIPF lands. Management planning applicable to these owners should possess the following elements:

- Tailored to the needs of each owner
- Understandable to the layperson
- Flexible and adaptive
- Must provide information on current forest conditions, projected growth and yield, cultural and management prescriptions, forest health, and environmental protection

There are many considerations that the forest owner and/or manager of NIPF lands must take into account. These are covered in greater detail in Chapter 6, but they include decision-making strategies, growth and yield projections, financial and economic factors, forest health considerations, and general management planning—including sale and harvest of forest products.

THE BIOLOGICAL POTENTIAL
OF CENTRAL HARDWOODS

Unmanaged Stands

In the central hardwood region there are a number of small stands of old-growth forest in scattered locations. Such stands provide clues about the forests that probably existed over much wider areas before European settlement. In addition, they offer a vision of what today's second-growth central hardwood forests might look like, if unmanaged, in about 200–300 years.

Many remnant old-growth forests have an oak component, often constituting part of the dominant overstory (Lafer and Wistendahl 1970; Choo and Boerner 1991; Nowacki and Trianosky 1993). An almost universal occurrence in these old-growth oak forests is the development of a maple understory. In some situations other shade tolerant species, such as American beech and eastern hemlock, are also prominent understory components. Another attribute common to unmanaged old-growth remnants is a lack of oak regeneration (Abrams et al. 1997). These facts all point to the likelihood of fire as a disturbance that promoted regeneration of oaks and shade intolerant species like yellow-poplar in the presettlement forest. It is also obvious that the lack of fire in the maturing forest stands has enabled maples to regenerate. The shade tolerant maples, beech, and hemlocks are gradually replacing the oaks and intolerant species (Cho and Boerner 1991). This successional process is not unexpected and, in fact, validates the principles that ecologists have espoused for many years. But observations from remnant old-growth stands give credibility to these principles and offer a vision of what many of our present-day second-growth central hardwood stands could look like in 100–200 years, if unmanaged. Certainly factors like heavy deer browsing and gypsy moth are different in today's forests and will play a role in the development of the future forests.

Managed Stands

Forest management is a relatively recent development in North America and the central hardwood region. Experimental forests, such as the Bent Creek, Fernow, Kane, and Sinkin Forests, offer evidence as to how managed forests could appear after 30–40 years of management. In general, managed stands in these forests are performing according to expectations, although there is a common problem of obtaining adequate oak regeneration on good sites, even when management is directed toward this goal. A difficulty with using experimental managed forests as indicators is the fact that most are managed to provide research results and not to simulate constraints under which most private landowners work. There are managed industrial forests within the central hardwood region to serve as examples, but here again, management has only been underway for a few decades at most, and the level of management is generally higher than most NIPF owners could afford. In addition, most NIPF owners have very different management goals than private industrial owners.

One of the most common occurrences in the central hardwood region is that of mismanaged stands. Such examples include stands of forestland that have been repeatedly high-graded, farm woodlots that have received heavy grazing in addition to high-grading, and a variety of forests that have been heavily damaged by insects, diseases, deer browsing, and other damaging agents. As has been stated many times throughout this book, repeated episodes of high-grading, diameter-limit cutting, and injurous agents can lead to impoverished stands that are no longer manageable since there are few good management choices that remain.

The problems of mismanagement are exacerbated on poor sites, where recovery time is greater and the diversity of species is smaller. But even on good sites repeated mismanagement can lead to a degradation in quality and the widespread practice of high-grading for an extended period of time could lead to disgenic selection where the **genotype** of the species is degraded.

THE FUTURE OF THE CENTRAL HARDWOODS: CONCLUSIONS AND RECOMMENDATIONS

The Stakeholders

There are three broad categories of **stakeholders** concerned with the resources of the central hardwood region. These groups are the *public*, the natural resource-using *industry*, and the *landowners*. The public includes a wide array of individuals from urban/suburban populations of large cities in or near the region to people living in smaller communities within the region. Public use includes recreation and tourism, hunting and fishing, gathering of wild harvested products, such as ginseng and mushrooms, or working in a job that is economically tied to the natural resources of the region.

Vocal outcries from public sectors, such as the environmental community, have had a significant impact on the management of public lands (National Forests, National Parks, state forests and parks, state game lands, etc.). Laws such as the Endangered Species Act and landmark court rulings like the Monongahela Decision have set important precedents. The move toward "ecosystem management" on public lands, with its strong social component, has come as a result of this political process. But most of the land within the central hardwood region is in the hands of small NIPF owners. The public has much less control over the NIPF sector, although substantial public funds are spent on NIPF lands through government assistance programs, such as the Stewardship Incentives Program, pest suppression, and various extension services. On the other hand, NIPF owners pay taxes themselves, including property taxes, severance taxes, inheritance taxes, or taxes on income generated from sales. In addition, forest resources provide jobs for many people and materials (paper, building materials, etc.) necessary for the functioning of the American social and economic system, which also benefit the public. It is a difficult task to try to sort out the tangible and intangible benefits and detractions of natural resource utilization to the public, and even more difficult to determine what role the public interest should play in management of private lands. But, given the prevailing situation of a maturing hardwood resource, strong demand for forest products, new technology to utilize hardwoods, a general reduction in available conifer resources, and increasing environmental concerns, the stage is set for conflicts over resource utilization.

The natural resource industries are important stakeholders and include those practicing consumptive uses, such as mineral extraction and utilization of renewable resources (timber, game), to non-consumptive uses, such as white-water rafting, enjoyment of scenic beauty, and birding. Indeed, the various industries that utilize the central hardwood resources are themselves frequently in conflict over how to best utilize them. In a few cases, using a resource for one purpose enhances another. For example, timber cutting usually enhances the habitat for certain game species, such as white-tailed deer, ruffed grouse, or wild turkey. But often one use diminishes another; for example, clearcutting reduces the aesthetic value of an area, and property development for housing may negatively affect fish and wildlife habitat and reduce the land available for timber production. Within the central hardwood region, some forest-products industries own their own land. But total industry ownership within the region is less than 10 percent. Thus, forest industries, as well as recreation and tourism industries, must rely heavily on private lands to supply their resources. The forest industry is rather diverse, ranging from logging contractors to large integrated companies. Types of material used range from pulpwood and rail wood to high-value sawtimber and veneer logs. A relatively recent industrial entrant is the composite industry which can use a range of sizes and quality of softer hardwood species (yellow-poplar, red maple, sweet birch, aspens, etc.).

In a free market economy, forest industries may be in direct or indirect competition for the same products. For example, if small-diameter yellow-poplar is usable for either glue-lam (a sawn laminate material), pulpwood, or decorative fence rails, companies who seek these raw materials may be in direct competition for stumpage. But if yellow-poplar is also in demand for plywood and high-quality lumber, when smaller-diameter stands are cut for the products mentioned earlier, trees will not be available for these large-diameter, high-value products. The obvious solution is to utilize topwood, silvicultural thinnings, or other intermediate management products for the small-diameter product while stimulating growth and improving quality of the residual stand in order to obtain the greatest yield of high-value products in the future. In such an approach, the landowner *and* the industry benefit. Unfortunately, the structure of the timber industry, with loggers and timber brokers often having interests opposing those of property owners or end users, creates short-sighted perspectives that often do not promote the best interests of either party.

Nontimber resources, such as recreation and tourism, are often given little regard by NIPF owners when doing timber harvesting, although timber extraction most certainly has an impact on the scenic qualities of the landscape. Property owners who own land within the viewsheds of scenic vistas and scenic highways could be targeted for actions aimed at preserving scenic qualities of the landscape. Perhaps some type of tax incentive and/or direct compensation could be implemented to offset the loss in income from timber sales that a private owner may incur by practicing "aesthetic forest management." Such programs could only be accomplished through governmental agencies and would be controversial.

Natural resource industries are certainly important stakeholders in the central hardwood region, and factors such as land ownership patterns, structure of the industry, conflicts over use, and public involvement will affect how these industries can function and will likely become more important in the future. Such issues fall under the general heading of **natural resource policy**, and the way they play out in the socioeconomic and political arenas will have as great an impact on how central hardwood forests are managed as ecological and biological principles.

Forest landowners are stakeholders who would appear to have the most to gain from proper forest management. But most NIPF owners (95 percent) do not have a management plan (Birch 1996), even though cost sharing and assistance programs are available for such planning. In fact, it is far more likely that landowners will spend more time posting their land to warn would-be trespassers or hunters than in learning about and planning forest management. Characteristics of NIPF owners nationwide are presented in Birch's (1996) report and summarized in Chapter 7 of this book. Perhaps these characteristics, interpreted in the context of social and ethnic demographics, can help explain what, to many natural resource managers, seems to be illogical behavior on the part of NIPF owners. First, 30 percent of the NIPF owners are retired. Second, 86 percent of the NIPF owners own less than 50 acres. Third, almost 74 percent have owned the land for less than 35 years. Only about 3 percent of the owners claim timber production as a primary reason for owning the land, although almost half say they have harvested timber during the last 20 years and almost one-third of them indicate that they expect to harvest timber in the next 10 years.

Further a significant proportion of owners are nonresident. Birch (1996) found that in the mid-Atlantic region, which includes a substantial part of the central hardwood region, 17.4 percent of the owners lived more than 15 miles from their forested tracts. Many of these are heirs who have inherited the land, and, although they may have grown up on a farm, which is now forested, they now have other careers and do not feel "connected" to the land. Along with those who have recently acquired their property by inheritance, there is a substantial proportion of recent purchasers of property who are speculating on the land as an investment (6–10 percent). Finally, a portion of the retired owners are recent retirees who have bought the property for a retirement home. Following are some extrapolations from this data, generalizations, admittedly speculative, concerning the attitudes and motivations of small landowners in the central hardwood region.

- *Farmers/residents:* In the central hardwood region, the primary groups to settle the land were European (especially Scotch-Irish and British). Land management, involving clearing for farming, was the traditional life-style, carried with the first settlers from their homelands. The notion of "cleaning up" the land by rooting out weeds, brush, and "filth" was inherent in their thinking, and this idea persists among rural people today. When pasture or cropland grows up to forest, the owner who has tended the land experiences a sense of loss — the land has reverted to a wild state. Thus, traditionally many resident farmers regard forestland as the leftover, unusable, or untamed land, and this attitude is manifest in their lack of motivation to manage forest resources. They also have a strong sense of independence and individuality, which comes into play in relation to issues like property rights and may also be a source of suspicion of outsiders, such as people from government agencies or universities who may advocate forest resource management.

- *Nonresident owners:* People in this category are often heirs to an estate. They have careers and often live far away from the family land. The land is usually a family farm that has been subdivided between several heirs. The current owners may remember the land as a farm, regardless of the fact that it is presently forested. Often these owners become aware of the timber value through a phone call from a timber buyer who has determined the ownership from the county clerk's records. Lack of motivation for forest resource management among such owners is a result of ignorance. Although many of these owners are middle-class and fairly well educated, they are generally disconnected from, and poorly informed about, their land.

- *Speculators/developers:* These owners, by definition, we might say, view the forest simply as an exploitable resource. They seldom keep land long enough to contemplate long-term forest management. Thus, they view timber and other natural resources as commodities to be sold rather than as resources to be managed.

- *Retirees:* According to Birch (1996) the population of retired owners is significant and growing. For example, in the United States in 1978, 26.8 percent of NIPF owners were retired; by 1994 that percentage was 29.4. Retirees have special needs and perspectives. From Birch's data on the mid-Atlantic region, 37 percent of retired landowners said they would harvest timber in the next 10 years, compared to 27 percent who indicated they never intended to harvest. For all "white collar" workers these same figures were 32 percent and 36 percent, respectively. Retirees may need income, and this often prompts them to sell timber. But, perhaps because of their age, retirees find it difficult to embrace the long-term view needed to initiate forest resource management.

As natural resource stakeholders in the central hardwood region, property owners (especially NIPF owners) seem to be poorly informed and disengaged from resource management. This is a problem that has concerned natural resource managers for many years, and it is especially so in the central hardwood region where almost 75 percent of the owners are NIPF owners.

There have been various attempts to organize forest landowners, through forest landowner associations and organizations like the American Tree Farm Association. But the participation rate in these is generally low. It may be due, in part, to the inherent independence of NIPF owners, which precludes them from organizing. But, in addition, the economic incentives that proved instrumental in the creation of farm organizations (Farm Bureau, Grange, etc.) are not as clearly evident in forest resource management, and the time span between harvests may be decades for forest stands. With smaller ownerships typical of the central hardwood region, these economics of scale become a factor in motivating owners to become organized or involved in management.

What is particularly frustrating from a resource manager's point of view is the fact that not only are NIPF owners often failing to practice forest management, but they are also easy targets for exploitation. NIPF owners seldom use a forester when harvesting timber; thus, they are at the mercy of the buyer. They are usually unable to discern between partial cuttings, such as high-grading, and silviculture-based methods like improvement cutting. Thus, with the appropriate knowledge, NIPF owners could accomplish silvicultural objectives with a timber harvest while obtaining almost as much money as they would get if the same stand was high-graded, and the future value of their forest would be greatly enhanced by proper management.

THE ROLE OF NATURAL RESOURCE MANAGERS

Technology Transfer

In the central hardwood region, as in other regions of the United States, the primary resource management players are foresters, wildlife biologists, and recreation and parks specialists. These managers have affiliations with government agencies, institutions (universities and private foundations), and industries. Some function as private consultants, but almost all have an interest in applying management to the landscape, and in the central hardwood region this means working with NIPF owners. The NIPF "problem" is not new and it is not from a lack of effort that it remains unsolved. The simple diagram in Figure 141 depicts the steps in the "technology transfer" process going from the resource management professionals to the landowners. It appears that the weak link in the process is between the "assistance" and "application" phases. Although our knowledge of ecosystem-level processes concerning central hardwoods is far from complete, most of the key relationships have been elucidated (Smith 1978). But much of the knowledge we now have is not being applied. Our ability to collect and disseminate information is almost unlimited. The USDA, Forest Service, U.S. Geologic Survey, and National Oceanic and Atmos-

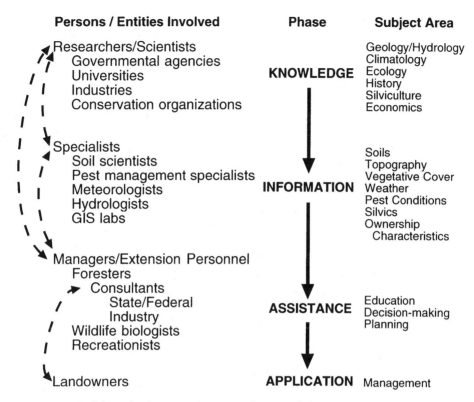

FIGURE 141   Diagram of the technology transfer process for natural resource management, from research to application.

pheric Association and other agencies have been collecting data concerning natural resources for many years, and with modern satellite technology and data processing and networking capabilities, our ability to collect, organize, and transmit information is highly advanced. Why then is the application of management lagging behind? Foresters have said for years: "If we could only *educate* landowners to the advantages of management...." Perhaps our problem is, in part, that we have not realized that communication is a two-way process and involves listening as well as telling. There may be several reasons that NIPF owners are not practicing forest management: lack of information, lack of available assistance, lack of incentive (economic, etc.), and lack of interest/ motivation.

Natural resource professionals have concentrated their efforts on the first two, and in the absence of incentives and motivation, these efforts have generally failed. Based on the developing economic picture for the central hardwoods, it appears likely that economic incentives for hardwood management will improve greatly over the next several decades. This should help motivate NIPF owners, and resource management professionals will have an opportunity to play a key role in the educational, motivational, and assistance activities. But before we can be effective, we must first look critically at ourselves. Foresters who work directly with property owners (consultants, industry procurement foresters, service foresters) all too often see a "timber sale" rather than a "management opportunity." This problem is aggravated by the fact that consultants are typically paid by commission and procurement foresters are given a bonus for bringing in timber below market value. Getting a fair price for a landowner's timber is a worthy objective, but in itself it is not enough. Management involves *planning*! Diameter-limit cuts or high-grading, even where timber is marked for sale, is not management if there is no objective other than marketing the timber. Occasionally, good regeneration results after a diameter-limit cut, which may help persons practicing such methods to rationalize them; but a practice is not management when it is not part of a plan. Sometimes foresters justify their actions by indicating that they are simply complying with the wishes of the landowner. This may be true at times, especially when dealing with land speculators, but one of the ethical responsibilities of the resource management professional is to *inform* property owners regarding the options available to them. There are several excellent publications and videos that can be referred to property owners. But what about property owners who do not use a forester? Egan (1997) observed that, although not *all* forest landowners come in contact with a forester when harvesting timber, all come in contact with a logger. He suggested that the loggers be required to distribute an informational packet to the property owner, and the owner must sign to indicate he/she received it. In West Virginia, such a packet was developed by the cooperative effort of several agencies and is available through the West Virginia Forestry Association. The packet entitled "Money Can Grow on Trees" contains information on topics such as silviculture, forest taxation, timber sales, management planning, and best management practices.

Professionalism and Ethics

Natural resource management professionals must go beyond what is good for business or what is good for themselves. The Society of American Foresters (1996) has published an Ethics Guide for Foresters and Natural Resource Professionals. This guide includes a "Code of Ethics" for members of the Society of American Foresters, which is reproduced opposite.

Note that the preamble to the ethical canons and the first canon discuss, as ethical standards, the "stewardship of the land, wise management of ecosystems and land management consistent with ecologically sound principles." Abiding by the code of ethics is a condition of membership in the Society, and it is in the setting and observing of such standards that true professionalism is achieved.

Code of Ethics

for Members of the Society of American Foresters

Preamble

Stewardship of the land is the cornerstone of the forestry profession. The purpose of these canons is to govern the professional conduct of members of the Society of American Foresters in their relations with the land, the public, their employers, including clients, and each other as provided in Article VII of the Society's Constitution. Compliance with these canons demonstrates our respect for the land and our commitment to the wise management of ecosystems, and ensures just and honorable professional and human relationships, mutual confidence and respect, and competent service to society.

These canons have been adopted by the membership of the Society and can only be amended by the membership. Procedures for processing charges of violation of these canons are contained in Bylaws established by the Council. The canons and procedures apply to all membership categories in all forestry-related disciplines, except Honorary Members.

All members upon joining the Society agree to abide by this Code as a condition of membership.

Canons

1. A member will advocate and practice land management consistent with ecologically sound principles.
2. A member's knowledge and skills will be utilized for the benefit of society. A member will strive for accurate, current, and increasing knowledge of forestry, will communicate such knowledge when not confidential, and will challenge and correct untrue statements about forestry.
3. A member will advertise only in a dignified and truthful manner, stating the services the member is qualified and prepared to perform. Such advertisements may include references to fees charged.
4. A member will base public comment on forestry matters on accurate knowledge and will not distort or withhold pertinent information to substantiate a point of view. Prior to making public statements on forest policies and practices, a member will indicate on whose behalf the statements are made.
5. A member will perform services consistent with the highest standards of quality and with loyalty to the employer.
6. A member will perform only those services for which the member is qualified by education or experience.
7. A member who is asked to participate in forestry operations which deviate from accepted professional standards must advise the employer in advance of the consequences of such deviation.
8. A member will not voluntarily disclose information concerning the affairs of the member's employer without the employer's express permission.
9. A member must avoid conflicts of interest or even the appearance of such conflicts. If, despite such precaution, a conflict of interest is discovered, it must be promptly and fully disclosed to the member's employer and the member must be prepared to act immediately to resolve the conflict.
10. A member will not accept compensation or expenses from more than one employer for the same service, unless the parties involved are informed and consent.
11. A member will engage, or advise the member's employer to engage, other experts and specialists in forestry or related fields whenever the employer's interest would be best served by such action, and a member will work cooperatively with other professionals.
12. A member will not by false statement or dishonest action injure the reputation or professional associations of another member.
13. A member will give credit for the methods, ideas, or assistance obtained from others.
14. A member in competition for supplying forestry services will encourage the prospective employer to base selection on comparison of qualifications and negotiation of fee or salary.
15. Information submitted by a member about a candidate for a prospective position, award, or elected office will be accurate, factual, and objective.
16. A member having evidence of violation of these canons by another member will present the information and charges to the Council in accordance with the Bylaws.

Adopted by the Society of American Foresters by Member Referendum, June 23, 1976, replacing the code adopted November 12, 1948, as amended December 4, 1971. The 1976 code was amended November 4, 1986, and November 2, 1992.

The SAF Ethics Guide also provides several "hypothetical ethics cases." Case number 2 involves "Diameter-Limit Cutting, Pay Me Now, or Cost Them Later" and was originally published by Nyland (1992). This case is included in Appendix 2, and it presents a provocative scenario that is all too familiar in the central hardwood region where diameter-limit cutting is frequently practiced and even sometimes recommended by foresters.

One of the defining characteristics of a profession, as compared to a job or career, is a set of ethical standards. The Society of American Foresters (1996) discusses the history of forestry ethics in America as far back as Gifford Pinchot, first Chief of the USDA, Forest Service and founder of the Society of American Foresters. Although Pinchot felt strongly about professional ethics, it was actually several years after his death that the first code of ethics for foresters was drafted in 1916 by T. S. Woolsey, Jr. Although leaders, such as H. H. Chapman, spoke strongly for adoption of a code of ethics in the 1920s, the first code was not ratified until 1948. The original canons have been added to and the most recent, a "land ethic" canon, was ratified in 1992. It is this canon, based primarily on the philosophical perspective of Aldo Leopold, that concerns the behavior of foresters relative to the management of land. The addition of the land ethic canon was championed by James E. Coufal, who first proposed its addition in 1989.

Resource management professionals are pulled in many directions, sometimes by their employers, their peers, as a backlash to what they perceive as "radical environmentalists," by tradition, or by belief and value systems that they have developed through their lives. It is not surprising that foresters often adopt a politically conservative philosophy, given the background of the profession. And it is also not surprising that some resource professionals see a land ethic canon as a threat to the profession; thus, the issue is not without controversy.

Perhaps then, an alternative way to look at land management ethics is to ask the question: "Is this an activity I would practice on my own land, which I plan to pass to my children?" If the answer to the question is "yes" then the manager has made a decision of conscience and should be able to proceed. Obviously, we do not all share the same ethical vision of the world, but people cannot be held to a higher standard than to treat others the way they would wish to be treated themselves.

The Future and the Role of Resource Professionals

As the twenty-first century approaches, the central hardwoods are reaching an important crossroad. Each locality within the region has its own history. The following examples can be given: the apple orchards of Green Ridge State Forest in western Maryland (Mash 1996); the railroad logging of Cass, West Virginia; copper smelting in Copper Basin, Tennessee; iron production in the Cumberland Plateau; chemical wood production in the Allegheny Plateau; and the Civilian Conservation Corps in many locations throughout the region (Carvell 1996). Each of these activities colors the resulting forest in some way, creating a mosaic on the landscape. But even with this varied historical backdrop, today there is a remarkably similar picture across the entire region. Almost everywhere, the forests of the central hardwood region have been heavily affected by human activities, dating back to the Indigenous People and continuing to the present. Fire, logging, and agriculture (and field abandonment) as well as pest introductions and white-tailed deer have spread across the central hardwood region like waves. And yet the forests persist, even flourish, as a testimony to the resilience of the ecosystem.

The crossroad we are approaching is one of a maturing resource that is owned primarily by private individuals. The last time the central hardwood region experienced a maturing forest, America was caught up in a frenzy of industrialization, world wars, the dust bowl, and eventually the Great Depression. Governmental structure and the scientific expertise were lacking to deal effectively with the exploitation that took place. But ignorance or naiveté is no longer a valid excuse. Some of the pieces that have always eluded resource managers are beginning to fall into place, for example, better markets for low-grade and smaller materials, a better educated public, results from long-term research, better resource information, and technological advances in data management. Some of the very same factors that could aid in better management might also have the opposite effect without proper planning. Thus, foresters and other resource professionals must learn to be better listeners; we must stop thinking in terms of timber sales and start seeing the management opportunities. We have to find innovative ways to involve NIPF owners in the process. Where economics of scale come into play, especially with regard to management of small parcels of privately owned land, landowner cooperatives may hold promise. As suggested by Wear et al. (1996), where federal and private land are intermingled, it may be wiser for the government to purchase or acquire by exchange the private lands that contain ''critical areas'' (areas that possess special ecosystem attributes) so that these could be managed to achieve societal objectives, rather than to regulate management of such resources on private land.

The Missouri Ozark Forest Ecosystem Project is an ambitious project involving state forestlands in the Missouri Ozarks (Missouri Conservation Commission of the State of Missouri 1994). Scientists from the Missouri Department of Conservation, University of Missouri and the USDA, Forest Service are studying the effects of even-age forest management, uneven-age management, and no management (control). The 60-year study involves 9200 acres and investigates the effects of forest management on birds, mammals, amphibians, and invertebrates, as well as vegetation and nutrient cycling.

This and other studies in the central hardwood region continue to supply information that will help to guide forest resource management into the future.

But we cannot wait for these long-term results in dealing with the needs of today. As reported in the earlier sections of this book, many studies have already been completed regarding management of central hardwoods, and we have developed a strong experiential basis for managing central hardwood ecosystems. The immediate challenge is to seek ways to apply this knowledge and experience to achieve better management and to learn from the mistakes of the past. As resource management professionals, we must become more effective communicators in order to meet the challenges in the central hardwood region during the years to come.

Glossary*

acceptable growing stock Trees occupying a site that are contributing substantially to future growth.

access Refers to having a means of entering a property for the purpose of conducting an operation such as a harvest/sale of timber. Access is one of the principal elements of planned management.

adaptive forest management An approach that combines economic opportunities with the most practical silvicultural treatments for forest property owners (see **adaptive silviculture**).

adaptive silviculture The application of silvicultural practices by foresters and knowledgeable experts who seek to develop scenarios in which silvicultural management can be accomplished within the overall framework of ownerships, taking into account not only what is biologically *best*, but what is *realistic*. Adaptive silviculture can be tailored to meet landowner objectives/constraints, take advantage of favorable market conditions, exploit good crops of desirable regeneration, and respond to insect or disease outbreaks. The adaptive silviculture approach becomes "management" when management plans are developed, goals are set, and work is done to implement these goals.

advance regeneration The persistence of shade tolerant species in a subordinate position awaiting an opportunity when a canopy tree dies allowing them adequate sunlight to grow. This is typical of highly shade tolerant species, such as maples, beech, hemlock, and basswood.

*Definitions differentiated by an asterisk are from *Terminology of Forest Science, Technology, Practice, and Products*, edited by F. C. Ford-Robertson (1983).

agricultural disturbance   A major occurrence in the central hardwood region; many of the native forests have, at some time, been cleared for agriculture. The abandonment of subsistence farms has led to the development of many of today's stands (Kalisz 1993). Many central hardwood stands were also subjected to livestock grazing.

allelochemicals   Chemicals, such as toxins, produced by a plant to kill or retard the growth of another competing plant.

allelopathic chemicals   A toxic substance produced in a plant's leaves, bark, nut husks, and/or roots that is antagonistic to competing vegetation.

allelopathy   A process whereby some plants (e.g., walnut trees) control their competitive interactions with many other species. One plant produces substances, toxic to other plants, that escape into the environment. The toxin kills or retards the growth of the other plant.

allowable cut   The amount that can be harvested over time without cutting more than the net growth.

American beech (*Fagus grandifolia*)   A widely distributed tree species found through-out the central hardwood region, having its greatest importance in association with the northern hardwoods where it often occurs with maples and birches.

angiosperms (Angiospermae)*   The botanical name for the group of vascular flowering plants that produce seeds enclosed in an ovary; includes the hardwoods.

anthropogenic   Human caused.

anticline   An upward fold in the earth's crustal material.

area control   A method of forest harvesting by which an area of the forest that represents the inverse of the rotation length (e.g., for a 100-year rotation — 1/100) can be harvested annually.

aspect*   The direction toward which a slope faces.

association   Term used by Lucy Braun (1950) to denote the so-called "climax unit" of a formation. For example, the mixed mesophytic forest is an association.

autotrophs   Green plants with the ability to use solar energy in photosynthesis. Autotrophic plants are the first step in collecting solar energy by converting oxidized carbon (CO_2) into a reduced form (carbohydrates), which can then be consumed and oxidized by heterotrophs to provide energy for their needs.

balanced uneven-age stand   A stand where an array of diameter classes are represented and possess an inverted J-shaped diameter distribution (Fig. 116).

biodiversity   Diversity of organisms that exists at different scales (micro, meso, macro, etc.), as measured by the number of different species (richness) and their distribution over the landscape (evenness).

biomass accumulation   A way of viewing the chronological progress of an ecosystem. Bormann and Likens (1979) proposed four phases of biomass accumulation following clearcutting of a northern hardwood ecosystem (Fig. 87). Biomass accumulation looks at total productivity without regard to commercial value.

biophysical   An approach to site quality evaluation that attempts to directly measure the environment associated with site productivity (Lee and Sypolt 1974).

black cherry (*Prunus serotina*)   A tree species with its commercially most important range in the Allegheny Plateau of Pennsylvania and at higher elevations in the Appalachians, from West Virginia through Virginia and North Carolina, where trees of good quality grow.

boreal spruce-fir   A unique community that is apparently a relic from the glaciation period. Spruce and spruce-fir forests cover the highest peaks in the southern Appalachians (normally above 4500 ft.), such as Mt. Mitchell in North Carolina.

capital gains tax   Tax on income based on the *increase* in value of a capital investment.

cardinal limits   The range of conditions that a given species can tolerate.

cation exchange capacity*   The total ionic charge of the adsorption complex active in the adsorption of ions, expressed as milliequivalents of cations or anions per 100 g of soil or other adsorbing material. *Note*: For soils, the cation exchange capacity (=T-value) is generally expressed in milliequivalents of NH_4 ions adsorbed by 100 g of soil from a large excess of N-ammonium acetate (CH_3COONH_4) at pH 7; the capacity varies considerably according to pH, particularly with organic materials.

cation exchange capacity   The ability to adsorb and release positively charged ions.

cedar breaks   Communities growing on soils derived from dolomite limestone parent material which weathers rapidly and produces a thin soil with relatively high pH. Communities on these thin soils do not progress past the cedar stage, hence the cedar break community (i.e., cedars of Lebanon State Park in the Nashville Basin area of Middle Tennessee and sites in the Bluegrass Region of Kentucky).

central hardwood region   The region south of the beech-maple forest, east of the Great Plains, and north and west of the southern pine forests of the Coastal

Plain and Piedmont (Fig. 1). The central hardwood forest covers an area of approximately 235,000 square miles and is centered along the axes of the Appalachian Mountains east of the Mississippi River and the Ouachita/Ozark Mountains west of the Mississippi.

chestnut blight   A disease caused by the fungus *Cryphonectria parasitica* introduced to the New York Botanical Gardens in 1904. By 1915 most states east of the Mississippi River reported infected trees. By the late 1930s, most mature chestnut trees were dead throughout the eastern states.

cleaning   A release operation that removes overtopping trees of similar age to favor trees of better species or quality. Cleanings are done during the *sapling stage*. Cleanings are appropriate to remove lower value species, such as black locust, sassafras, eastern redcedar, sourwood, blackgum, aspens, and Virginia pine, which become established at the same time as do more valuable oaks, yellow-poplar, ash, and black cherry in old-field stands. These stands may also contain limby or poorly formed dominants of the desired species as well. Cleanings to remove undesirable trees can be effective using either cutting, girdling, or herbicide injection.

clearcutting   An even-age silvicultural system that involves the felling of *all* trees in a stand in one operation. This system relies on natural regeneration, but in some cases artificial regeneration (planting or direct seeding) is used as a primary or supplemental source of regeneration. The basic ecological premise behind clearcutting is to redistribute the resources of the site to a new crop by removing the existing stand. This type of cutting is designed to mimic natural disturbances, such as fire, windstorms, and catastrophic insect and disease outbreaks, which promote regeneration of species that have evolved to exploit these conditions.

climatic climax   An association that, due to the specific climate of a region, would ultimately predominate in the region in the absence of disturbance or other limitations.

climax   Term used by Clements (1916) to refer to a vegetative community that perpetuates itself. The climax is theoretically the end product of a process of succession.

codominant crown class   Tree crowns that receive direct light from above and some direct light from the sides. The crowns of codominant trees generally form the main canopy level.

community   An assemblage of organisms living in an environment and interacting with each other and with the environment. Each community possesses species that are adapted to the specific conditions that exist at the time.

competitive advantage   In regard to a limited resource can be gained by having either an advantage due to position or an advantage due to vigor and may occur above ground (crown competition) or below ground (root competition). Where competition occurs for limited resources, once one individual gains a competitive advantage over its neighbors, the advantage usually compounds.

competition (for resources)   The interaction between plants that is a result of limited resources. For example, species such as oaks and hickories, with their better developed root systems, are more competitive on sites where moisture is limited.

conservative   Term used with respect to a species' ecological role. Conservative species typically occur later in the successional sequence and are shade tolerant and slow growing, have heavy seed, and are more efficient at using low-intensity light. They have determinate growth and a higher root/shoot ratio.

continental position   Position within a land mass (continent) relative to the ocean.

control of harvest   An activity that is performed to ensure a sustained yield of products and is constrained by the site productivity and the size of material needed to grow. In general, two methods are used: area control and volume control.

coppice method   Silvicultural system that relies on sprout reproduction.

crop-tree assessment   A phase of crop-tree management that includes crop-tree selection, timber management, wildlife management, and aesthetic and water management goals. Perkey et al. (1993) provide guidelines for this assessment.

crop-tree enhancement   A phase of crop-tree management usually accomplished by *releasing* them from competition, although Houston et al. (1995) include fertilization as a possible enhancement activity and pruning of butt logs could also be added to the list.

crop-tree management   A type of intermediate management that focuses on individual trees of potentially high value. In most cases, these operations direct the allocation of above- and below-ground resources by making more resources available to the residual crop trees. It is a promising technique to facilitate the production of high-value grade sawlogs or veneer logs.

crown class   The relative position of tree crowns in a stand used to indicate the degree of competition experienced by individual trees. Four crown classes are generally recognized (Fig. 96): see **dominant**, **codominant**, **intermediate**, and **overtopped**.

crown thinning   A technique that removes trees from the middle and upper strata of the canopy to favor good trees in the same canopy range (Smith 1986).

crown-touching   The recommended method for releasing crop trees (Lamson et al. 1988). To employ this method, the crop-tree crown is divided into quadrants and these "sides" are evaluated as to whether or not the tree is free-to-grow (Fig. 106). The recommendation is to release potential crop trees on at least three sides (Lamson et al. 1990, Wilkins 1994).

cruise   An inventory of the timber growing on a site conducted by measuring a sample or all the trees (100 percent cruise). A sample cruise would be conducted for larger tracts, greater than 50 acres (Wiant 1989).

culls   Trees occupying a site but which are not producing a harvestable yield.

cutting cycle   The interval between successive harvests. Use of yield projections, in combination with stand size and site quality, can be useful in establishing the interval between cutting cycles, as well as the allowable cut. The length of the cutting cycle may be dependent, in part, on the length of time required to produce a viable sale from the stand.

cutting/felling   A method to accomplish a release operation by removing the competing vegetation and leaving the desired trees standing.

deciduous*   Refers to perennial plants that are normally leafless for some time during the year. Deciduous species have the advantage of concentrating their growth and bioproductivity during a portion of the year when environmental conditions, such as soil moisture and temperature, are most favorable.

deciduous angiosperms   See **angiosperms**.

decline   A syndrome involving an accumulation of stress by the tree. As stress accumulates, different secondary organisms become involved. The decline process is illustrated in Figure 92.

deer, impact of   The consumption of tree regeneration by deer is an ecological factor that will continue to have a significant impact on future stands in the central hardwood region.

deferment cutting   A modification of the shelterwood method, the goal of which is to obtain regeneration of shade intolerant species by reducing the overstory basal area to approximately 20 ft.^{2} per acre and to defer harvest of the overwood through a complete rotation of the regeneration (Miller and Schuler 1995). This technique carries a sparse stand of large residual trees for an extended period, thus mitigating some of the adverse aesthetic effects of clearcutting while maintaining some of the qualities of mature forests, such as hard mast production. Retention of the overwood ensures a seed source for a long period in case there is difficulty in obtaining regeneration immediately after cutting. Because a high percentage of the stand basal area is removed in the initial cut, such harvests can be commercially attractive.

designation of timber   Identification of trees to be harvested. One of the principal elements of a planned harvest/sale.

determinate   A morphologic growth type in woody plants related to the extension of apical meristems. Buds that are formed at the end of a growing season contain primordia representing all the tissues that will expand in the next growing season. In other words, the enclosed winter bud contains a miniature of the next year's stem and leaf tissue, which expands the following spring. The determinate species base their next year's growth on the current year's conditions. For example, a sugar maple, once the leaf and twig tissue is fully expanded in the spring, cannot extend any more during the growing season. See **indeterminate**.

direct factors   Components needed for the functioning of autotrophic plants. These are heat, light, water, carbon dioxide, oxygen, and mineral nutrients. These factors are obtained from the environment (site) and, at this level, are referred to as the site resources.

direct suppression   An action taken in reference to pest (e.g., gypsy moth) management. At such time as populations reach a threshold where damage is likely to occur (500–1000 egg masses per acre), the manager may want to intervene with direct suppression. In a forestry context, this usually equates to some type of spray program using biological or chemical pesticides.

disturbance   Forest disturbance is a component of the process of secondary succession and has played a significant part in the development of the central hardwood forest. Disturbances can come from within (endogenous), such as tree fall, native insect, or disease outbreak, or may come from outside agents (exogenous), such as fire, windstorm, and ice.

diversity   The occurrence in a forest of a variety of life forms, species, age classes, and so on. Most managers would agree that maintaining a degree of diversity is desirable in hardwood stands to ensure market flexibility, reduce vulnerability to pests, and provide habitat for wildlife.

division   The second unit in a hierarchical order that depicts ecosystems on a regional basis (compiled by R. G. Bailey, *Ecoregions of the United States*, 1994). Successively smaller ecosystems (units) are defined within larger ecosystems. Different ecosystem attributes are given prime importance in each unit. Divisions are based mainly on the large ecological climate zones. The units from larger to smaller include domains, divisions, provinces, and sections.

domain   The largest unit in a hierarchical order that depicts ecosystems on a regional basis (compiled by R. G. Bailey, *Ecoregions of the United States, 1994*). Successively smaller ecosystems (units) are defined within larger ecosystems. Different ecosystem attributes are given prime importance in each unit. Domains are based mainly on the large ecological climate zones. The units from larger to smaller include domains, divisions, provinces, and sections.

dominant crown class   Crowns of trees in the dominant crown class receive direct light from above and the sides. Trees are clearly emergent above the general canopy.

ecological classification   An attempt to group sites with similar ecological characteristics. Johnson (1992) states: "Rigorously developed ecological classification systems can be important silvicultural tools for identifying potential problems and opportunities." Several researchers (Smalley 1987; Bailey 1996) have provided in-depth regional classifications of ecosystems within the central hardwood region. They have synthesized this and other work to provide a national framework for classifying forest ecosystems.

ecological optimum   A concept that is useful when looking at competitive interactions among species. If the cardinal limits represent the range of conditions that a given species can tolerate, the optimum represents the point within that range at which the species functions best. When examining the adaptation of a species to a site, the optimum is usually more significant than the cardinal limit because, even though a particular species' cardinal limits may not be exceeded on a given site, the species usually lives in a competitive association with other species. Other things being equal, when several species are in competition, the ones that are closest to their optimum will prevail.

ecological role   Silvical characteristics define a tree species' ecological role (or niche) and are therefore key to managing the species. Each species' ecological role is a product of its inherent genetic makeup and interactions with the environment and other species. Survival growth, and regeneration are all a function of these adaptations and interactions. Important information regarding the management of a species is gained by observing the natural conditions under which it grows best, such as geographic distribution, soils, topographic conditions and associated species.

ecosystem   An association in which the prevailing vegetation creates habitat for animal forms that are adapted to it, and they, in turn, have a biofeedback relationship with the vegetation (Odum 1971). An assemblage of organisms that function in a particular environment, having interactions with their environment and with each other.

ecosystem management   "Ecosystem management defines a paradigm that weaves biophysical and social threads into a tapestry of beauty, health and sustainability. It embraces both social and ecological dynamics in a flexible and adaptive process" (Cornett 1994). Wiant (1995) describes ecosystem management as a "retreat from reality."

edaphic climax   A condition occurring when soil factors impose limitations on plant communities preventing them from reaching a "true" successional climax. Communities that occur in the shale barrens of the Ridge and Valley Province and in the Missouri Ozarks are a good example.

endogenous   A term that refers to forest disturbances that come from within, such as native insect or disease outbreaks or tree fall.

ethical responsibility   The obligation of professionals in a field to use their knowledge and skills responsibly. For example, foresters have an obligation to promote the concepts of planned management versus forest exploitation in dealing with other professionals as well as landowners.

even-aged (single-cohort) stand   A stand that developed as a result of a single disturbance event, such as the logging/fire scenario that occurred around the turn of the century. The dynamics of most central hardwood stands appears to fit this case.

even-age management   A silvicultural system designed to totally remove an existing stand and create a new single-cohort stand. Even-age management methods include clearcutting, seed-tree, and shelterwood systems. It is best suited to regenerate shade intolerant species.

exclusion/protection   A reactive approach to management of a deer over-population problem in forests. Two primary devices are utilized for exclusion/ protection—fences and tree shelters. When fences (either electric or non-electric) are well constructed and maintained, they work well but are very difficult to build and maintain. They need to be at least 8 ft. tall and must conform closely to the ground to prevent deer from going over or under. Tree shelters must be at least 5 ft. tall in order to allow trees to get above the reach of deer. On an operational scale, tree shelters might be used in limited areas where regeneration is particularly problematic but are not likely to be used in large-scale operations because of expense and difficulty of installation.

exogenous   A term that refers to forest disturbances that come from outside agents, such as fire.

exploitive   A term used with respect to a species' ecological role. Exploitive species are those that are capable of rapidly taking advantage of temporary enrichment of a site. They are analogous to pioneer species in plant succession terms. Such species characteristically have certain attributes, such as mechanisms that disseminate seed widely (light weight, animals, etc.). They are shade intolerant, grow fast, use high-intensity light efficiently, and have indeterminate growth and a lower root/shoot ratio.

federal assistance programs   Governmental programs enacted to create incentives for forest management and other activities on non industrial private land.

fertilization   The addition of supplementary fertilizer to a site. It is a possible crop-tree enhancement activity recommended by Houston et al. (1995).

fire   An important type of disturbance that may not only kill vegetation but may alter the site by converting biomass into ash. Fire is one of the most influential factors in the development of the current forest that occupies much of the central hardwood region.

fire climax   A condition occurring when periodic fire events impose limitations on plant communities preventing them from reaching a "true" successional climax. The pine barrens that occur along the coastal plains from Massachusetts to New Jersey are maintained by periodic fires and are, therefore, a "fire climax" community.

flexible diameter-limit cut   A partial cutting system where trees are harvested when they drop below an acceptable rate of return, based on growth and quality.

food chain or food web   The association of organisms through energy flow forming the energy base for all life.

forest harvesting   An anthropogenic disturbance to the forest in which trees are removed for commercial products. Harvesting may or may not be done in association with silvicultural operations.

forest hazard rating   A key component of forest health management. In the context of forest health, hazard describes a condition of the forest under which a damaging event is likely to occur. Hazard rating, therefore, involves the ranking of stands as to their hazard.

forest health   "A dynamic condition that relates to the vigor of trees and associated organisms and is reflected in their physiological performance (survival, growth, reproduction, etc.) as compared to accepted norms for similar environments" (Hicks and Mudrick 1994).

forest management plan   A strategy for implementing forest management practices. See **forest resource management**. A good plan is one that is effective in accomplishing the forest manager's or landowner's objectives.

forest resource management*   The practical application of scientific, economic, and social principles to the administration and working of a forest estate for specified objectives (Society of American Foresters, Ford-Robertson 1983). Traditional forest management focuses on commodity yield and economic gain.

forest roads   See **roads**.

forest stand dynamics   Functional relationships in stands, such as how trees function where they compete with other trees in varying mixtures of species on different sites, at a variety of stocking densities and in varying age structures (Oliver and Larson 1996).

forestry*   A profession embracing the science, business, and art of creating, conserving, and managing forests and forestland for the continuing use of their resources, material, or other forest produce.

formation   The "major vegetation unit" of an area as termed by Lucy Braun (1950), for example, deciduous forest.

geosyncline   A downward fold in the earth's crustal material.

genotype*   The entire genetic constitution (expressed or latent) of one individual.

Geographic Information System (GIS)   A powerful computer-mediated tool to assist in organizing information to facilitate forest management, including forest health management (Fleischer et al. 1992b). GIS databases for forest hazard conditions, insect and disease populations, resource management zones, topography, roads, and so on can be overlain in various combinations (Hutchinson and DeLost 1993).

grassy balds   A type of grassland occurring on scattered high mountain tops along the Appalachian highlands. Examples are Bald Knob (West Virginia), Stratton's Meadows (Tennessee), and Brasstown Bald (Georgia). The reason for the occurrence of these balds is still unexplained, since they do not occur above a true timberline. Several hypotheses have been put forward, including historic agricultural clearing and fire.

grassy meadows   The occurrence of these natural grasslands is related to soil and parent material with the added influence of climate and periodic grass fires (i.e., Nashville Basin of Tennessee, Bluegrass Region of Kentucky, and the Ozark glades of Missouri).

group   A uniform inclusion of trees within a stand (Smith 1986). The recognition of a group has more to do with the management perspective than with biological factors that characterize similar groups.

group selection   Group selection is an uneven-age management method in which larger openings are created than in single-tree selection. It has the advantage of permitting regeneration of intermediate and shade intolerant species. Smith (1986) sets an upper limit on group selection openings as having a diameter smaller than twice the height of the tallest trees. In central hardwood stands, a half-acre circular opening is a reasonable approximation.

growth model   Mathematical model that simulates forest growth and provides outputs in terms of commercial forest products.

growth factor method   A method for estimating annual growth rates (Ashley 1989). Using this procedure, diameter growth (in rings per inch) and average dbh are the only variables needed.

gypsy moth (*Lymantria dispar*)   A defoliating insect that was inadvertently introduced into the Boston area in 1869. Deciduous species, especially oaks, are their preferred hosts. Gypsy moth is an ecological factor that will most likely have a significant impact on future stands in the central hardwood region.

hazard   In the context of forest health management, hazard identifies the potential for a damaging agent to attack the forest. Hazard is often associated with characteristics of the host plant, such as age, species, and vigor. See **risk**.

herbicides   Chemicals that either selectively or non-selectively kill plants. They can be used effectively in release operations to remove competing vegetation.

heterotrophs*   Organisms that cannot live without an external source of organic food.

hickories (Carya spp.)   A tree genus commonly associated with upland oaks. Although several authors have identified an oak-hickory forest cover type, others have questioned this ranking. The greatest distribution of hickory growing stock occurs in a band including most of Tennessee and Kentucky extending into southern West Virginia.

high-forest method   Silvicultural system that relies on reproduction from seed. Within high-forest methods, there are two subdivisions—even-age systems and uneven-age systems.

high-grading   The practice of cutting the best and most desirable trees and leaving scattered poor quality and unmerchantable residual trees. If repeated over and over, high-grading will essentially leave stands of trees with low growth potential and can ultimately lead to an impoverished condition where few good management alternatives remain.

ice age   Perhaps the most recent (and most significant) episode of climatic and vegetational change to affect the distribution of present-day vegetation in the central hardwood region. The Wisconsin ice sheet began to retreat northward about 14,000 years ago. Prior to that, continental glaciers extended down through North America into southern Illinois, and a zone of tundra extended well down into much of the present central hardwood region.

impoverished stands   Poorly managed stands in the central hardwood region that have become depleted either through repeated high-grading, fires, animal damage, or pest attacks.

improvement cutting*   The elimination or suppression of less valuable in favor of more valuable tree growth, typically of mixed or uneven-aged forests. It is an intermediate management technique designed to improve stand quality (Smith 1986).

indeterminate   A morphologic growth type in woody plants related to the extension of apical meristems. Primordia enclosed in the winter bud represent only a portion of the potential leaf and stem tissue that can be produced the following year. Leaves found in the winter bud of indeterminate species are morphologically distinct from those produced later in the growing season. Indeterminate species have the capacity to exploit temporary episodes of abundant resources. See **determinate**.

industrial private property owners   In reference to forestland, these owners are generally timber companies whose primary goal is to make a profit.

Industrial private land represents a relatively small part (less than 20 percent) of the overall forestland in the central hardwood region.

initial floristics model　In this pattern of vegetation development, proposed by Egler (1954) (Fig. 88), most of the species have already regenerated or regenerate immediately after the disturbance event, but some short-lived species may dominate early and drop out quickly. The regeneration in present-day central hardwood stands that developed synchronously after a disturbance event with different species, assuming dominance through time, exemplify this model.

intermediate crown class　Tree crowns receiving direct light from the top only. Their crowns are usually narrow and slightly below the main canopy level.

intermediate cuttings　A silvicultural treatment, or combination of treatments, performed simultaneously that are used when stands are *not* ready for a final harvest, which, in general, is currently the case for most central hardwood stands.

intermediate management　A regimen of silvicultural treatments that are used to "tend" existing stands.

inventory　See **cruise**. An inventory involves the gathering and organizing of information relating to a forest property. Such information may include soils, timber volume, species composition, and wildlife habitat.

killing　A method to accomplish a release operation by deadening the competing vegetation and leaving them standing. Killing of competing vegetation is usually accomplished by stem girdling, herbicide treatment, or both.

landscape level　Viewing the ecosystem at a scale of tens to hundreds of square miles. At this mesoscale level, macroclimate is relatively homogeneous, but elements such as soils, topography, and drainage basins may vary.

liberation　A release operation done to remove *older trees* that are overtopping sapling-size or younger trees. Liberation operations are often appropriate in stands that have been high-graded and cut over.

light, quality of　Light quality refers to the specific wavelengths of light. In addition to reduced light quantity, the quality of light reaching the forest floor is depleted in the most photosynthetically active wavelengths, those in the red and blue portions of the spectrum. Many shrub or small-tree species are obligate understory dwellers and therefore must be shade tolerant or possess some mechanism for coping with their altered light environment.

light compensation point　The light intensity at which the rates of photosynthesis and respiration are equal. That is, the point at which food production and consumption rates are equal. Other things being equal, species that are shade tolerant have a lower light compensation point than shade intolerant species.

limiting factor concept   The idea that the rate of a process (e.g., photosynthesis) is limited by the factor that is in shortest supply. For example, bioproductivity of a boreal forest may be limited by temperature whereas an oak forest growing on a shaley soil may be limited more by mineral nutrients and/or by available water.

live crown ratio   The percent of the tree's bole occupied by live branches; a crown characteristic that foresters have traditionally used to assess the severity of competition among trees. The lower the live crown ratio, the greater the degree of competition. It should also be noted that shade tolerant species have inherently deeper crowns than intolerant species; thus, live crown ratio is only useful as a relative measure and for a given species.

low thinning   Also known as thinning from below, this thinning method involves removal of trees from lower canopy positions (overtopped, intermediate, and weak codominant). The primary objective of a low thinning is to salvage anticipated losses. This technique is most applicable to pure, even-aged stands of relatively shade intolerant species and would remove trees that have become, or are becoming, overtopped.

macroclimate   An overriding condition that affects the type of vegetation that predominates in a given area. The direct influence of climate on vegetation includes the supply of resources (water, heat, light) and the limitations imposed on plant functions (coldest temperature, most severe drought, seasonal cycles, wind, etc.). Indirect impacts are also important in that soils and landforms are themselves a product of a given climate (Bailey 1996). In addition to influencing the amount of heat, light, and water, climate has had a major effect on soil formation; thus, macroclimate is indirectly associated with mineral nutrient levels as well.

macroscale   Level at which species' adaptations are in response to latitude, elevation, and continental position.

management   A philosophical approach to forestry based on the human desire to control or organize. Management involves planning and implementation phases.

marking guides   When trees are designated for cutting in conjunction with silvicultural operations, it may be necessary to develop marking guides, the purpose of which is to facilitate the application of the silvicultural and management operations on the ground.

marking priority   When more than one silvicultural operation (crop-tree release, improvement cutting, etc.) are performed simultaneously, the objectives may best be achieved using a marking priority approach. Establishment of a residual basal area target can be a useful supplement to this procedure.

mean annual increment   Average annual volume growth over the length of the rotation.

mesoscale   A level of ecosystem classification at which species adaptations are in response to general landform differences. For example, in West Virginia, the geology, soils, climate, and vegetation all differ between the Ridge and Valley and Appalachian Plateau Provinces, and these differences translate into different resource management constraints.

microscale   A level of ecosystem classification at which adaptations of vegetation occur in response to soil and topographic differences. The mosaic of species that typically occurs throughout the mesophytic forest areas where oaks predominate on tops of ridges and southwest slopes and mesic species, such as yellow-poplar, predominate in coves and northeast slopes is a good example of a microscale adaptation. At this scale microclimate is controlled by factors such as slope inclination, slope position, aspect, and soil color (if exposed) or vegetative cover.

mineral nutrient cycling   A fundamental process in ecosystems in which terrestrial forest systems typically conserve mineral elements needed for growth by returning them in biomass and releasing them slowly via the decomposition process (Bormann and Likens 1979).

monadnock   A protruding feature on the landscape composed of resistant geologic material that has been extruded to the surface. Examples are Stone Mountain in Georgia and Mt. Katahdin in Maine.

Monongahela Decision   Controversy over the use of clearcutting on public land resulted in the 1970 landmark decision that recommended uneven-age management as a primary management policy for the Monongahela National Forest in West Virginia. This decision was more significant as a precedent than it was to the Monongahela National Forest.

multiple use*   Any practice of forestry fulfilling two or more objectives of management.

mycorrhizae   Fungal roots that develop when plant roots are infected by certain fungi, and, as a result, the absorbing surface of the root is increased. This is one of the most important symbiotic relationships. The fungi are benefited by deriving nutrition from the tree.

natural pruning   A response by which trees grown in dense stands tend to lose their lower branches, which die and fall off at a point in the canopy below which the light compensation point is reached. It is more dramatic for shade intolerant species than for tolerant species.

natural resource policy   Statement of strategy for handling natural resource issues.

natural thinning   An inherent process occurring in stands by which the thousands of seedlings and/or sprouts per acre in the original stand are gradually reduced in number over time as a result of competition.

niche The specific ecological role that each species or type occupies in the ecosystem. Species adaptations at this scale are usually related to very precise environmental differences. For example, several woody species, due to their small mature height, are relegated to understory positions in the forest. These species (e.g., flowering dogwood, black haw, spice bush, witch hazel, rhododendron) are, of necessity, highly shade tolerant.

non-industrial private property owner The NIPF (usually individual) owner owns the largest portion of forestland of any ownership group in the central hardwood region. These owners hold land in small tracts (generally less than 500 acres) for a wide variety of reasons including farming, residence, investment, and recreational use.

northern hardwoods The beech/maple/birch type consists of various mixtures of shade tolerant species, usually found at higher elevations in the central hardwood region.

oak-chestnut Prior to the chestnut blight, American chestnut occupied an important position in the hardwood ecosystem. Braun (1950) used the name "oak-chestnut" to identify one of the most significant forest associations in the eastern deciduous forest.

oaks (*Quercus* spp.) The oaks are a diverse group of species that are consistently identified with the central hardwood region. Common oaks include white, black, northern red, chestnut, and scarlet oaks. They are often found on harsher environments such as south-facing slopes.

overtopped crown class Tree crowns receiving only indirect sunlight. Their crowns are clearly below the main canopy.

parasitism An association in which organisms (parasites) reside in or on another organism and derive nourishment from their host. Many parasites are detrimental to their host. For example, diseases, such as chestnut blight, anthracnoses, *Nectria*, and Dutch elm disease, are all parasitic. Examples of parasitic plants include mistletoe, which sometimes becomes problematic to individual trees. Diseases, such as *Armillaria* root disease, can function as parasites, when they attack living trees, or, on dead hosts, as **saprophytes**.

partial cutting The removal of some trees while leaving others. It may take the form of a silvicultural operation such as an improvement cutting or single-tree selection, or may simply be a timber harvest such as high-grading or diameter-limit cutting.

photosynthesis The ability of plants to capture solar energy in the form of chemical energy using readily available compounds (carbon dioxide, water) to produce carbohydrates has been the key to all biological development on earth.

physiographic regions Regions of the Earth's surface that have certain structural features in common.

phytophagy   The eating of plants by animals and/or insects, e.g., the browsing of white-tailed deer on tree seedlings.

pine barrens   Areas that are generally vegetated by drought tolerant species, such as pitch pine and scrub oaks, and other so-called pygmy forests, which usually occur on droughty soils. The community also owes its existence to the periodic occurrence of fire, which cleans out the competing vegetation and prepares the seedbed for germination of pine seeds.

pioneer community   The first community of a secondary succession that follows a disturbance event. This community presumably gives way to what is termed the climax.

pioneer species   The early invaders of a site that are typically shade intolerant, fast growing, short-lived, and capable of rapid and wide dissemination of propagules. Classically, pioneer plant species are herbaceous, but some tree species, such as Virginia pine, pin cherry, and eastern redcedar, also function in this role.

position advantage   A strictly physical advantage. For example, a species that arrives at the site first may achieve a position advantage relative to light by growing up more quickly and shading out potential competitors.

preparatory cutting   A shelterwood cutting treatment designed to remove poor-quality trees and increase vigor and seed production among the residuals.

presalvage   A type of salvage cutting that removes trees that are highly vulnerable to mortality *before* they die. The advantages of presalvage are to obtain live-timber value for material sold and to avoid depressed markets that often accompany large-scale pest outbreaks.

presalvage harvest   A silvicultural treatment for gypsy moth management recommended for mature, oak-dominated stands; proposed by Gottschalk (1987).

presalvage shelterwood   A silvicultural treatment for gypsy moth management, which is a modification of the presalvage harvest where residual oaks are left to serve as a seed source to initiate a new stand; proposed by Gottschalk (1987).

presalvage thinning   A silvicultural treatment for gypsy moth management recommended for immature, predominantly oak stands that are fully stocked; proposed by Gottschalk (1987).

primary pollutants   Air pollutants that are emitted directly into the atmosphere from their sources. Included in this group are sulfur dioxide, nitrogen oxides, volatile organic compounds, and heavy metals. USDA, Forest Service (1990) classification.

primary succession   A theoretical sequence of vegetational changes that begins with a previously unoccupied site (sand dune, rock outcrop, lava flow, etc.). The sequence might start with a geologic event, for example, sediment deposition into a basin or volcanic activity. This geologic material is exposed to a particular climatic sequence (erosion cycles, freeze-thaw, etc.), which begins to weather the material, forming soil. Propagules of plant species adapted to the conditions (soil, climate) become established and their roots and organic matter further change the site in a feedback relationship. As the site changes, it becomes suitable for other plants and animals, which may indeed also have an effect on the local climate (Clements 1916).

proactive   An approach to forest management where problems are anticipated and actions taken before they become critical. Deer herd management through hunting regulation is an example in which overpopulation by deer can be avoided. See **reactive**.

profitability of forestry   As a strict economic venture, one should compare the income that could be generated (e.g., through interest) from the money that would be derived from the sale of land and timber to the income that can be generated from sale of forest products that result from management.

province   The third unit in a hierarchical order of land classification that depicts ecosystems on a regional basis (compiled by R. G. Bailey, *Ecoregions of the United States*, 1994). Successively smaller ecosystems (units) are defined within larger ecosystems. A different ecosystem component is given prime importance in each unit. Provinces are based mainly on macrofeatures of the vegetation. The units from largest to smallest include domains, divisions, provinces, and sections.

pruning   A crop-tree enhancement activity that involves removing living or dead branches from butt logs in order to improve the quality of products produced.

public property owners   Public land is held in the public trust and usually managed by a public agency. The USDA Forest Service, Bureau of Land Management, National Park Service, and a variety of state forestry and wildlife agencies are involved in management of forests, parks, and wildlife areas.

reactive   An approach to management of forests where critical problems are dealt with after they occur. Reactive approaches to overpopulation by deer include (1) exclusion/protection, (2) inhibition, (3) satiation, and (4) repelling. Two primary devices are utilized for exclusion/protection—fences and tree shelters. Inhibition is attempted by making it difficult for deer to reach seedlings, usually by scattering and/or piling brush from logging so as to discourage browsing. Satiation is achieved by creating so much food (undergrowth) that deer cannot eat it all. Therefore, some portion is able to achieve sufficient height to get above the reach of deer. Repelling is attempted when some material (repellent) is placed on the trees or in the forest that discourages deer from browsing.

red maple (*Acer rubrum*)   A tree species with a broad distribution throughout the central hardwood region. It is most abundant in the northeastern part of the region, extending down the higher Appalachians into western North Carolina.

regeneration systems   Silvicultural treatments that are designed to regenerate new stands.

release operation   Intermediate cutting designed to improve stand quality (Smith 1986). It is important to leave those trees that are *capable of responding* to release. This may be a function of the tree's species, age, condition, site, and so on and generally relates to its silvical characteristics.

removal cutting   A shelterwood cutting treatment that is done after regeneration is established to remove the overstory and allow the new stand to grow.

reproduction method   "A procedure by which a stand is established or renewed; the process is accomplished during the regeneration period by artificial or natural reproduction" (Smith 1986).

reproductive strategies   Ways in which plants propagate. Some species rely primarily on vegetative propagation to spread and colonize an area (e.g., ferns). Taller-growing trees become established by advance reproduction (seedlings already present, usually of tolerant species) or new individuals. The latter can be from vegetative sources (root or stump sprouts) or from seed.

reserve shelterwood   A shelterwood method, similar to deferment cutting, in which the overwood is maintained for more than 20 percent of the length of the rotation. This technique carries a sparse stand of large residual trees for an extended period, thus mitigating some of the adverse aesthetic effects of clearcutting while maintaining some of the qualities of mature forests, such as hard mast production. Because of the low density of the residual stand, regeneration under such cutting is usually composed of shade intolerant species, similar to those that regenerate under clearcutting. See *deferment cutting*.

residuum   Soils formed in place on ridges and certain slope segments from resident parent material types.

resources, site   Factors, including heat, light, water, carbon dioxide, oxygen, and mineral nutrients, obtained from the environment (site) and needed for the functioning of autotrophic plants.

risk   In the context of forest health, risk takes into account the *probability* of a damaging event occurring. Risk may be influenced by pest population status as well as host plant conditions. See **hazard**.

sale contract   See **timber sale contract**.

salvage cutting   An intermediate management technique designed to cope with pest problems. They are cuttings "made for the purpose of removing trees that have been or are in imminent danger of being killed or damaged by injurious agencies other than competition between trees" (Smith 1986). The philosophy behind salvage cutting is more *reactive*. It is often done after damage has occurred, and the primary objective would be to recover value from trees that have died or are expected to die.

sanitation conversion   A silvicultural treatment for gypsy moth management recommended to convert high-hazard oak stands to less susceptible species, such as pines or other mixed hardwoods; proposed by Gottschalk (1987).

sanitation cutting   Removal of trees to "reduce the spread of damaging organisms to the residual stand." It is an intermediate management technique designed to cope with pest problems (Smith 1986). The philosophy behind sanitation cutting is *proactive*.

sanitation thinning   A silvicultural treatment proposed by Gottschalk (1987) for gypsy moth management. The sanitation thinning is recommended for stands having a more diverse species mixture, including non-oaks.

saprophytes   Organisms that obtain nutrition from dead tissue after the affected organism (e.g., tree) has died. Leaf and wood decaying organisms are saprophytic.

satiation   An approach used to mitigate deer damage to regenerating forests. Satiation is achieved by creating so much food (undergrowth) that deer cannot eat it all. Therefore, some portion of tree stock is able to achieve sufficient height to get above the reach of deer.

secondary pollutants   Air pollutants that form when primary pollutants (those emitted directly into the atmosphere from their sources) undergo chemical changes. Included in this group are photochemical oxidants and acid deposition. USDA, Forest Service (1990) classification.

secondary succession   Succession that begins with the interruption of a community, resulting from some type of disturbance (forest harvesting, fire, wind or ice storms, insect or disease outbreaks). This type of process is typical in the forests of the central hardwood region, most of them being even-aged forests that regrew after logging, fires, or agricultural abandonment within the past 100 years.

section   The fourth unit in an hierarchical order that is used to classify ecosystems on a regional basis (compiled by R. G. Bailey, *Ecoregions of the United States*, 1994). Successively smaller ecosystems (units) are defined within larger ecosystems. A different ecosystem component is given prime importance in each unit. Sections are based mainly on physiography, which exerts the major control over ecosystems within climatic-vegetation zones. The units from larger to smaller include domains, divisions, provinces, and sections.

seed cutting   A shelterwood cutting treatment performed to open the stand sufficiently to encourage the development of regeneration.

seed tree*   A tree selected, and often reserved, for seed production and/or collection.

seed-tree cutting*   An even-age silvicultural system involving removal in one cut of the mature timber from an area, save for a small number of seed bearers left singly or in small groups. A seed-tree cutting treatment is performed to open the stand sufficiently to encourage the development of regeneration.

selection thinning   A technique that removes poorly formed dominants to favor crop trees in the upper canopy stratum (Smith 1986).

severance tax*   A tax on a fixed natural resource (e.g., on timber), following its removal from the natural site and therefore severance from the natural state. When timber is cut, in some states, such as West Virginia, the owner of the timber at that time must pay a severance tax.

shade tolerance   A silvical characteristic referring to the ability of a plant to subsist in the shade of other plants. This attribute is important in defining the ecological role of a species. The stereotypic "pioneer" community would be dominated by fast-growing, shade intolerant species, which is frequently the case. But, depending on the type of disturbance that initiates a secondary succession, many representatives of shade tolerant species may be present as well.

shale barrens   Sites occurring in shale outcrop areas along the Ridge and Valley from Pennsylvania into West Virginia; dominated by widely spaced scrub oaks and other drought tolerant species, such as Virginia and pitch pines. The shale soils are generally thin, low in fertility, and droughty.

shelterwood   An even-age management system where the objective is to develop a standing crop of advance regeneration through a series of partial removal cuttings of the overstory (Smith 1986). This system relies on natural regeneration, but in some cases artificial regeneration (planting or direct seeding) is used as a primary or supplemental source of regeneration. The shelterwood method appears to be especially well suited to regenerating species that are intermediate in shade tolerance and have slower initial growth. In the central hardwood region, this fits the description of the oaks. Where advance oak regeneration is inadequate, the shelterwood system of regeneration has been used successfully, especially on the poorer sites. The shelterwood method involves cutting treatments extended over a 15- to 30-year period.

shredders   Organisms that enhance the decay process of forest litter by processing the leaf tissue. Mites in the suborders Oribatei, Mesostigmata, and Prostigmata, as well as Collembola, are important shredders of organic residues in hardwood forests.

silvical characteristics   Literally means "tree characteristics." The inherent characteristics of trees that determine their ecological role. They define the ecological requirements of a species, determine if individuals of a species can tolerate excesses or deficiencies in resources of a given site, and determine how efficient members of a species might be in utilizing the resources of the site. They determine how trees function in stands where they compete with other trees in varying mixtures of species on different sites, at a variety of stocking densities and in varying age structures. Silvical characteristics include shade tolerance, growth rate, site requirements, regeneration strategies, and injurious agents, such as insects and diseases.

silvics*   The study of the life history and general characteristics of forest trees and stands, with particular reference to locality factors, as a basis for the practice of silviculture. Silvics deals with the principles underlying the growth and development of single trees and of the forest as a biological unit (Smith 1986).

silvicultural management   Forest management that utilizes silvicultural principles to achieve management goals. There are three cardinal rules that must be observed in any silvicultural management. First, cuttings should never be conducted in a manner that amounts to high-grading — that is, never cut only good trees and leave poor ones. Second, regeneration should always be a consideration of management, either implicit or explicit. Finally, the harvest should be designed to remove only the growth, which is a basic tenet of sustained yield management.

silvicultural system   "A planned program of silvicultural treatment during the whole life of the stand; it not only includes reproduction cuttings but also any tending operations or intermediate cuttings" (Smith 1986).

silviculture*   The theory and practice of controlling the establishment, composition, constitution, and growth of forests.

single-cohort stand   See **even-aged stand**.

single-tree selection   An uneven-age management method based on the removal of single mature trees. This technique simulates the natural gap dynamics that occurs in mature unmanaged natural stands. This technique theoretically results in a "balanced" uneven-age stand, where an array of diameter classes are represented and possess an inverted J-shaped diameter distribution (Fig. 116). This method leaves relatively small canopy gaps that can close fairly rapidly due to crown expansion of residual trees. Thus, it promotes the regeneration of shade tolerant species. It is difficult to apply and requires long-term commitment on the part of the owner/manager.

site   The sum total of all environmental factors affecting the functioning of a forest community in a given locale (soil, climatic, and biotic factors).

site index   Use of tree height growth as a measure of site quality. Height of dominant or codominant trees of a certain species (usually oaks) at 50 years of age is generally used in central hardwoods.

skid trails (roads)* Any surface, more or less prepared, over which logs are dragged.

slope inclination Degree of rise for a particular unit of land. It is often expressed as slope percent (vertical rise divided by horizontal distance × 100). Slope inclination is one of several factors that control climate at the microscale.

slope position A term used to indicate the location on a slope relative to the ridge top or valley bottom.

soil A complex of mineral and organic fragments that supplies minerals, water, and support for plant growth (Wilde 1946). The soil, as a medium for plant growth, serves as a reservoir for water, mineral nutrients, and oxygen (for root growth).

stakeholders (natural resource) People who have a claim in the present and future use of natural resources. They include the public, the natural resource-using industry, and the landowners.

stand A spatially continuous group of trees and associated vegetation having similar structures and growing under similar soil and climatic conditions (Oliver and Larson 1996). It is analogous to the ecological concept of "community" but focuses more on the trees and vegetation. A stand is a group of trees with similar age structure, species composition, site quality, and condition so as to be recognizable from adjacent stands. It is the basic unit of the forest to which a silvicultural treatment is applied.

stand dynamics A term analogous to succession but focusing on the "changes in forest stand structure with time, including stand behavior during and after disturbances" (Oliver and Larson 1996).

stocking Refers to the occupancy of a site (number of trees or basal area per unit area) relative to the optimum the site can carry.

stocking guides Guides developed for upland hardwoods in the central hardwood region that are based on the concept that there is a theoretical "full stocking" condition that exists for a given stand, considering its age, species composition, and site quality (Gingrich 1967; Roach 1977). Stocking can be expressed in terms of number of stems per unit area, basal area per unit area, or percent of full stocking. These guides can be displayed graphically as overstocked, fully stocked, or understocked (Fig. 96).

subclimax communities Interim successional communities between pioneer and climax communities. Sometimes a community may, due to some site factor or other environmental condition, be prevented from reaching the "climatic climax" for the region and will be maintained in a subclimax state. Many of the existing oak-dominated forests of the central hardwood region appear to be a subclimax and an artifact of the disturbance history of the region.

succession An orderly change in community species composition over time, with each community possessing species that are adapted to the specific conditions that exist at the time. According to successional theory, ecosystems not subjected to strong exogenous disturbances change in a progressive and directional way (Margalef 1963).

sugar maple (*Acer saccharum*) A tree species distributed across the central hardwood region. Sugar maple is typically associated with the northern hardwoods (American beech and birches) at higher elevations in the Appalachians and occurring broadly in New England and the Lake States.

superorganism This term can be applied to Clements's (1916) view that the ecosystem is a "complex organism," and for any given site (environment), there is a particular climax community that represents the "mature" form of that ecosystem. In a sense, this view of the ecosystem has at its core the view that the whole is greater than the sum of the parts.

survival strategies Methods, such as seed dissemination mechanisms, seed dormancy, sprouting ability, and allelopathy, that plants have evolved, in addition to energy relationships, to ensure their perpetuation.

susceptible Trees likely to be attacked by damaging agents, for example, gypsy moth.

sustainable forestry Stout (1995) cites the World Commission on Environment and Development definition of sustainable forestry as forestry that "meets the needs of the present without compromising the ability of future generations to meet their own needs." Sustainable forestry focuses on both commodity and non-commodity production.

sustained yield* The yield that a forest can produce continuously at a given intensity of management. As forestry in America evolved as a profession, the term "yield" has been broadened to include more than one commodity or use, although retaining a commodity focus.

symbiosis A type of interaction that occurs in forests where two organisms interact for their mutual benefit (e.g., formation of mycorrhizae on plant roots). Symbiotic nitrogen fixation is another example of symbiosis that is important in central hardwood forests, especially in early successional stages and in fire-dominated ecosystems.

thinning Thinning is an intermediate cutting aimed at controlling stand density (Smith 1986). The primary purpose of thinning is to redirect the resources of the site to the residual trees in order to improve their vigor and growth.

timber sale contract   An agreement negotiated between the forest landowner and the logging contractor and/or timber buyer. A sale contract is a legal document and, at a minimum, should provide a description of the timber for sale, indicate a sale price, stipulate a duration of the agreement, provide for adherence to BMPs, and require a performance bond to be held by the landowner to secure the proper performance of the job. One of the principal elements of a planned harvest/sale.

two-aged management   A silvicultural system that provides for regeneration of shade intolerant species while carrying a sparse overstory of mature trees to offset the negative aesthetic impact of clearcutting (Miller and Schuler 1995). A silvicultural treatment in which the residual stand density is reduced to 20–25 ft.^{2} of basal area per acre, and a second age class develops as regeneration. Both cohorts are then tended as a two-aged stand. Some of the adverse aesthetic effects of clearcutting are mitigated while maintaining some of the qualities of mature forests, such as hard mast production.

unacceptable growing stock (UGS)   Trees occupying a site but which are not expected to produce a harvestable yield.

uneven-age management   Silvicultural systems that produce uneven-age (multi-cohort) stands of shade tolerant species. In actuality, uneven-age stands are aggregations of many small even-age stands. They usually possess an inverted J-shaped diameter distribution that is related to tree *age* rather than differential growth rates of various species in mixed stands (Fig. 116). Uneven-age stands managed by single-tree methods ultimately become dominated by shade tolerant species, since the size of stand openings is often too small to permit the successful regeneration of intolerant species.

vegetation frequency   An approach to site evaluation that uses occurrence of vegetation to predict site quality.

vegetative cover*   A broad term comprising all the vegetation occupying an area.

vigor   The condition of an organism that relates to its physiological functioning. A vigorous organism (e.g., tree) is healthy and functioning within the normal range expected for the particular conditions that exist.

vigor advantage   A competitive advantage that one plant holds over another due to its greater vigor. Plants that are situated near a resource (e.g., stream) or in an above-average microsite may gain an advantage initially due to vigor and later, as it becomes proportionately larger than its competitors, due to position. Genotype may enable one plant to gain an advantage due to vigor over another, as well.

visual crown rating (VCR)   Systems developed and applied by researchers (Belanger et al. 1991; Kelly et al. 1992) to facilitate monitoring of forest and tree health. These ratings are based on assessments of crown density, depth, foliage color, and amount of dieback. These ratings are useful as subjective measures when applied by an experienced evaluator.

volume control   A method of setting the allowable cut in a managed forest based on the annual volume growth or mean annual increment. This method is suited to situations where a good inventory of the forest is available, site productivity is known, and the type of product desired is clearly understood. This approach has some advantages over area-based systems in that it is unaffected by variations in site productivity, and it is much more applicable to silvicultural systems that involve uneven-age management or partial cutting.

vulnerable   Term applied to trees likely to be damaged by a pest agent (e.g., gypsy moth).

weeding   A release operation performed in seedling stands to remove herbaceous plants and shrubs that overtop desirable tree seedlings. Weeding is appropriate to remove a fern, bramble, or herbaceous understory, which develops during the regeneration phase following harvest of the overstory. Plants, seeds, or other propagules of competing vegetation are usually already present and are capable of rapid response to the additional resources.

Wilderness Act   1964 environmental policy legislation whereby areas in the National Forests could be set aside for "wilderness," and timber cutting in these areas would be prohibited.

yellow-poplar (*Liriodendron tulipifera*)   A major component of stands from the Allegheny Plateau of central Pennsylvania through West Virginia and into Kentucky and Tennessee. Yellow-poplar is usually associated with the so-called mixed mesophytic hardwood type and is often found in pure stands on north-facing slopes.

yield model   Forest growth model that provides outputs in terms of commercial forest products.

Literature Cited

Abrams, M. D. 1992. Fire and the development of oak forests. *Bioscience* 42(5):346–353.

Abrams, M. D., Orwig, D. A., and Dockry, M. J. 1997. Dendroecology and successional status of two contrasting oak forests in the Blue Ridge Mountains, U.S.A. *Can. J. For. Res.* 27:994–1002.

Ahern, G. P. 1928. *Deforested America.* Washington, D.C (no reference given).

Alban, L. M., Thomasma, S. A., and Twery, M. J. 1995. Forest Stewardship Planning Guide. *USDA For. Serv. Gen. Tech. Rep.* NE 203, 15 pp.

Alerich, C. L. 1993. Forest statistics for Pennsylvania — 1978 and 1989. *USDA For. Serv. NEFES Res. Bull.* NE-126, 244 pp.

Alig, R. J., Hohewsten, W. G., Murray, B. C., and Haight, R. G. 1990. Changes in area of timberland in the United States, 1952–2040, by ownership, forest type, region and state. *USDA For. Serv. SEFES Gen. Tech. Rep.* SE-64, 34 pp.

Anderson, R. G. and Loucks, O. L. 1979. White-tailed deer (*Odocoileus virginianus*) influence on structure and composition of *Tsuga canadensis* forests. *J. Appl. Ecol.* 16:855–861.

Anderson, R. L. 1994. How people, pests and the environment have changed, and continue to change the southern Appalachian, forest landscape. In: *Threats to Forest Health in the Southern Appalachians,* C. Ferguson and P. Bowman (eds.). Southern Appalachian Management and the Biosphere Cooperative, Gatlinburg, TN, pp. 1–4.

Andresen, J. W. 1957. Precocity of *Pinus rigida* Mill. *Castanea* 22:130–134.

Arends, E. and McCormick, J. F. 1992. Replacement of oak-chestnut forests in the Great Smoky Mountains. In: *Proceedings: 6th Central Hardwood Forest Conference* R. Hay, F. W. Woods, and H. DeSelm (eds.). University of Tennessee Press, Knoxville, TN, pp. 305–316.

Argon, K. A. 1996. This land is their land. *J. For.* 94(2):30–33.

Ashby, W. C. 1962. Germination capacity in American basswood. *Trans. Ill. Acad. Sci.* 55:120–123.

Ashley, B. S. 1989. *Reference Handbook for Foresters.* USDA, Forest Service, Northeastern Area State and Private Forestry, NA-FR-15, 37 pp.

Ashley, B. S. 1991. *Simplified Point-Sample Cruising.* USDA, Forest Service, Northeastern Area State and Private Forestry, NA-UP-01-91, 51 pp.

Atkins, J. M. 1988. West Virginia in flames. *WV Div. For. Tech. Rep.* 88-1, 16 pp.

Atkins, J. M. and Wimer, G. 1990. Severe soil erosion following forest fires in southern West Virginia. *WV Div. For. Tech. Rep.* 90-1, 14 pp.

Auchmoody, L. R. and Smith, H. C. 1979. Oak soil-site relationships in northwestern West Virginia. *USDA For. Serv. Res. Pap.* NE-434, 27 pp.

Auchmoody, L. R., Smith, H. C., and Walters, R. S. 1993. Acorn production in northern red oak stands in northwestern Pennsylvania. *USDA For. Serv. Res. Pap.* NE-680. 5 pp.

Bailey, R. G. 1988. Ecogeographic analysis. A guide to the ecological division of land for resource management. *USDA For. Serv. Misc. Publ.* 1465, 16 pp.

Bailey, R. G. 1994. *Ecoregions of the United States* (map), Rev. USDA Forest Service, Washington, DC.

Bailey, R. G. 1996. *Ecosystem Geography.* Springer, New York, 204 pp.

Bailey, R. G., Avers, D. E., King, T., and McNab, H. 1994. *Ecoregions and Subregions of the United States* (map). U.S. Geologic Survey, Washington, DC.

Baker, F. S. 1949. A revised tolerance table. *J. For.* 47:179–181.

Bartram, J. 1751. *Observations in His Travels from Pennsylvania to Onondaga, Oswego and Lake Ontario.* J. Whieston and B. White, London.

Bartram, W. 1791. *Travels Through North and South Carolina, Georgia, East and West Florida, the Cherokee Country, the Extensive Territories of the Muscogulges or Creek Confederacy and the Country of the Chactaws.* Reprinted 1988, Penguin Press with introduction by James Dickey, 414 pp.

Baumgras, J. E. and Luppold, W. G. 1993. Relative price trends for hardwood stumpage, sawlogs and lumber in Ohio. In: *Proceedings: 9th Central Hardwood Forest Conference. USDA For. Serv. Gen. Tech. Rep.* NC-161:381–389.

Baumgras, J. E., Hassler, C. C., and Le Doux, C. B. 1993. Estimating and validating harvesting system production through computer simulation. *For. Prod. J.* 43(11/12):656–671.

Baumgras, J. E., Miller, G. W., and Le Doux, C. B. 1995. Economics of hardwood silviculture using skyline and conventional logging. *Hardwood Symp. Proc.* 5–14.

Beck, D. E. 1962. Yellow-poplar site index curves. *USDA For. Serv. Res. Note* SE-180, 2 pp.

Beck, D. E. 1977. Twelve-year acorn yield in southern Appalachian oaks. *USDA For. Serv. Res. Note* SE-244, 8 pp.

Beck, D. E. 1986. Thinning Appalachian pole and small sawtimber stands. In: *Proceedings: Guidelines for Managing Immature Appalachian Hardwood Stands,* H. C. Smith and M. C. Eye (eds.). West Virginia University Books, Morgantown, pp. 85–98.

Beck, D. E. 1988. Regenerating cove hardwood stands. In: *Proceedings: Guidelines for Regenerating Appalachian Hardwood Stands,* H. C. Smith, A. W. Perkey, and W. A. Kidd, Jr. (eds.). SAF Publ. 88-03:156–166.

Beck, D. E. 1990. *Liriodendron tulipifera* L., yellow-poplar. In: *Silvics of North America. Vol. 2: Hardwoods,* R. M. Burns and B. H. Honkala (eds.). *USDA For. Serv. Agric. Handb.* 654:406–416.

Beck, D. E. and Della-Bianca, L. 1970. Yield of unthinned yellow-poplar. *USDA For. Serv. Res. Pap.* SE-58, 20 pp.

Beck, D. E. and Della-Bianca, L. 1972. Growth and yield of thinned yellow-poplar. *USDA For. Serv. SEFES Res. Pap.* SE-101, 20 pp.

Beck, D. E. and Della-Bianca, L. 1981. Yellow-poplar: characteristics and management. *USDA For. Serv. Agric. Handb.* 583, 91 pp.

Beck, D. E. and Hooper, R. M. 1986. Development of a southern Appalachian hardwood stand after clearcutting. *South. J. Appl. For.* 10:168–172.

Behan, R. W. 1967. The succotash syndrome, or multiple use: a heartfelt approach to forest land management. *Nat. Res. J. Univ. New Mexico School of Law* 7(4):473–484.

Beilmann, A. P. and Brenner, L. G. 1951. The recent intrusion of forest in the Ozarks. *Ann Mo. Bot. Gard.* 38:261–281.

Belanger, R. P. 1990. *Quercus falcata* Michx. var. *falcata*. Southern red oak (typical). In: *Silvics of North America. Vol. 2*, Hardwoods, R. M. Burns and B. H. Honkala (eds.). *USDA For. Serv. Agric. Handb.* 654:640–644.

Belanger, R. P., Zarnoch, S. J., Anderson, R. L., and Cost, N. D. 1991. Relation between visual crown characteristics and periodic radial growth in loblolly pine trees. Southwide Disease Workshops, Durham, NC, 1 p.

Belcher, D. M. 1982. TWIGS: the woodsman's ideal growth projection system. In: *Microcomputers: A New Tool for Foresters*, J. W. Moser Jr. (ed.). SAF Publ. 82-05:70–95.

Belcher, D. M., Holdaway, M. R., and Brand, G. J. 1982. A description of STEMS, the stand and tree evaluation and modelling system. *USDA For. Serv. Gen. Tech. Rep.* NC-79, 18 pp.

Beltz, R. C., Cost, N. D., Kingsley, N. P., and Peters, J. R. 1992. Timber volume distribution maps for the eastern United States. *USDA For. Serv. Gen. Tech. Rep.* WO-60, 59 pp.

Berglund, J. V. 1969. Silvics. Unpubl., Silviculture Dept., State University College of Environmental Science and Forestry, Syracuse, NY, 328 pp.

Bess, H. A., Spurr, S. H., and Littlefield, E. W. 1947. Forest site conditions and the gypsy moth. *Harvard For. Bull.* 22, 56 pp.

Birch, T. W. 1996. Private forest-land owners of the United States, 1994. *USDA For. Serv. NEFES Res. Bull.* NE-134, 183 pp.

Birch, T. W. 1997. Private forest owners of the central hardwood forest. In: *Proceedings: 11th Central Hardwood Forest Conference*, S. G. Pallardy, R. A. Cecich, G. H. Garrett, and P. S. Johnson (eds.). *USDA For. Serv. Gen. Tech. Rep.* NC-188:89–97.

Birch, T. W. and Kingsley, N. P. 1978. The forest-land owners of West Virginia. *USDA For. Serv. NEFES For. Res. Bull.* NE-58, 76 pp.

Birch, T. W. and Moulton, R. J. 1997. Northern forest landowners: a profile. *Nat. Woodlands* Jan. 1997, 3 pp.

Birch, T. W. and Stelter, C. M. 1993. Trends in owner attitudes. In: *Penn's Woods, Change and Challenge: Proceedings of the 1993 Penn State Forest Research Issues Conference*, J. C. Finley and S. B. Jones (eds.). State College, PA, pp. 50–60.

Birch, T. W., Lewis, D. G., and Kaiser, H. F. 1982. The private forest-land owners of the United States. *USDA For. Serv. Res. Bull.* WO-1, 64 pp.

Blackhurst, W. E. 1954. *Riders of the Flood*. Vantage Press, New York, 198 pp.

Blethen, T. and Wood, C. A. 1985. A process begun: the settlement era. In: *The Great Forest: An Appalachian Story*, B. N. Buxton. and M. L. Crutchfield (eds.). Appalachian Consortium Press, Boone, NC. pp. 5–14.

Bode, H. R. 1958. Beiträge ur Kenntnis allelopathischer Erscheinungen bei einigen Jaglandaceen. *Planta* 51:440–480.

Boring, L. R. and Swank, W. T. 1984. The role of black locust (*Robinia pseudoacacia*) in forest succession. *J. Ecol.* 72:749–766.

Boring, L. R., Monk, C. D., and Swank, W. T. 1981. Early regeneration of a clear-cut southern Appalachian forest. *Ecology* 62(5):1244–1253.

Bormann, F. H. and Likens, G. E. 1979. *Pattern and Process in a Forested Ecosystem.* Springer-Verlag, New York, 253 pp.

Botkin, D. B., Janak, J. F., and Wallis, J. R. 1972. Some ecological consequences of a computer model on forest growth. *J. Ecol.* 60:849–872.

Brantley, E. A., Anderson, R. L. and Smith, G. 1994. How to identify ozone injury on eastern forest bioindicator plants. *USDA For. Serv. Prot. Rep.* R8-PR25, 78 pp.

Braun, E. L. 1950. *Deciduous Forests of Eastern North America.* Blakiston Co., Philadelphia, PA, 596 pp.

Brenneman, B. B. 1995. Westvaco wildlife and ecosystem research forest. In: *Ecosystem Management: Translating Concepts to Practice.* 1995 Penn State School of Forest Resources Issues Conference, State College, PA, pp. 59–60.

Brock, S. M., Hollenhorst, S., and Freimund, W. 1990. Effects of gypsy moth infestation on aesthetic preferences and behavior intentions (Abstract). *Proc. USDA For. Serv. Interagency Gypsy Moth Res. Rev. Gen. Tech. Rep.* NE-146.

Brooks, A. B. 1910. *Forestry and Wood Industries.* West Virginia Geol. Survey, Vol. 5. Acme Publishers, Morgantown, WV, 481 pp.

Brooks, M. 1965. *The Appalachians.* Houghton Mifflin, Boston, 346 pp.

Brown, N. C. 1923. *The American Lumber Industry.* John Wiley & Sons, New York, 279 pp.

Buckner, E. 1989. Evolution of forest types in the southeast. In: *Proceedings: Pine Hardwood Mixtures: A Symposium on Management and Ecology of the Type. USDA For. Serv. SEFES Gen. Tech. Rep.* SE-58.

Buckner, E. 1992. The changing landscape of eastern North America. In: *Conference Proceedings: 100 Years of Professional Forestry, Seventy-First Annual Meeting of Appalachian SAF,* R. Thatcher and T. McLintock (eds.). pp. 55–59.

Budyko, M. I. 1974. *Climate and Life.* Academic Press, New York, 508 pp.

Butler, L. and Kondo, V. 1994. Macrolepidopterous moths collected by blacklight trap at Cooper's Rock State Forest, West Virginia: a baseline study. *WV Agric. For. Exp. Stn. Bull.* 705, Morgantown, WV, 25 pp.

Butler, L. and Wood, P. S. 1985. Native hardwood defoliating caterpillars. Establishing baseline data prior to gypsy moth invasion. In: *Proceedings of the 1984 National Gypsy Moth Review,* Charleston, WV, pp. 117–119.

Buxton, B. M. and Crutchfield, M. L. 1985. *The Great Forest: An Appalachian Story.* Appalachian Consortium Press, Boone, NC. 30 pp.

Campbell, R. W. and Sloan, R. J. 1977. Forest stand responses to defoliation by gypsy moth. *For. Sci. Monogr.* 19, 34 pp.

Campbell, S. M. and Kittredge, D. B. 1996. Ecosystem-based management on multiple NIPF ownerships. *J. For.* 94(2):24–29.

Carmean, W. H. 1965. Yellow buckeye (*Aesculus octandra* Marsh.). In: *Silvics of Forest Trees of the United States,* H. A. Fowells (ed.). *USDA Agric. Handb.* 271:78–81.

Carmean, W. H. 1971. Site index curves for black, white, scarlet and chestnut oaks in the Central States. *USDA For. Serv. Res. Pap.* NC-62, 8 pp.

Carmean, W. H., Hahn, J. T., and Jacobs, R. D. 1989. Site index curves for forest tree species in the eastern United States. *USDA For. Serv. Gen. Tech. Rep.* NC-128, 142 pp.

Carter, K. K. and Snow, A. G., Jr. 1990. *Pinus virginiana*, Virginia pine. In: *Silvics of North America, Vol. 1, Conifers*, R. M. Burns and B. H. Honkala (eds.). *USDA, For. Serv. Agric. Handb.* 654:513–519.

Carvell, K. L. 1967. The response of understory oak seedlings to release after partial cutting. *WV Univ. Agric. Exp. Stn. Bull.* 553, 20 pp.

Carvell, K. L. 1973. Effects of improvement cuttings and thinnings on the development of cove and mixed oak stands. *WV Agric. For. Exp. Stn. Tech. Bull.* 616T, 20 pp.

Carvell, K. L. 1983. Intermediate cuttings in Appalachian mixed oak stands. In: *Proceedings: 11th Annual Hardwood Symposium.* Hardwood Research Council, Cashiers, NC, pp. 40–43.

Carvell, K. L. 1987. Effects of past history on present stand composition and condition. In: *Proceedings: Guidelines for Managing Immature Appalachian Hardwood Stands,* H. C. Smith and M. C. Eye (eds.). SAF Publ. 86-02, Morgantown, WV.

Carvell, K. L. 1996. Camp Rhododendron, Civilian Conservation Corps Camp at Cooper's Rock State Forest (1936-42). Monongalia Historical Society, Morgantown, WV.

Carvell, K. L. and Tryon, E. H. 1961. The effect of environmental factors on the abundance of oak regeneration beneath mature oak stands. *For. Sci.* 7(2):98–105.

Chang, S. J. 1996. US forest property taxation systems and their effects. In: *Symposium on Nonindustrial Private Forests: Learning from the Past, Prospects for the Future,* Baughman, M. J. and Goodman, N. (eds.). Extension Special Programs. Minnesota Extension Service, St. Paul, MN, pp. 318–325.

Chapman, H. H. 1950. *Forest Management.* Hildreth Press Publishers, Bristol, CT, 582 pp.

Cho, D. and Boerner, R. E. J. 1991. Structure, dynamics and composition of Sears Woods and Carmean Woods State Nature Preserves, North-Central Ohio. *Castanea* 56(2):77–89.

Clark, B. F. and Boyce, S. G. 1964. Yellow-Poplar Seed Remains Viable in the Forest Litter. *J. For.* 62:100–102.

Clarkson, R. B. 1964. *Tumult on the Mountains: Lumbering in West Virginia, 1770–1920.* McClain Printing Co., Parsons, WV, 410 pp.

Clatterbuck, W. K. 1990. Forest development following disturbances by fire and timber cutting for charcoal production. In: *Fire and the Environment: Ecological and Cultural Perspectives,* S. C. Nodvin and T. A. Waldrop (eds.). Knoxville, TN. *USDA For. Serv. Gen. Tech. Rep.* SE-69:60–65.

Clatterbuck, W. K. 1993. Are overtopped white oak good candidates for management? In: *Proceedings: Seventh Biennial Southern Silvicultural Research Conference,* J. C. Brissette (ed.). *USDA For. Serv. Gen. Tech. Rep.* SO-93:497–500.

Clements, F. E. 1916. *Plant Succession.* Carnegie Institute, Washington, DC, Pub. 242.

Clinton, B. D. and Vose, J. M. 1996. Effects of *Rhododendron maximun* L. on *Acer rubrum* L. seedling establishment. *Castanea* 61(1):38–45.

Clinton, B. D., Boring, L. R., and Swank, W. T. 1994. Regeneration patterns in canopy gaps of mixed-oak forests of the southern Appalachians: influences of topographic position and evergreen understory. *Am. Midl. Nat.* 132:308–319.

Coder, K. D., Wray, P. H., and Countryman, D. W. 1987. Group shelterwood system for regenerating oak in eastern Iowa. In: *Proceedings: 6th Central Hardwood Forest Conference*, R. L. Hay, F. W. Woods, and H. DeSelm (eds.). University of Tennessee, Knoxville, TN, pp. 83–90.

Colbert, J. J. and Racin, G. 1995. User's guide to the stand damage model: a component of the gypsy moth life system model. *USDA For. Serv. Gen. Tech. Rep.* NE-207, 38 pp.

Cornett, Z. J. 1994. Ecosystem management: why now? *Ecosystem Management News*, 3:14.

Critchfield, W. B. 1960. Leaf dimorphism in *Poplulus trichocarpa. Am. J. Bot.* 47:699–711.

Cronon, W. 1983. *Changes in the Land: Indians, Colonists and the Ecology of New England.* Hill and Wang, New York, 241 pp.

Crow, G. R. and Hicks, R. R. Jr. 1990. Predicting mortality in mixed oak stands following spring insect defoliation. *For. Sci.* 36(3):831–841.

Crow, T. R. 1990. *Tilia americana* L., American basswood. In: *Silvics of North America, Vol. 2: Hardwoods*, R. M. Burns and B. H. Honkala (eds.). *USDA For. Serv. Agric. Handb.* 654:784–791.

Crow, T. R. and Metzger, F. T. 1987. Regeneration under selection cutting. In: *Managing Northern Hardwoods*, R. D. Nyland (ed.). SAF Publ. 87-03: 81–94.

Dale, M. E., Smith, H. C., and Pearcy, J. N. 1994. Size of clearcut opening affects species composition, growth rate and stand characteristics. *USDA For. Serv. Res. Pap.* NE-698, 21 pp.

Dana, S. T. and Fairfax, S. K. 1980. *Forest and Range Policy, Its Development in the United States*, 2nd. ed. McGraw-Hill, New York, 458 pp.

Daubenmire, R. 1959. *Plants and Environment.* John Wiley & Sons, New York, 422 pp.

Davidson, W. H. 1988. Potential for planting hardwoods in the Appalachians. In: *Proceedings: Guidlines for Regenerating Appalachian Hardwood Stands*, H. C. Smith, A. W. Perkey, and W. E. Kidd, Jr. (eds.). SAF Publ. 88-03:255–268.

Davis, K. P. 1954. *American Forest Management.* McGraw-Hill, New York, 482 pp.

Davis, M. B. 1983. Holocene vegetational history of the eastern United States. In: *Late Quaternary Environments of the United States, Vol II: The Holocene*, H. E. Wright (ed.). University of Minnesota Press, Minneapolis, pp. 166–181.

Davis, R. P. S. Jr. 1978. Final report: a cultural resource overview of the Monongahela National Forest, West Virginia. West Virginia Geological and Economic Survey, Morgantown, WV.

Dawson, J. O., McCarthy, J., Roush, J. A., and Stengier, D. M. 1989. Oak regeneration by clearcutting after a series of partial cuts. In: *Proceedings: 7th Central Hardwood Conference*, G. Rink and C. A. Budelsky (eds.). *USDA For. Serv. Gen. Tech. Rep.* NC-132:181–184.

Day, F. P. and Monk, C. D. 1977. Net primary production and phenology on a southern Appalachian watershed. *Am. J. Bot.* 64(9):1117–1125.

Della-Bianca, L. and Beck, D. E. 1985. Selection management in southern Appalachian hardwoods. *South. J. Appl. For.* 9(3):191–196.

Dey, D. 1992. Predicting quantity and quality of reproduction in the uplands. In: *Proceedings: Oak Regeneration: Serious Problems, Practical Recommendations*, D. L. Loftis and C. E. McGee (eds.). *USDA For. Serv. Gen. Tech. Rep.* SE-84:138–145.

Dickson, R. E. 1994. Height growth and episodic flushing in northern red oak. In: *Proceedings: Biology and Silviculture of Northern Red Oak in the North Central Region. A Synopsis. USDA For. Serv. Gen. Tech. Rep.* NC-173:1–9.

DiGiovanni, D. M. 1990. Forest Statistics for West Virginia 1975 and 1989. *USDA, Forest Service, Northeast Forest Exp. Stn. Res. Bull.* NE-114.

Doolittle, W. T. 1958. Site index comparisons for several species in the southern Appalachians. *Soil Sci. Soc. Am. Proc.* 22:455–458.

Dosser, R. C. and Hicks, R. R. Jr. 1975. Ortet and season of collection significantly affect rooting of river birch cuttings. *Tree Planter's Notes* 26(2):11–12.

Duerr, W. A., Teeguarden, D. E., Christiansen, N. B., and Guttenberg, S. 1979. *Forest Resource Management, Decision-making Principles and Cases.* W. B. Saunders, Philadelphia, 612 pp.

Dunlap, R. E. 1991. Trends in public opinion toward environmental issues, 1965–1990. *Soc. Nat. Resources* 4:285–312.

Eagar, C., Miller-Weeks, M., Gillespie, A. J. R. and Burkman, W. 1992. *Summary Report: Forest Health Monitoring in the Northeast, 1991.* National Association of State Foresters, US EPA and USDA Forest Service, NEFES, 13 pp.

Eagle, T. R. Jr. 1993. The effects of gypsy moth defoliation on soil water nutrient concentration in two forest cover types in West Virginia. Unpubl. M.S. Thesis. West Virginia University, 149 pp.

Eardley, A. J. 1951. *Structural geology of North America.* Harper and Brothers, New York, 624 pp.

Egan, A. F. 1996. Harvesting nonindustrial private forests: Who's in charge? In: *Symposium on Nonindustrial Private Forests: Learning from the Past, Prospects for the Future*, M. J. Baughman and N. Goodman (eds.). Extension Special Programs, University of Minnesota, St. Paul, pp. 276–284.

Egan, A. F. 1997. Personal communication. Division of Forestry, West Virginia University, Morgantown.

Egan, A. and Jones, S. 1993. Do landowner practices reflect beliefs? *J. For.* 91(10):39–45.

Egler, F. E. 1954. Vegetation science concepts: I. Initial floristic composition: a factor in old-field vegetation development. *Vegetatio* 4:412–417.

Eller, R. 1985. Land as commodity: industrialization of the Appalachian forests, 1880–1940. In: *The Great Forest: An Appalachian Story*, B. M. Buxton and M. L. Crutchfield (eds.). Boone, NC, pp. 15–22.

Elliott, K. J. and Swank, W. T. 1994. Changes in tree species diversity after successive clearcuts in the southern Appalachians. *Vegetatio* 115:11–18.

Erdmann, G. G. 1990. *Betula alleghaniensis* Britton, yellow birch. In: *Silvics of North America, Vol. 2: Hardwoods*, R. M. Burns and B. H. Honkala (eds.). *USDA Agric. Handb.* 654:133–147.

Espenshade, E. B. Jr. (ed.). 1995. *Goode's World Atlas*, 19th ed. Rand McNally, New York, 372 pp.

Etgen, R. J. and Hicks, R. R. Jr. 1987. Impact of looper defoliation: a case study in West Virginia. *North. J. Appl. For.* 4:201–204.

Evans, L. T. 1980. The natural history of crop yield. *Am. Scientist* 68:388–397.

Eyre, S. R. 1963. *Vegetation and Soils: A World Picture.* Aldine Publishing Co., Chicago, 323 pp.

Fenneman, N. M. 1938. *Physiography of the Eastern United States.* McGraw-Hill, New York.

Fleischer, S., Carter, J., Reardon, R., and Ravlin, W. F. 1992a. Sequential sampling plans for estimating gypsy moth egg mass density. *USDA For. Serv. AIPM Technol. Transfer* NA-TP-07-92, 14 pp.

Fleischer, S., Roberts, A., Young, J., Mahoney, P., Ravlin, F. W., and Reardon, R. 1992b. Development of geographic information system technology for gypsy moth management within a county: an overview. *USDA For. Serv. AIPM Technol Transfer* NA-TP-01-93, 24 pp.

Ford-Robertson, F. C. 1983. *Terminology of forest science, technology, practice and products.* Society of American Foresters, Washington DC, 370 pp.

Fosbroke, D. E. and Hicks, R. R. Jr. 1989. Gypsy moth-induced tree mortality in southwestern Pennsylvania: a preliminary report. *Proc. Nat. Gypsy Moth Rev. (Dearborn, MI),* 167–178.

Fosbroke, D. E. and Meyers, J. R. 1995. Logging safety in forest management education. In: *Proceedings: 10th Central Hardwood Forest Conference,* K. W. Gottschalk and S. L. C. Fosbroke (eds.). *USDA For. Serv. Gen. Tech. Rep.* NE-197:442–453.

Fosbroke, S. and Carvell, K. L. 1989. Managing Appalachian oak: a literature review. *WV Agric. For. Exp. Stn. Circ.* 149, 49 pp.

Fountain, M. S. 1980a. Relating understory vegetation to site quality in north-central West Virginia. *Castanea* 45:1–5.

Fountain, M. S. 1980b. Prediction of oak site index from stand composition. *WV. For. Notes,* No 5:5-6.

Franklin, J. F., Shugart, H. H., and Harmon, M. E. 1987. Tree death as an ecological process. *Bioscience* 37(8):550–556.

Frederick, K. D. and Sedjo, R. A. 1991. *America's Renewable Resources: Historical Trends and Current Challenges.* Resources for the Future, Washington, DC.

Fries, R. F. 1951. *Empire in Pine, the Story of Lumbering in Wisconsin, 1830–1900.* State Historical Society of Wisconsin, Madison, 285 pp.

Gaddis, D. A. 1996. Accomplishments and program evaluations of forestry financial assistance programs. In: *Symposium on Nonindustrial Private Forests: Learning from the Past, Prospects for the Future,* M. J. Baughman and N. Goodman (eds.). Extension Special Programs, Minnesota Extension Service, St. Paul, pp. 357–366.

Gansner, D. A. and Herrick, O. W. 1987. Impact of gypsy moth on the timber resource. In: *Proceedings: Coping with the Gypsy Moth in the New Frontier,* S. Fosbroke and R. R. Hicks, Jr. (eds.). West Virginia University Books, Morgantown, pp. 11–19.

Gansner, D. A., Herrick, O. W., Mason, G. N., and Gottschalk, K. W. 1987. Coping with the gypsy moth on new frontiers of infestation. *South J. Appl. For.* 11(4):201–209.

Geiger, R. 1965. *The Climate Near the Ground.* Harvard University Press, Cambridge, MA.

George, D. W. and Fischer, B. C. 1989. The effect of site and age on tree regeneration in young upland hardwood clearcuts. In: *Proceedings of the Seventh Central Hardwood Conference,* G. Rink and C. A. Budelsky (eds.). *USDA For. Serv. Gen. Tech. Rep.* NC-132: 40–47.

Geraghty, J. J., Miller, D. W., van der Leeden, F., and Troise, F. 1973. *Water Atlas of the United States.* Water Information Center, Mahassett Isle, Port Washington, NY.

Ghent, J. 1994. The gypsy moth. In: *Threats to Forest Health in the Southern Appalachians,* C. Ferguson and P. Bowman (eds.). Southern Appalachian Management and Biosphere Cooperative, Gatlinburg, TN pp 13–16.

Giddings, N. J. 1912. The chestnut bark disease. *WV Agric. Exp. Stn. Dept. Plant Pathol. Bull.* 137:207–225.

Gillespie, W. H., Miller, A., Wimer, G., Circle, J., Upton, S., and Dameron, B. 1992. Damages sustained and accruing from forest fires in southern West Virginia, Oct.

26–Nov. 10, 1991. For. Mgt. Rev. Commission, Guide Book. West Virginia Division of Forestry, MP-92-1, 17 pp.

Gingrich, S. F. 1967. Measuring and evaluating stocking and stand density in upland hardwood forests in the central states. *For. Sci.* 13:38–53.

Gleason, H. A. 1926. The individualistic concept of plant association. *Bull. Torrey Bot. Club* 53:7–26.

Godman, R. M. and Lancaster, K. 1990. *Tsuga canadensis* (L.) Carr. eastern hemlock. In: *Silvics of North America, Vol. 1: Conifers*, R. M. Burns and B. H. Honkala (eds.). *USDA, For. Serv. Agric. Handb.* 654:604–612.

Godman, R. M. and Mattson, G. A. 1976. Seed crops and regeneration problems of 19 species in northeastern Wisconsin. *USDA For. Serv. Res. Pap.* NC-123, 5 pp.

Godman, R. M., Yawney, H. W., and Tubbs, C. H. 1990. *Acer saccharum* Marsh., sugar maple. In: *Silvics of North America, Vol. 2: Hardwoods*, R. M. Burns and B. H. Honkala (eds.). *USDA For. Serv. Agric. Handb.* 654:78–87.

Goff, F. G. and West, D. 1975. Canopy-understory interaction effects on forest populations. *For. Sci.* 21:98–108.

Gottschalk, K. W. 1982. Silvicultural alternatives for coping with the gypsy moth. In: *Conference Proceedings: Coping with the Gypsy Moth.* 1982 Penn State Forest Issues Conference, University Park, PA, pp. 137–156.

Gottschalk, K. W. 1987. Prevention: the silvicultural alternative. In: *Conference Proceedings: Coping with the Gypsy Moth in the New Frontier*, S. Fosbroke and R. Hicks, Jr. (eds.). West Virginia University Books, Morgantown, pp. 92–102.

Gottschalk, K. W. 1993. Silvicultural guidelines for forest stands threatened by the gypsy moth. *USDA For. Serv. Gen. Tech. Rep.* NE-171, 49 pp.

Gottschalk, K. W. and MacFarlane, W. R. 1992. Photographic guide to crown condition of oaks: use for gypsy moth silvicultural treatments. *USDA For. Serv. Gen. Tech. Rep.* NE-168, 8 pp.

Graney, D. L. 1987. Ten-year growth of red and white oak crop trees following thinning and fertilization in the Boston Mountains of Arkansas. In: *Proceedings: 4th Biennial Southern Silviculture Research Conference. USDA For. Serv. Gen. Tech. Rep.* SE-42.

Graney, D. L. 1990. *Carya ovata* (Mill.) K. Koch, shagbark hickory. In: *Silvics of North America, Vol. 2: Hardwoods*, R. M. Burns and B. H. Honkala (eds.). *USDA For. Serv. Agric. Handb.* 654:219–225.

Graves, D. H. 1986. A landowner's guide, measuring farm timber. *Univ. Ky. Coll. Agric. Coop. Ext. Serv.* FOR9, 22 pp.

Grelen, H. E. 1990. *Betula nigra* L., river birch. In: *Silvics of North America*, Vol. 2: *Hardwoods*. R. M. Burns and B. H. Honkala (eds.). *USDA For. Serv. Agric. Handb.* 654:153–157.

Gribko, L. 1997. Personal communication. Division of Forestry, West Virginia University, Morgantown.

Gribko, L. and Hix, D. M. 1993. Effect of small rodents on northern red oak acorns in north-central West Virginia. In: *Proceedings: 9th Central Hardwood Conference*, A. P. Gillespie, G. R. Parker, and P. E. Pope (eds.). *USDA For. Serv. Gen. Tech. Rep.* NC-161.

Grime, J. P. 1979. *Plant Strategies and Vegetation Process.* John Wiley & Sons, New York, 222 pp.

Grisez, T. J. 1975. Flowering and seed production in seven hardwood species. *USDA For. Serv. Res. Pap.* NE-315, 8 pp.

Grumbine, R. E. 1994. What is ecosystem management? *Conservation Biol.* 8(1):27–38.

Guthrie, R. L. 1989. Xylem structure and ecological dominance in a forest community. *Am. J. Bot.* 76:1216–1228.

Hamel, D. R. 1983. Forest management chemicals. A guide to use when considering pesticides for forest management. *USDA For. Serv. Agric. Handb.* 585, 645 pp.

Hammond, E. H. 1954. Small-scale continental landform maps. *Annals Association of American Geographers* 44: 33–42.

Haney, H. L. Jr. and Siegel, W. C. 1993. Estate planning for forest landowners. What will become of your timberland? *USDA For. Serv. Gen. Tech. Rep.* SO-97, 186 pp.

Hannah, P. R. 1968. Topography and soil relations for white and black oak in southern Indiana. *USDA For. Serv. Res. Pap.* NC-25, 7 pp.

Hansen, G. D. and Nyland, R. D. 1986. Effects of diameter distribution on the growth of simulated uneven-aged sugar maple stands. *Can. J. For. Res.* 17:1–8.

Harshberger, J. W. 1911. *Phytogeographic Survey of North America.* G. W. Stexhert & Co., New York, 427 pp.

Hawley, R. C. and Smith, D. M. 1954. *The Practices of Silviculture.* 6th ed. John Wiley & Sons, New York, 525 pp.

Healy, W. M. 1988. Effects of seed-eating birds and mammals on Appalachian hardwood regeneration. In: *Conference Proceedings: Guidelines for Regenerating Appalachian Hardwood Stands,* H. C. Smith, A. W. Perky, and W. E. Kidd, Jr. (eds.). West Virginia University Books. Morgantown, WV, pp 104–111.

Healy, W. M. and Lyons, P. T. 1987. Deer and forests on Boston's municipal watershed after 50 years as a wildlife sanctuary. In: *Proceedings: Deer, Forestry and Agriculture: Interactions and Strategies for Management.* SAF, Warren, PA, pp. 3–21.

Hedden, R. L. 1981. Hazard-rating system development and validation: an overview. In: *Proceedings: Hazard-Rating Systems in Forest Insect Pest Management. USDA For. Serv. Gen. Tech. Rep.* WO-27:9–21.

Heidman, L. J. 1963. Deer repellents are effective on ponderosa pine in the Southwest. *J. For.* 61:53–54.

Hemphill, D. C. 1986. Shovel logging. *N. Engl. Logging Industry Res. Assoc. Tech. Release* 8(1), 5 pp.

Hepting, G. H. 1971. Diseases of forest and shade trees of the United States. *USDA For. Serv. Agric. Handb.* 386, 658 pp.

Herrick, D. W. and Gansner, D. A. 1987. Gypsy moth on a new frontier: forest tree defoliation and mortality. *North. J. Appl. For.* 4(3):128–133.

Herrick, D. W. and Gansner, D. A. 1988. Changes in forest composition associated with gypsy moth on new frontiers of infestation. *North. J. Appl. For.* 5(1):59–61.

Hicks, R. R. Jr. 1981. Climatic, site and stand factors. In: *The Southern Pine Beetle,* R. C. Thatcher, J. Searcy, J. Coster, and G. Hertel (eds.). *USDA For. Serv. Tech. Bull.* 1631:55–68.

Hicks, R. R. Jr. 1992. Nutrient fluxes for two small watersheds: seven-year results from the West Virginia University Forest. *WV Agric. For. Exp. Stn. Bull.* 707, 29 pp.

Hicks, R. R. Jr. 1993. Applying gypsy moth research to forest management, a case study at the West Virginia University Forest. *WV For. Notes* 15:3–6.

Hicks, R. R. Jr. and Fosbroke, D. E. 1987a. Can gypsy moth damage be predicted? In: *Conference Proceedings: Coping with the Gypsy Moth in the New Frontier.* West Virginia University Press, Morgantown, pp. 73–80.

Hicks, R. R. Jr. and Fosbroke, D. E. 1987b. Mortality following gypsy moth defoliation in the central Appalachians. In: *Proceedings: 6th Central Hardwood Forest Conference.* R. L. Hay, F. W. Woods, and H. DeSelm (eds.). University of Tennessee, Knoxville, TN, pp 423–426.

Hicks, R. R. Jr. and Frank, P. S. Jr. 1984. Relationship of aspect to soil nutrients, species importance and biomass in a forested watershed in West Virginia. *For. Ecol. Mgt.* 8:281–291.

Hicks, R. R. Jr. and Mudrick, D. A. 1994. *Forest Health, 1993: A Status Report for West Virginia.* West Virginia Department of Agriculture, Charleston, 68 pp.

Hicks, R. R. Jr. and Reines, M. 1967. Sweetgum phenology. In: *Proceedings: 15th Northeast Forest Tree Improvement Conference.* Morgantown, WV.

Hicks, R. R. Jr. and Stephenson, G. K. 1978. *Woody plants of the Western Gulf Region.* Kendall/Hunt Publishing Co., Dubuque, IA, 339 pp.

Hicks, R. R. Jr., Coster, J. E., and Mason, G. N. 1987. Forest insect hazard rating. *J. For.* 85(10):20–26.

Hicks, R. R. Jr., Frank, P. S. Jr., Wiant, H. V. Jr., and Carvell, K. L. 1982. Biomass productivity related to soil-site factors on a small watershed. *WV For. Notes* No. 9:9–12.

Hicks, R. R. Jr., Riddle, K. S., and Brock, S. M. 1989a. Benefit/cost analysis of spraying to prevent timber value loss. *Proc. Nat. Gypsy Moth Rev.* (*Dearborne, MI*), pp. 38–54.

Hicks, R. R. Jr., Riddle, K. S., and Brock, S. M. 1989b. Direct control of insect defoliation in oak stands is economically feasible in preventing timber value loss. In: *Proceedings: 7th Central Hardwood Forest Conference,* G. Rink and C. A. Budelsky (eds.). *USDA For. Serv. Gen. Tech. Rep.* NC-132:86–94.

Hilt, D. E. 1979. Diameter growth of upland oaks after thinning. *USDA For. Serv. NEFES Res. Pap.* NE-437, 12 pp.

Hilt, D. E. 1985. OAKSIM: an individual-tree growth and yield simulator for managed, even-aged, upland oak stands. *USDA For. Serv. Res. Pap.* NE-562. 21 pp.

Hix, D. M., Fosbroke, D. E., and Hicks, R. R. Jr. 1990. Effects of gypsy moth defoliation on regeneration of Appalachian Plateau and Ridge and Valley hardwood stands. In: *Proceedings of the 1990 SAF Convention,* Washington, DC.

Holcomb, C. J. and Bickford, C. A. 1952. Growth of yellow-poplar and associated species in West Virginia, as a guide to selective cutting. *USDA For. Serv. NEFES Res. Pap.* 52, 29 pp.

Hook, D. D., Kormanik, P. P., and Brown, C. L. 1970. Early development of sweetgum root sprouts in Coastal South Carolina. *USDA For. Serv. Gen. Tech. Res. Pap.* SE–62. 6 pp.

Hoosein, M. 1991. *Relationships of Aspect, Season and Leaf Substrate to Microarthropod Densities in Decomposing Leaf Litter.* M.S. Thesis. West Virginia University, Morgantown, 63 pp.

Hopkins, A. D. 1938. Bioclimatics: a science of life and climate relations. *USDA Misc. Pub.* 280, Washington, D.C, 188 pp.

Horsley, S. B. 1977. Allelopathic inhibition of black cherry by fern, grass, goldenrod and aster. *Can. J. For. Res.* 7:205–216.

Horsley, S. B. 1981. Control of herbaceous weeds in Allegheny hardwood forests with herbicides. *Weed Sci.* 29:655–662.

Horsley, S. B. 1988. How vegetation can influence regeneration. In: *Proceedings: Guidelines for Regenerating Appalachian Hardwood Stands,* H. C. Smith, A. W. Perkey, and W. E. Kidd, Jr. (eds.). SAF Publ. 88-03:38–55, Morgantown, WV.

Houck, L. 1908. *History of Missouri.* University of Chicago Press, Chicago, 350 pp.

Houston, A. E., Buckner, E. R., and Meadows, J. S. 1995. Romancing the crop tree. *For. Farmer* Sept./Oct.:32–35.

Houston, D. R. 1992. A host-stress-saprogen model for forest dieback-decline diseases. In: *Forest Decline Concepts,* P. D. Manion and D. Lachance (eds.) American Phytopathology Society, St. Paul, MN, 249 pp.

Houston, D. R. and Valentine, H. T. 1985. Classifying forest susceptibility to gypsy moth defoliation. *USDA For. Serv. Agric. Handb.* 542, 19 pp.

Hubler, C. 1995. *America's Mountains: An Exploration of their Origins and Influence from the Alaska Range to the Appalachians.* Facts on File Books, New York, 196 pp.

Huntley, J. C. 1990. *Robinia pseudoacacia* L., black locust. In: *Silvics North America, Vol. 2: Hardwoods,* R. M. Burns and B. H. Honkala (eds.). *USDA For. Serv. Agric. Handb.* 654:755–761.

Hutchinson, J. and DeLost, S. 1993. Cooper's Rock demonstration project: a decision support system for gypsy moth managers. *USDA For. Serv. AIPM Technol. Transfer* NA-TP-09-93, 13 pp.

Huyler, N. and Le Doux, C. B. 1991. A comparison of small tractors for thinning central hardwoods. In: *Proceedings: 8th Central Hardwood Forest Conference,* L. H. McCormick and K. W. Gottschalk (eds.) *USDA For. Serv. Gen. Tech. Rep.* 148:92–104.

Ike, A. F. Jr. and Huppuch, C. D. 1968. Predicting tree height growth from soil and topographic site factors in the Georgia Blue Ridge Mountains. *Georgia For. Res. Pap.* 54, 11 pp.

Illick, J. S. and Aughanbaugh. 1930. Pitch pine in Pennsylvania. *Penn. Dept. For. Waters Res. Bull.* 2, 108 pp.

Jenkins, L. C. 1993. Timber sales and the federal income tax. *Penn. State Coll. Agric. Sci. Ext. Circ.* 406, 43 pp.

Johnson, G. C. and Ware, S. 1982. Post-chestnut forests in the central Blue Ridge of Virginia. Castanea 47(4):329–343.

Johnson, J. E., Bollig, J. J., and Rathfon, R. A. 1997. Growth response of young yellow-poplar to release and fertilization. *South. J. Appl. For.* 2(4):175–179.

Johnson, P. S. 1977. Predicting oak stump sprouting and sprout development in the Missouri Ozarks. *USDA For. Serv. Res. Pap.* NC-149, 11 pp.

Johnson, P. S. 1990. *Quercus coccinea* Muenchh., scarlet oak. In: *Silvics of North America, Vol. 2: Hardwoods,* R. M. Burns and B. H. Honkala (eds.). *USDA For. Serv. Agric. Handb.* 654:625–630.

Johnson, P. S. 1992. Perspectives on ecology and silviculture of oak-dominated forests in the Central and Eastern States. *USDA For. Serv. Gen. Tech. Rep.* NC-153, 28 pp.

Johnson, P. S. 1993. Sources of oak regeneration. In: *Proceedings: Oak Regeneration: Serious Problems, Practical Recommendations,* D. Loftis. and C. E. McGee (eds.). *USDA For. Serv. Gen. Tech. Rep.* SE-84:112–133.

Jones, J. R. 1969. Review and comparison of site evaluation methods. *USDA For. Serv. Res. Pap.* RM-51, 27 pp.

Jones, S. B. 1995. SAF's committee on forest health and productivity: the profession wrestles with ecosystem management. In: *Ecosystem management: Translating Concepts to Practice.* 1995 Penn State School of Forestry Resources Issues Conference, J. C. Finley and S. B. Jones (eds.). State College, PA, pp. 29–34.

Jones, S. B., deCalesta, D., and Chunko, S. E. 1993. Whitetails are changing our woodlands. *Am. For.* Dec.:20–25.

Kalisz, P. J. 1993. Forest disturbance by historic farming practices in eastern Kentucky. In: *Forest Disturbance. A Special Issue.* Natural Resources Newsletter, Univiversity of Kentucky, pp. 7–11.

Keese, K. C. 1991. *Facing the Stewardship Challenge in the 90s.* The Forum, West Virginia Forestry Association, Ripley, pp. 3–4.

Kelly, R. S., Smith, E. L., and Cox, S. M. 1992. *Vermont Hardwood Tree Health in 1991 Compared to 1996.* Vermont Dept. of Forests, Parks and Recreation, Morrisville, 25 pp.

Kennedy, H. E. Jr. 1990. *Fraxinus pennsylvanica* Marsh., green ash. In: *Silvics of North America, Vol. 2: Hardwoods*, R. M. Burns and B. H. Honkala (eds.). *USDA For. Serv. Agric. Handb.* 654:348–354.

Kesner, A. L. 1986. Characteristics of, and management for, veneer quality hardwood logs. In: *Proceedings: Guidelines for Managing Immature Appalachian Hardwood Stands.* West Virginia University Books, Morgantown, pp. 22–32.

King. P. B. 1950. Tectonic framework of the southeastern states. *Bull. Am. Assoc. Petroleum Geol.* 34 pp.

Kingsley, N. P. 1985. A forester's atlas of the northeast. *USDA For. Serv. Gen. Tech. Rep.* NE-95, 96 pp.

Kirkham, K. B. and Carvell, K. L. 1980. Effect of improvement cuttings and thinnings on the understories of mixed oak and cove hardwood stands. *WV Univ. Agric. For. Exp. Stn. Bull.* 67, 16 pp.

Klopatek, J. M., Olson, R. J., Emerson, C. J., and Jones, J. L. 1979. Land-use conflicts with natural vegetation in the United States. *Env. Sci. Div. Pub.* 1333. Oak Ridge Nat. Lab., Oak Ridge, TN, 19 pp.

Kochenderfer, J. N. 1970. Erosion control on logging roads in the Appalachians. *USDA For. Serv. Res. Pap.* NE-158, 8 pp.

Kochenderfer, J. N., Wendel, G. W., and Smith, H. C. 1984. Cost and soil loss on "minimum standard" forest truck roads constructed in the central Appalachians. *USDA For. Serv. Res. Pap.* NE-544, 8 pp.

Kolb, T. E. and Steiner, K. C. 1990. Growth and biomass partitioning of northern red oak and yellow-poplar seedlings: effects of shading and grass root competition. *For. Sci.* 36(1):34–44.

Kolb, T. E., Steiner, K. C., McCormick, L. H., and Bowersox, T. W. 1990. Growth response of northern red-oak and yellow-poplar seedlings to light, soil moisture and nutrients in relation to ecological strategy. *For. Ecol. Mgt.* 38:65–78.

Köppen, W. 1931. *Grundriss der Klimakunde.* Walter de Gruyter, Berlin, 388 pp.

Kormanik, P. P. 1990. *Liquidambar styraciflua* L., sweetgum. In: *Silvics of North America, Vol. 2, Hardwoods*, R. M. Burns and B. H. Honkala (eds.). *USDA For. Serv. Agric. Handb.* 654:400–405.

Labosky, P. Jr. 1987. Salvaging dead hardwoods. In: *Coping with the Gypsy Moth in the New Frontier,* S. Fosbroke and R. R. Hicks, Jr. (eds.) West Virginia University Books, Morgantown, pp. 118–123.

Lacey, S. E. 1991. The Forest Stewardship and Stewardship Incentives Programs: New programs for technical assistance and cost-sharing for private non-industrial forestland. USDA Forest Service State and Private Forestry. *For. Mgt. Update* 13:10–15.

Lafer, N. G. and Wistendahl, W. A. 1970. Tree composition of Dysart Woods, Belmont County, Ohio. *Castanea* 35(4):302–307.

Lamson, N. I. 1980. Site index prediction tables for oak in northwestern West Virginia. *USDA For. Serv. NEFES Res. Pap.* NE-462, 5 pp.

Lamson, N. I. 1985. Thinning increases growth of 60-year-old cherry-maple stands in West Virginia. *USDA For. Serv. NEFES Res. Pap.* NE-571, 8 pp.

Lamson, N. I. 1987. Estimating northern red oak site-index class from total height and diameter of dominant and codominant trees in central Appalachian hardwood stands. *USDA For. Serv. Res. Pap.* NE-605, 3 pp.

Lamson, N. I. 1990. *Betula lenta,* sweet birch. In: *Silvics of North America, Vol. 2: Hardwoods,* R. M. Burns and B. H. Honkala (eds.). *USDA For. Serv. Agric. Handb.* 654:149–152.

Lamson, N. I. and Smith, H. C. 1989. Crop-tree release increases growth of 12-year-old yellow-poplar and black cherry. *USDA For. Serv. NEFES Res. Pap.* NE-622, 7 pp.

Lamson, N. I., Smith, H. C., and Miller, G. W. 1984. Residual stocking not seriously reduced by logging damage from thinning of West Virginia cherry-maple stands. *USDA For. Serv. Res. Pap.* NE-541, 7 pp.

Lamson, N. I., Smith, H. C., Perkey, A. W., and Brock, S. M. 1990. Crown release increases growth of crop trees. *USDA For. Serv. NEFES Res. Pap.* NE-635, 8 pp.

Lamson, N. I., Smith, H. C., Perkey, A. W., and Wilkins, B. L. 1988. How to release crop trees in precommercial hardwood stands. *USDA For. Serv. NEFES* NE-INF-80–88. 2 pp.

Lawson, E. R. 1990a. *Juniperus virginiana* L., eastern redcedar. In: *Silvics of North America,* Vol. 1: *Conifers,* R. M. Burns and B. H. Honkala (eds.). *USDA For. Serv. Agric. Handb.* 654:131–140.

Lawson, E. R. 1990b. *Pinus echinata* Mill., shortleaf pine. In: *Silvics of North America, Vol. 1: Conifers,* R. M. Burns and B. H. Honkala (eds.). *USDA For. Serv. Agric. Handb.* 654:316–326.

Leak, W. B. and Filip, S. M. 1977. Thirty-eight years of group selection in New England northern hardwoods. *J. For.* 75:641–643.

Le Doux, C. B. 1988. Importance of timber production and transport costs on stand management. *USDA For. Serv. Res. Pap.* NE-612, 5 pp.

Le Doux, C. B., Baumgras, J. E., and Selbe, R. B. 1989. PROFIT-PC: a program for estimating maximum net revenue from multiproduct harvests in Appalachian hardwoods. *The Compiler* 7(4):27–32.

Lee, R. and Sypolt, C. R. 1974. Toward biophysical evaluation of forest site potential. *For. Sci.* 20(2):145–154.

Leopold, D. J. and Parker, G. R. 1985. Vegetation patterns on a southern Appalachian watershed after successive clearcuts. *Castanea* 50(3):164–186.

Lesser, W. H. 1993. Prehistoric human settlement in the upland forest region. In: *Upland Forests of West Virginia,* S. L. Stephenson (ed.). pp. 231–260.

Leuschner, W. A. 1992. *Introduction to Forest Resource Management.* Krieger Publishing Co., Malabar, FL. 297 pp.

Lewis, R. L. 1995. Railroads, deforestation and transformation of agriculture in West Virginia back counties, 1880–1920. In: *Appalachia in the Making,* M. B. Pudup, D. B. Billings, and A. L. Walker (eds.). University of North Carolina Press, Chapel Hill, pp. 297–321.

Liebhold, A. 1989. Etienne Leopold Trouvelot, perpetrator of our problem. Gypsy Moth News. *USDA For. Serv. Northeastern Area State & Private For.* No. 20:8–9.

Liebhold, A., Gottschalk, K. W., Muzika, R. M., Montgomery, M. E., Young, R., O'Day,

K., and Kelley, B. 1995. Suitability of North American tree species to the gypsy moth: a summary of field and laboratory tests. *USDA For. Serv. NEFES Gen. Tech. Rep.* NE-211, 34 pp.

Lilly, C. K., Tajchman, S. J., Frank, P. S. Jr., and Eckhard, D. A. 1980. On the microclimates of light and dark strip mine surfaces. *WV For. Notes Circ.* 121(8):13–16.

Little, C. E. *The Dying of the Trees: The Pandemic in America's Forests.* Viking Press, Publ. by Penguin Books USA, Inc., New York, 221 pp.

Little, S. and Garrett, P. W. 1990. *Pinus rigida* Mill., pitch pine. In: *Silvics of North America, Vol. 1, Conifers,* R. M. Burns and B. H. Honkala (eds.). *USDA For. Serv. Agric. Handb.* 654:456–462.

Loewenstein, E. F., Garrett, H. E., Johnson, P. S., and Dwyer, J. P. 1995. Changes in a Missouri Ozark oak-hickory forest during 40 years of uneven-aged management. In: *Proceedings: 10th Central Hardwood Forest Conference,* K. W. Gottschalk and S. L. C. Fosbroke (eds.). *USDA For. Serv. Gen. Tech. Rep.* NE-197:159–164.

Loftis, D. L. 1985. Preharvest herbicide treatment improves regeneration in southern Appalachian hardwoods. *South. J. Appl. For.* 9(3):177–180.

Loftis, D. L. 1988a. Regenerating oaks on high-quality sites, an update. In: *Proceedings: Guidelines for Regenerating Appalachian Hardwood Stands,* H. C. Smith, A. W. Perkey, and W. E. Kidd (eds.). *SAF Publ.* 88-03:199–209.

Loftis, D. L. 1988b. Species composition of regeneration after clearcutting southern Appalachian hardwoods. In: *Proceedings: Biennial Southern Silviculture Research Conference,* Memphis, TN, *USDA, Forest Service, Gen. Tech. Rep.* 50–74: pp. 253–257.

Loftis, D. L. 1989. Regeneration of southern hardwoods: some ecological concepts. 1989 National Silviculture Workshop, Petersburg, Alaska.

Loftis, D. L. 1990a. Predicting post-harvest performance of advance red oak reproduction in the southern Appalachians. *For. Sci.* 36(4):908–916.

Loftis, D. L. 1990b. A shelterwood method for regenerating red oak in the southern Appalachians. *For. Sci.* 36(4):917–929.

Loftis, D. L. 1993. Predicting oak regeneration—state of the art. In: *Proceedings: Oak Regeneration: Serious Problems, Practical Recommendations,* D. L. Loftis and C. E. McGee (eds.). *USDA For. Serv. Gen. Tech. Rep.* SE-84:134–137.

Lorimer, C. G. 1976. *Stand History and Dynamics of a Southern Appalachian Virgin Forest.* Ph.D. Dissertation, Duke University, Durham, NC, 201 pp.

Lorimer, C. G. 1984. Development of red maple understories in northeastern oak forests. *For. Sci.* 30(1):3–22.

Lorimer, C. G. 1993. Causes of the oak regeneration problem. In: *Proceedings: Oak Regeneration: Serious Problems, Practical Recommendations,* D. L. Loftis and C. E. McGee (eds.). *USDA For. Serv. SEFES Gen. Tech. Rep.* SE-84:14–39.

Love, R. 1795. Letter *in* Old Buncombe Co. Heritage, NC. Old Buncombe Co. Genealogical Society, D. C. Wood and C. D. Biddix (eds.), 1981.

Luther, E. T. 1977. *Our Restless Earth.* University of Tennessee Press, Knoxville, 94 pp.

Lutz, H. and Chandler, R. F. 1946. *Forest Soils.* John Wiley & Sons, New York, 514 pp.

Lutz, H. J. 1959. Forest ecology, the biological basis of silviculture. *Publ. Univ. Brit. Col.* 8 pp.

Lynch, J. A., Bowersox, V. C., and Simmons, C. 1995. Precipitation chemistry trends in the United States: 1980–1993. National Atmospheric Deposition Program. Summary Report, 103 pp.

MacCleery, D. W. 1990. Brief overview of timber conditions and trends of US forests. USDA Forest Service Timber Management Staff, Washington, DC, 5 pp.

MacCleery, D. W. 1992. American forests: a history of resiliency and recovery. *USDA For. Serv.* FS-540, 59 pp.

Manion, P. D. 1981. *Tree Disease Concepts.* Prentice-Hall, Englewood Cliffs, NJ, 399 pp.

Margalef, R. 1963. On certain unifying principles in ecology. *Am. Nat.* 97:357–374.

Marks, P. L. 1975. On the relation of extension growth and successional status of deciduous trees of the northeastern United States. *Bull. Torrey Bot. Club* 102:172–177.

Marquis, D. A. 1979a. Ecological aspects of shelterwood cutting. In: *Proceedings: National Silviculture Workshop*, Charleston, SC, pp. 40–56.

Marquis, D. A. 1979b. Shelterwood cutting for Allegheny hardwoods. *J. For.* 77(3):140–144.

Marquis, D. A. 1981a. Effect of deer browsing on timber production in Allegheny hardwood forests of northwestern Pennsylvania. *USDA For. Serv. Res. Pap.* NE-475, 10 pp.

Marquis, D. A. 1981b. Management of Allegheny hardwoods for timber and wildlife. In: *Proceedings: XVII IUFRO World Congress, Forest Environment and Silviculture*, Kyoto, Japan, pp. 369–380.

Marquis, D. A. 1982. Effect of advance seedling size and vigor on survival after clearcutting. *USDA For. Serv. Res. Pap.* NE-498, 7 pp.

Marquis, D. A. 1987. Silvicultural techniques for circumventing deer damage. In: *Proceedings: Deer, Forestry, and Agriculture: Interactions and Strategies for Management.* SAF, Warren, PA, pp. 125–136.

Marquis, D. A. 1990. *Prunus serotina* Ehrh., black cherry. In: *Silvics of North America, Vol. 2, Hardwoods*, R. M. Burns and B. H. Honkala (eds.). *USDA For. Serv. Agric. Handb.* 654:594–604.

Marquis, D. A. and Grisez, T. J. 1978. The effect of deer exclosures on the recovery of vegetation in failed clearcuts on the Allegheny Plateau. *USDA For. Serv. Res. Note* NE-270, 5 pp.

Marquis, D. A. and Twery, M. J. 1992. Decision-making for natural regeneration in the northern forest ecosystem. In: *Proceedings: Oak Regeneration: Serious Problems, Practical Recommendations*, D. L. Loftis and C. E. McGee (eds.). *USDA For. Serv. Gen. Tech. Rep.* SE-84:156–173.

Marquis, D. A., Ernst, R. L., and Stout, S. L. 1984. Prescribing silvicultural treatments in hardwood stands in the Alleghenies. *USDA For. Serv. NEFES Gen. Tech. Rep.* NE-96. 90 pp.

Marquis, D. A., Ernst, R. L., and Stout, S. L. 1992. Prescribing silvicultural treatments in hardwood stands in the Alleghenies (Revised). *USDA For. Serv. Gen. Tech. Rep.* NE-96. 101 pp.

Martin, P. and Houf, G. F. 1993. Glade grasslands in southwest Missouri. *Rangelands* 15(2):70–73.

Martinat, P. J., Jennings, D. T., and Whitmore, R. C. 1993. Effects of diflubenzuron on litter spider and orthopteroid community in a central Appalachian forest infested with gypsy moth (Lepidoptera: Lymantridae). *Environ. Entomol.* 22(5):1003–1008.

Marx, D. H. and Beattie, D. J. 1977. Mycorrhizae—promising aid to timber growers. *For. Farmer* 36:6–9.

Maser, C. 1994. *Sustainable Forestry: Philosophy, Science and Economics.* St. Lucie Press, Delray Beach, FL, 435 pp.

Mash, J. 1996. *The Land of the Living. The Story of Maryland's Green Ridge Forest.* Living History Foundation, Cumberland Press, Cumberland, MD, 895 pp.

Maxwell, J. A. and Davis, M. B. 1972. Pollen evidence of Pleistocene and Holocene vegetation in the Allegheny Plateau, MD. *Quat. Res.* 2:506–530.

McGee, C. E. 1975. Regeneration alternatives in mixed oak stands. *USDA For. Serv. Res. Pap.* SE-125, 8 pp.

McGee, C. E. 1990. *Nyssa sylvatica* Marsh. var. *sylvatica*, black tupelo (typical). In: *Silvics of North America, Vol. 2, Hardwoods,* R. M. Burns and B. H. Honkala (eds.). *USDA For. Serv. Agric. Handb.* 654:482–485.

McGee, G. G., Leopold, D. J., and Nyland, R. D. 1995. Understory response to springtime prescribed fire in two New York transition oak forests. *For. Ecol. Mgt.* 76:149–168.

McGill, D. W., Jones, S. B., and Nowak, C. A. 1995. Identification of canopy strata in Allegheny hardwood stands (abstract). In: *Proceedings: 10th Central Hardwood Forest Conference,* Morgantown, WV. K. W. Gottschalk and S. L. C. Fosbroke (eds.) *USDA For. Serv. NEFES Gen. Tech. Rep.* NE-197:547–548.

McNab, W. H. 1987. Yellow-poplar site quality related to slope type in mountainous terrain. *North. J. Appl. For.* 4:189–192.

McNab, W. H. 1991. Land classification in the Blue Ridge Province: state-of-the-science report. In: *Proceedings: Symposium on Ecological Land Classification: Application to Identify the Productive Potential of Southern Forests,* Charlotte, NC, pp. 37–47.

McQuilkin, R. A. 1990. *Quercus prinus* L., chestnut oak. In: *Silvics of North America, Vol. 2: Hardwoods,* R. M. Burns and B. H. Honkala (eds.). *USDA For. Serv. Agric. Handb.* 654:721–726.

McWilliams, W. H., Stout, S. L., Bowersox, T. W., and McCormick, L. H. 1995. Adequacy of advance tree-seedling regeneration in Pennsylvania's forests. *North. J. Appl. For.* 12(4):187–192.

Mellilo, J. M., Aber, J. D., and Muratore, J. F. 1982. Nitrogen and lignin control of hardwood leaf litter decomposition dynamics. *Ecology* 63(3):621–626.

Michael, E. D. 1988. Effects of white-tailed deer on Appalachian hardwood regeneration. In: *Conferences Proceedings: Guidelines for Regenerating Appalachian Hardwood Stands,* H. C. Smith, A. W. Perkey, and W. E. Kidd, Jr. (eds.). West Virginia University Books, Morgantown, WV, pp. 89–96.

Michaux, F. A. 1805. *Travels to the Westward of the Allegheny Mountains in the States of the Ohio, Kentucky and Tennessee in the Year of 1802.* Barnard and Shulzer, London.

Miller, C. 1994. Sawdust memories, Pinchot and the making of forest history. *J. For.* 92(2):8–12.

Miller, G. W. 1984. Releasing young hardwood crop trees—use of a chain saw costs less than herbicides. *USDA For. Serv. NEFES Res. Pap.* NE-550, 5 pp.

Miller, G. W. 1993. Financial aspects of partial cutting practices in central Appalachian hardwoods. *USDA For. Serv. Res. Pap.* NE-673, 9 pp.

Miller, G. W. 1996. Epicormic branching on central Appalachian hardwoods 10 years after deferment cutting. *USDA For. Serv. Res. Pap.* NE-702. 9 pp.

Miller, G. W. and Schuler, T. M. 1995. Development and quality of reproduction in two-age central Appalachian hardwoods—10-year results. In: *Proceedings: 10th Central Hardwood Forest Conference,* K. W. Gottschalk and S. L. C. Fosbroke (eds.). *USDA For. Serv. Gen. Tech. Rep.* NE-197:364–374.

Miller, G. W. and Smith, H. C. 1991. Applying group selection in upland hardwoods. Uneven-Aged Silviculture of Upland Hardwood Stands, Workshop Notes, Virginia

Cooperative Extension Serv. Virginia Polytechnic Institute and State University, Blacksburg VA, pp. 20–25.

Miller, J. H. 1991. Application methods for forest herbicide research. In: *Standard Methods for Forest Herbicide Research*, J. H. Miller and G. H. Glover (eds.). South Weed Science Society Auburn University Herbicide Cooperative, pp. 45–60.

Minkler, L. S. 1967. Release and pruning can improve growth and quality of white oak. *J. For.* 65(9):654–655.

Missouri Conservation Commission of the State of Missouri. 1994. Missouri Ozark Forest Ecosystem Project. Missouri Department of Conservation, 12 pp.

Mudrick, D. A., Hoosein, M., and Hicks, R. R. Jr. 1994. Decomposition of leaf litter in an Appalachian forest: effects of leaf species, aspect, slope position and time. *For. Ecol. Mgt.* 68:231–250.

Munn, L. C. and Vimmerstedt, J. P. 1980. Predicting height growth of yellow-poplar from soil and topography in southeastern Ohio. *Soil Sci. Soc. Am. J.* 44:384–387.

Murphy, P. A., Shelton, M. G., and Graney, D. C. 1993. Group selection — problem and possibilities for the more shade-intolerant species. In: *Proceedings: 9th Central Hardwood Forest Conference*, A. R. Gillespie, G. R. Parker, and P. E. Pope (eds.). *USDA For. Serv. Gen. Tech. Rep.* NC-161:229–247.

Murrill, W. A. 1904. A serious chestnut disease. *J. N.Y. Bot. Gar.* 7:143–153.

National Oceanic and Atmospheric Administration. 1980. *Climates of the States*, Vol. 2. Gale Research Co., Detroit, MI, 1175 pp.

Nelson, T. C. 1959. Silvical characteristics of mockernut hickory. *USDA For. Serv. Stn. Pap.* 105. SEFES, Ashville, NC, 10 pp.

Nelson, T. C. 1965. Silvical characteristics of the commercial hickories. *USDA For. Serv. Task Force Rep.* 10. SEFES, Ashville, NC, 16 pp.

Noble, I. R. and Slayter, R. O. 1980. The use of vital attributes to predict successional changes in plant communities subject to recurrent disturbances. *Vegetatio* 43:5–21.

Nowacki, G. J. and Trianosky, P. A. 1993. Literature on old-growth forests of eastern North America. *Nat. Areas J.* 13(2):87–107.

Nyland, R. D. 1986. Logging damage during thinning in even-age hardwood stands. In: *Proceedings: Guidelines for Managing Immature Appalachian Hardwood Stands*, H. C. Smith and M. C. Eye (eds.). *SAF Bull.* 86-02:150–166.

Nyland, R. D. 1990. Logging damage during conventional harvesting in northern hardwood stands. In: *Colloque sur la forêt feuillue. Twenty-first Semaine des Sciences forestières*, G. Van der Kelen (ed.). Universitè Laval, Quêbec, pp. 53–58.

Nyland, R. D. 1992. Exploitation and greed in the eastern forests. *J. For.* 90(1):33–37.

Nyland, R. D. 1996. *Silviculture, Concepts and Applications*. McGraw-Hill, New York, 633 pp.

Nyland, R. D., Alban, L. N., and Nissen, R. L. Jr. 1993. Greed or sustention: silviculture or not. In: *Proceedings: Nurturing the Northeastern Forest*, R. D. Briggs and W. B. Krohn (eds.). *Maine Agric. For. Exp. Stn. Misc. Rep.* 382:37–52.

Nyland, R. D., Larson, C. C., and Shirley, H. L. 1983. *Forestry and Its Career Opportunities*, 4th ed. McGraw-Hill, New York, 150 pp.

Oak, S. W. 1993. Insects and diseases affecting oak regeneration success. In: *Proceedings: Oak Regeneration: Serious Problems, Practical Recommendations*, D. L. Loftis and C. E. McGee (eds.). *USDA For. Ser. Gen. Tech. Rep.* SE-84:105–111.

Odum, E. P. 1971. *Fundamentals of Ecology*, 3rd. ed. Saunders, Philadelphia, 574 pp.

Oefinger, S. W. and Halls, L. K. 1974. Identifying woody plants valuable to wildlife in southern forests. *USDA For. Serv. Res. Pap.* SO-92, 76 pp.

Oliver, C. D. 1981. Forest development in North America following major disturbances. *For. Ecol. Mgt.* 3:153–168.

Oliver, C. D. and Larson, B. C. 1996. *Forest Stand Dynamics.* John Wiley & Sons, New York, 520 pp.

Oosting, H. J. 1956. *The Study of Plant Communities.* W. H. Freeman & Co., San Francisco, 440 pp.

Overstreet, J. 1993. Kentucky's forest fires: a variable disturbance. In: *Forest Disturbance, A Special Issue.* Natural Resources Newsletter, Department of Forestry University of Kentucky, pp. 23–25.

Paff, W. D. 1982. Use of best management practices on logging operations in West Virginia. West Virginia Div. of Forestry, Charleston, WV. 53 p.

Paine, T. D., Stephen, F. M., and Mason, G. N. 1983. A risk population level. In: *The Role of the Host in Population Dynamics of Forest Insects*, L. Safranyck (ed.). Canadian Forest Service, Banft, Alberta.

Parker, G. R. and Swank, W. T. 1982. Tree response to clearcutting in a southern Appalachian watershed. *Am. Midl. Nat.* 108(2):304–310.

Patric, J. H. and Schell, K. F. 1990. Why does the clearcutting controversy persist? American Pulpwood Assoc., Inc., 23 pp.

Perkey, A. W. 1991. A comparison of crop tree management to uneven-aged management. *USDA For. Serv. Northeastern Area State and Private For. For. Mgt. Update* 13:22–27.

Perkey, A. W. 1992. Managing red oak crop trees to produce financial benefits. *USDA For. Serv. Northeastern Area State and Private For. For. Mgt. Update* 14:23–29.

Perkey, A. W. and Powell, D. S. 1988. Regenerating Appalachian hardwoods: the current situation? In: *Conference Proceedings: Guidelines for Regenerating Appalachian Hardwood Stands*, H. C. Smith, A. W. Perkey, and W. E. Kidd, Jr. (eds.). Morgantown, WV, pp. 5–16

Perkey, A. W., Wilkins, B. L., and Smith, H. C. 1993. Crop tree management in eastern hardwoods. *USDA For. Serv. Northeast Area State and Private For.* NA-TP-19-93, 54 p.

Perry, D. A. 1996. *Forest Ecosystems.* The Johns Hopkins University Press, Baltimore, MD, 649 pp.

Petersen, M. S., Rigby, K. J., and Hintze, L. F. 1980. *Historical Geology of North America*, 2nd ed. Wm. C. Brown Co., Dubuque, IA, 232 pp.

Philips, J. F. and Ward, W. W. 1971. Basal area growth of black cherry trees following cutting. *Penn. State Univ. Res. Briefs* 5:9–12.

Phillips, J. J. 1966. Site index of yellow-poplar related to soil and topography in southern New Jersey. *USDA For. Serv. NEFES Res. Pap.* NE-52, 10 pp.

Pope, P. C. 1993. Oak regeneration via seedling planting: Historical perspective and current status. *Purdue Univ. Dept. For. Nat. Resour. Stn. Bull.* No. 6760, 36 pp.

Powell, D. S., Faulkner, J. L., Darr, D. R., Zhu, Z., and MacCleery, D. W. 1993. Forest resources of the United States, 1992. *USDA For. Serv. Rocky Mount. For. Range Exp. Stn. Gen. Tech. Rep.* RM–234, 133 pp.

Quimby, J. 1987. Impact of gypsy moth defoliation on forest stands. In: *Conference Proceedings: Coping with Gypsy Moth in the New Frontier*, S. Fosbroke and P. Hicks (eds.). West Virginia University Books, Morgantown, WV, pp. 21–38.

Redding, J. 1995. History of deer population trends and forest cutting on the Allegheny National Forest. In: *Proceedings: 10th Central Hardwood Forest Conference*, K. W. Gottschalk and S. L. C. Fosbroke (eds.). *USDA For. Serv. Gen. Tech. Rep.* NE-197:214–224.

Rentch, J. S. and Fortney, R. H. 1997. The vegetation of West Virginia grass bald communities. *Castanea* 62(3):147–160.

Rexrode, K. R. and Carvell, K. L. 1981. The effects of late planting on survival, height growth, and vigor of eastern white pine. *Tree Planters Notes* 32:30–32.

Rice, E. L. 1974. *Allelopathy.* Academic Press, New York, 353 pp.

Roach, B. A. 1968. Is clear cutting good or bad? *KTG J.* 8(4). Keep Tennessee Green Assoc., Sewanee, TN.

Roach, B. A. 1974. Selection cutting and group selection. *SUNY Coll. Environ. Sci. For. Appl. For. Res. Inst. AFRI Misc. Publ.* No. 5, 9 pp.

Roach, B. A. 1977. A stocking guide for Allegheny hardwoods and its use in controlling intermediate cuttings. *USDA For. Serv. Res. Pap.* NE-373, 30 pp.

Roach, B. A. and Gingrich, S. F. 1968. Even-aged silviculture for upland central hardwoods. *USDA For. Serv. Agric. Handb.* 355, 39 pp.

Rogers, L. L. and Lindquist, E. L. 1992. Supercanopy white pine and wildlife. In: *White Pine Symposium Proceedings*, R. A. Stine and M. J. Baughman (eds.). Department of Forest Research, University of Minnesota, St. Paul, pp. 39–43.

Rogers, R. 1990. *Quercus alba* L., white oak. In: *Silvics of North America, Vol. 2: Hardwoods*, R. M. Burns and B. H. Honkala (eds.). *USDA For. Serv. Agric. Handb.* 654:605–613.

Rogers, R., Johnson, P. S., and Loftis, D. L. 1993. An overview of oak silviculture in the United States: the past, present and future. *Ann. Sci. For.* 50:535–542.

Sander, I. L. 1972. Size of oak advance reproduction: key to growth following harvest cutting. *USDA For. Serv. Res. Pap.* NC-79, 6 pp.

Sander, I. L. 1977. Manager's handbook for oaks in the North Central States. *USDA For. Serv. NCFES Gen. Tech. Rep.* NC-37, 35 pp.

Sander, I. L. 1990a. *Quercus rubra* L., northern red oak. In: *Silvics of North America, Vol. 2: Hardwoods*, R. M. Burns and B. H. Honkala (eds.). *USDA For. Serv. Agric. Handb.* 654:727–733.

Sander, I. L. 1990b. *Quercus velutina.*, black oak. In: *Silvics of North America, Vol. 2: Hardwoods*, R. M. Burns and B. H. Honkala (eds.). *USDA For. Serv. Agric. Handb.* 654:744–750.

Sander, I. L. 1992. Regenerating oaks in the Central States. In: *Proceedings: Oak Regeneration: Serious Problems, Practical Recommendations*, D. L. Loftis and C. E. McGee (eds.). *USDA For. Serv. Gen. Tech. Rep.* SE-8:174–183.

Sander, I. L. and Clark, F. B. 1971. Reproduction of upland hardwood forests in the central states. *USDA Agric. Handb.* 405, 25 pp.

Sander, I. L. and Graney, D. 1993. Regenerating oaks in the central states. In: *Proceedings: Oak Regeneration: Serious Problems, Practical Recommendations.* D. L. Loftis, and C. E. McGee (eds.). *USDA For. Serv. Gen. Tech. Rep.* SE-84:174–183.

Sander, I. L., Johnson, P. S., and Rogers, R. 1984. Evaluating oak advance reproduction in the Missouri Ozarks. *USDA For. Serv. Res. Pap.* NC-251, 16 pp.

Schier, G. A. 1987. Germination and early growth of four pine species on soil treated with simulated acid rain. *Can. J. For. Res.* 17(10):1190–1196.

Schlaegel, B. E., Kulow, D. L., and Baughman, R. N. 1969. Empirical yield tables for West

Virginia yellow-poplar. *WV Agric. Exp. Stn. Bull.* 574T. Morgantown, WV, 24 pp.

Schlesinger, R. C. 1990. *Fraxinus americana* L., white ash. In: *Silvics of North America, Vol. 2, Hardwoods*, R. M. Burns and B. H. Honkala (eds.). *USDA For. Serv. Agric. Handb.* 654:333–338.

Schlesinger, R. C. and Funk, D. T. 1977. Manager's handbook for black walnut. *USDA For. Serv. Gen. Tech. Rep.* NC-38. 22 pp.

Schnur, G. L. 1937. Yield, stand and volume tables for even-aged upland oak forests. *USDA Tech. Bull.* 560, 88 pp.

Schuler, T. M. and Miller, G. W. 1995. Shelterwood treatments fail to establish oak reproduction on mesic forest sites in West Virginia — 10-year results. In: *Proceedings: 10th Central Hardwood Forest Conference*, K. W. Gottschalk and S. L. C. Fosbroke (eds.). *USDA For. Serv. Gen. Tech. Rep.* NE-197:375–387.

Seymour, R. S., Hannah, P. R., Grace, J., and Marquis, D. A. 1986. Silviculture: the past 30 years, the next 30 years. Part IV. The Northeast. *J. For.* 84:31–38.

Shands, W. E. 1991. Problems and prospects at the urban-rural interface. *J. For.* 89(6):23–26.

Shigo, A. L. 1966. Decay and discoloration following logging wounds in northern hardwoods. *USDA For. Serv. Res. Pap.* NE-47, 43 pp.

Shugart, H. H. and West, D. C. 1980. Forest succession models. *Bioscience* 30:308–313.

Shumway, D. L., Steiner, K. C., and Kolb, T. E. 1993. Variation in seedling hydraulic architecture as a function of species and environment. *Tree Physiol.* 12:41–54.

Siegal, W. C., Hoover, W. L., Haney, H. L. Jr., and Liu, K. 1995. Forest owners' guide to the federal income tax. *USDA For. Serv. Agric. Handb.* 708. 138 pp.

Sims, D. H. 1992. The two-aged stand, a management alternative. *USDA For. Serv. Coop. For. Mgt. Bull.* RG-MB61, 2 pp.

Sims, D. H. and Loftis, D. L. 1990. Regenerating northern red oak on high quality sites. *For. Farmer* 49(1):12–13.

Skelly, J. M., Davis, D. D., Merrill, W., Cameron, E. A., Brown, H. D., Drummond, D. B., and Dochinger, L. S. (eds.). 1987. *Diagnosing Injury to Eastern Forest Trees: A Manual for Identifying Damage Caused by Air Pollution, Pathogens, Insects and Abiotic Stresses.* National Acid Precipitation Assessment Program, Forest Response Program, Vegetation Survey Research Cooperative, University Park, PA. College of Agriculture, Department of Pathology, Pennsylvania State University, 122 pp.

Smalley, G. W. 1987. Site classification and evaluation for the interior uplands. Forest sites of the Cumberland Plateau and Highland Rim/Pennyroyal. *USDA For. Serv. SFES Tech. Pub.* R8-TP9.

Smalley, G. W. 1990. *Carya glabra* (Mill.), sweet, pignut hickory. In: *Silvics of North America, Vol. 2: Hardwoods*, R. M. Burns and B. H. Honkala (eds.). *USDA For. Serv. Agric. Handb.* 654:198–204.

Smith, D. M. 1978. Implications for silvicultural management in the impacts of regeneration systems on soils and environment. In: *Proceedings: 5th North American Forest Soils Conference, Forest Soils and Land Use*, C. T. Youngberg (ed.). Department of Forest and Wood Sciences, Colorado State University, Ft. Collins CO. pp. 536–545.

Smith, D. M. 1986. *The Practice of Silviculture.* John Wiley & Sons, New York, 570 pp.

Smith D. W. 1993. Oak regeneration: the scope of the problem. In: *Oak Regeneration: Serious Problem, Practical Recommendations*, D. L. Loftis and C. E. McGee (eds.). *USDA For. Serv. Gen. Tech. Rep.* SE-84:40–52.

Smith D. W. 1995. The southern Appalachian hardwood region. In: *Regional Silviculture of the United States,* 3rd ed, J. W. Barrett (ed.). John Wiley & Sons, New York, pp. 173–225.

Smith, H. C. 1988. Possible alternatives to clearcutting and selection harvesting practices. In: *Proceedings: Guidelines for Regenerating Appalachian Hardwood Stands,* H. C. Smith, A. W. Perkey, and W. E. Kidd, Jr. (eds.). *SAF Publ.* 88-03:276–289.

Smith, H. C. 1990a. *Carya cordiformis* (Wangenh.) K. Koch, bitternut hickory. In: *Silvics of North America,* Vol. 2: *Hardwoods,* R. M. Burns and B. H. Honkala (eds.). *USDA For. Serv. Agric. Handb.* 654:190–197.

Smith, H. C. 1990b. *Carya tomentosa* (Poir) Nutt., mockernut hickory. In: *Silvics of North America, Vol. 2: Hardwoods,* R. M. Burns and B. H. Honkala (eds.). *USDA For. Serv. Agric. Handb.* 654:226–233.

Smith, H. C. 1992. Regenerating oaks in the central Appalachians. In: *Proceedings: Oak Regeneration: Serious Problems, Practical Recommendations,* D. L. Loftis and C. E. McGee (eds.). *USDA For. Serv. Gen. Tech. Rep.* SE-84:211–223.

Smith, H. C. and DeBald, P. S. 1975. Economics of even-aged and uneven-aged silviculture and management in eastern hardwoods. In: *Proceedings: Symposium on Uneven-Aged Silviculture and Management in the Eastern United States. USDA For. Serv. Gen. Tech. Rep.* WO-24:121–137.

Smith, H. C. and Lamson, N. I. 1982. Number of residual trees: a guide to selection cutting. *USDA For. Serv. Gen. Tech. Rep.* NE-80, 33 pp.

Smith, H. C. and Lamson, N. I. 1986a. Cultural practices in Appalachian hardwood sapling stands — if done, how to do them. In: *Proceedings: Guidelines for Managing Immature Appalachian Hardwood Stands,* H. C. Smith and M. C. Eye (eds.). West Virginia University Books, Morgantown, pp. 46–61.

Smith, H. C. and Lamson, N. I. 1986b. Wild grapevines — a special problem in immature Appalachian hardwood stands. In: *Proceedings: Guidelines for Managing Immature Appalachian Hardwood Stands,* H. C. Smith and M. C. Eye (eds.). West Virginia University Books, Morgantown, pp. 228–239.

Smith, H. C., Miller, G. W., and Lamson, N. I. 1994a. Crop-tree release thinning in 65-year-old commercial cherry-maple stands (5-year results). *USDA For. Serv. NEFES Res. Pap.* NE-694, 11 pp.

Smith, H. C., Miller, G. W., and Schuler, T. M. 1994b. Closure of logging wounds after 10 years. *USDA For. Serv. NEFES Res Pap.* NE-692. 6 pp.

Smith, H. C., Trimble, G. R., and DeBald, P. S. 1979. Raise cutting diameters for increased returns. *USDA For. Serv. Res. Pap.* NE-455, 33 pp.

Smith, R. L. 1966. Wildlife and forest problems in Appalachia. *Trans. Thirty-First N. Am. Wildlife Mgt. Inst.* 212–226.

Smith, R. L. 1993. Wildlife of the upland forest. In: *The Upland Forests of West Virginia,* S. Stephenson (ed.). McClain Printing Co., Parsons, WV, pp. 211–229.

Smith, W. H. 1987. Future of the hardwood forest: some problems with declines and air quality. In: *Proceedings: 5th Central Hardwood Forest Conference,* R. L. Hay, F. W. Woods, and H. DeSelm (eds.). Univesity of Tennessee, Knoxville, TN, pp. 3–14.

Society of American Foresters. 1993. *Task Force Report on Sustaining Long-Term Forest Health and Productivity.* 83 pp.

Society of American Foresters. 1996. *Ethics Guide for Foresters and Other Natural Resource Professionals.* 56 pp.

Sonderman, D. L. and Rast, E. D. 1988. Effect of thinning on mixed-oak stem quality. *USDA For. Serv. NEFES Res. Pap.* NE-618, 7 pp.

Spalding, H. M. and Fernow, B. E. 1899. The white pine. *USDA Bur. For. Bull.* 22.

Spurr, S. H. and Barnes, B. V. 1980. *Forest Ecology.* John Wiley & Sons, New York, 687 pp.

Steiner, K. C. 1995. Autumn predation of northern red oak seed crops. In: *Proceedings: 10th Central Hardwood Forest Conference. USDA For. Serv. Gen. Tech. Rep.* NE-197:489–494.

Stephens, G. R. 1987. Effect of gypsy moth defoliation on conifers. In: *Proceedings: Coping with Gypsy Moth in the New Frontier,* S. Fosbroke and R. R. Hicks, Jr. (eds.). West Virginia University Books, Morgantown, pp. 30–38.

Stephens, G. R. 1988. Mortality, dieback and growth of defoliated hemlock and white pine. *N. J. Appl. For.* 5(2):93–96.

Stephenson, S. L. 1974. Ecological composition of some former oak-chestnut communities in western Virginia. *Castanea* 39(3):278–286.

Stephenson, S. L. (ed.). 1993. *Upland Forests of West Virginia.* McClain Printing Co., Parsons, WV, 295 pp.

Stephenson, S. L. and Adams, H. S. 1993. Threats to the upland forests. In: *Upland Forests of West Virginia.* S. L. Stephenson (ed.). McClain Printing Co., Parsons, WV, pp. 261–273.

Steyermark, J. A. 1959. *Vegetational History of the Ozark Forest.* The University of Missouri Studies, Columbia, 138 pp.

Stout, S. L. 1986. Twenty-two-year growth of four planted hardwoods. *N. J. Appl. For.* 3(2):69–72.

Stout, S. L. 1995. Overview of ecosystem management and its importance. In: *Proceedings: Ecosystem Management: Translating Concepts to Practice.* Pennsylvania State School of Forest Resource Issues Conference, J. C. Finley and S. B. Jones (eds.). State College, PA, pp. 1–13.

Stout, S. L., Nowak, C. A., Redding, J. A., White, R., and McWilliams, W. 1995. Allegheny National Forest health. In: *Proceedings: Forest Health Through Silviculture.* 1995 National Silviculture Workshop, Mescalero, NM. *USDA For. Serv.* RM-GTR-267:79–86.

Stransky, J. J. 1990. *Quercus stellata,* Wangenh., post oak. In: *Silvics of North America, Vol. 2, Hardwoods,* R. M. Burns and B. H. Honkala (eds.). *USDA For. Serv. Agric. Handb.* 654:738–743.

Stringer, J. W., Miller, G. W., and Wittwer, R. F. 1988. Applying a crop-tree release in small-sawtimber white oak stands. *USDA For. Serv. NEFES Res. Pap.* NE-620, 5 pp.

Swank, W. T. and Vose, J. M. 1988. Effects of cutting practices in microenvironment in relation to hardwood regeneration. In: *Proceedings: Guidelines for Regenerating Appalachian Hardwood Stands,* H. C. Smith, A. W. Perkey, and W.E. Kidd (eds.). *SAF Publ.* 88-03:71–88.

Swanton, J. R. 1979. *The Indians of the Southeastern United States.* Smithsonian Institution Press, Washington, DC.

Tajchman, S. J. 1982. A biogeophysical approach to the site quality problem. *WV For. Notes Circ.* 121(9):3–6.

Tajchman, S. J. and Lacey, C. J. 1986. Bioclimatic factors in forest site potential. *For. Ecol. Mgt.* 14:211–218.

Tilghman, N. G. 1987. Maximum deer population compatible with forest regeneration, an estimate from deer enclosure studies in Pennsylvania. In: *Proceedings: Symposium on Deer, Forestry and Agriculture.* Allegheny Society of American Foresters, Warren, PA.

Tilghman, N. G. 1989. Impacts of white-tailed deer on forest regeneration in north-western Pennsylvania. *J. Wildlife Mgt.* 53(3):524–532.

Torbert, J. and Burger, J. 1995. Achieving a productive forestry postmining land-use to benefit land owners and coal operators. In: *Proceedings: Reclamation of Surface-Mined Forest Land.* Twin Falls State Park, Mullens, WV, pp. 9–14.

Trewartha, G. T. 1968. *An Introduction to Climate*, 4th ed. McGraw-Hill, New York, 408 pp.

Trimble, G. R. Jr. 1969. Diameter growth of individual hardwood trees. *USDA For. Serv. Res. Pap.* NE-145, 25 pp.

Trimble, G. R. Jr. 1973. The regeneration of central Appalachian hardwoods with emphasis on the effects of site quality and harvesting practice. *USDA For. Serv. Res. Pap.* NE-282, 14 pp.

Trimble, G. R. Jr. 1975. Summaries of some silvical characteristics of several Appalachian hardwood trees. *USDA For. Serv. Gen. Tech. Rep.* NE-16, 5 pp.

Trimble, G. R. Jr. and Weitzman, S. 1956. Site index studies of upland oaks in the northern Appalachians. *For. Sci.* 2(3):162–173.

Trimble, G. R. Jr., Mendel, J. J., and Kennell, R. A. 1974. A procedure for selection marking in hardwoods — combining silvicultural considerations with economic guidelines. USDA For. Serv. Res. Pap. NE-292, 13 pp.

Tryon, E. H. and Carvell, K. L. 1958. Regeneration under oak stands. *WV Univ. Agric. Exp. Stn. Bull.* 424T, 22 pp.

Tubbs, C. H. and Houston, D. R. 1990. *Fagus grandifolia* Ehrh., American beech. In: *Silvics of North America. Vol. 2: Hardwoods*, R. M. Burns and B. H. Honkala (eds.). *USDA For. Serv. Agric. Handb.* 654:325–332.

Twardus, D., Miller-Weeks, M., and Gillespie, A. 1995. Forest health assessment for the northeastern area. *USDA For. Serv.* NA-TP-01-95, 61 pp.

Twery, M. J., Thomas, S. J., Twardus, D. B., Selmon, L. A., and Ghent, J. 1994. In: *Proceedings: Interagency Gypsy Moth Forum.* Annapolis, MD, pp. 80–81.

USDA Forest Service. 1970. *Fifty Year History of the Monongahela National Forest.* USDA For. Serv. Washington, DC, 66 pp.

USDA Forest Service. 1987. Forest statistics of the United States. *Resour. Bull.* PNW-RB-168, 140 pp.

USDA Forest Service. 1990. Air pollution and forest decline: Is there a link? *Agric. Inform. Bull.* 595, 13 pp.

USDA Forest Service. 1993. Health forests for America's future. A strategic plan. MP-1513, 58 pp.

USDA Forest Service. 1995. Shifts in stocking reveal forest health problems. *USDA For. Serv. NEFES Northeastern Area* NA-TP-07-95, 9 pp.

USDA Soil Conservation Service. 1981. Land resources and major land resource areas of the United States. *Agric. Handb.* 296, 156 pp.

U.S. Department of Commerce. 1983. *Climate Atlas of the United States.* Environmental Data Service, National Climatic Data Center, Ashville, NC. 80 pp.

Valentine, H. T. and Houston, D. R. 1979. A discriminant function for identifying mixed-oak stand susceptibility to gypsy moth defoliation. *For. Sci.* 25(3):468–474.

Van Hassent, D. 1994. Maryland's Forest Stewardship Program. USDA Forest Service, State and Private Forestry. *For. Mgt. Update* 15:24–25.

Van Lear, D. H. and Waldrop, T. A. 1989. History, uses and effects of fire in the Appalachians. *USDA For. Serv. SEFES Gen. Tech. Rept.* SE-54, 20 pp.

Van Lear, D. H. and Watt, J. M. 1993. The role of fire in oak regeneration. In: *Proceedings*: *Oak Regeneration: Serious Problems*, Practical Recommendations, D. Loftis and C. E. McGee (eds.). *USDA For. Serv. SEFES Gen. Tech. Rept.* SE-84:66–78.

Vankat, J. L. 1979. *The Natural Vegetation of North America: An Introduction*. John Wiley & Sons, New York, 261 pp.

Vose, J. M. and Swank, W. T. 1990. A conceptual model of forest growth emphasizing stand leaf area. In: *Process Modeling of Forest Growth Responses to Environmental Stress*, R. L. Dixon, R. S. Mendahl, G. A. Ruark, and W. G. Warren (eds.). Timber Press, Portland, OR 278–287.

Wagner, D. L., Henry, J. J., Peacock, J. W., McManus, M. L., and Reardon, R. C. 1995. Common caterpillars of eastern deciduous forests. *USDA For. Serv. Natl. Cent. For. Health Mgt.* FHM-NC-04-95, 31 pp.

Walters, R. S. 1993. Protecting red oak seedlings with tree shelters in northwestern Pennsylvania. *USDA For. Serv. Res. Pap.* NE-679, 5 pp.

Walters, R. S. and Yawney, H. W. 1990. *Acer rubrum*, red maple. In: *Silvics of North America, Vol. 2: Hardwoods*, R. M. Burns and B. H. Honkala (eds.). *USDA For. Serv. Agric. Handb.* 654:60–67.

Ware, S. 1992. Where are all the hickories in the Piedmont oak-hickory forest? *Castanea* 57(1):4–12.

Waring, R. H. 1987. Characteristics of trees predisposed to die. *Bioscience* 37(8):569–573.

Wear, D. N., Turner, M. G., and Flamm, R. O. 1996. Ecosystem management with multiple owners: landscape dynamics in a southern Appalachian watershed. *Ecol. Applications* 6(4):1173–1188.

Weigel, D. R. and Parker, G. R. 1995. Tree regeneration following group selection harvesting. In: *Proceedings: 10th Central Hardwood Forest Conference*, K. W. Gottschalk and S. L. C. Fosbroke (eds.). *USDA For. Serv. Gen. Tech. Rept.* NE-197:316–325.

Wells, O. O. and Schmidtling, R. C. 1990. *Platanus occidentalis* L., sycamore. In: *Silvics of North America, Vol. 2: Hardwoods*, R. M. Burns and B. H. Honkala (eds.). *USDA For. Serv. Agric. Handb.* 654:511–517.

Wendel, G. W. 1975. Stump sprout growth and quality of several Appalachian hardwood species after clearcutting. *USDA For. Serv. Res. Pap.* NE-329. 9 pp.

Wendel, G. W. and Smith, H. C. 1990. *Pinus strobus* L., eastern white pine. In: *Silvics of North America: Vol. 1, Conifers*, R. M. Burns and B. H. Honkala (eds.). *USDA For. Serv. Agric. Handb.* 654:476–488.

West Virginia Division of Forestry. 1993. Forest Stewardship Incentive Program. Helping forest landowners invest in their forest resources. WVDOF-MP-93-2, 2 pp.

West Virginia Forestry 2nd. Committee. 1958. *West Virginia Forest Facts*. American Forest Product Industries, Inc., Washington DC, 15 pp.

Whipkey, R. D. and Bennett, L. B. (eds.). 1989. *West Virginia Silvicultural Water Quality Management Plan*. West Virginia Division of Forestry, Tech. Rept. 89-6, 97 pp.

Wiant, H. V. Jr. 1989. How to estimate the value of timber in your woodlot. *WV Agric. For. Exp. Stn. Circ.* 148, 55 pp.

Wiant, H. V. Jr. 1995. Ecosystem management: retreat from reality. In: *Ecosystem Management: Translating Concepts to Practice*. 1995 Pennsylvania State School of Forest Resources Issues Conference, J. C. Finley and S. B. Jones (eds.). State College, PA, pp. 73–76.

Wiant, H. V., Jr. and Fountain, M. S. 1980. Oak site index and biomass yield in upland oak and cove hardwood timber types in West Virginia. *USDA Forest Service Res. Note* NE-291. 2 p.

Wilde, S. A. 1946. *Forest Soils and Forest Growth*. Chronica Botanica Co., Waltham, MA.

Wilkins, B. L. 1994. *Crop Tree Management Quick Reference*. USDA, Forest Service, Northeastern Area State and Private Forestry, Morgantown, WV, 10 pp.

Williams, M. 1989. *Americans and Their Forests. A Historical Geography*. Cambridge University Press, Cambridge, UK, 599 pp.

Williams, R. A., Voth, D. E., and Hitt, C. 1996. Arkansas' NIPF landowners' opinions and attitudes regarding management and use of forested property. In: *Symposium on Nonindustrial Private Forests: Learning from the Past, Prospects for the Future*, M. J. Baughman and N. Goodman (eds.). Extension Special Programs, Minnesota Extension Service, St. Paul, pp. 230–237.

Williams, R. D. 1990a. *Aesculus octandra* Marsh., yellow buckeye. In: *Silvics of North America: Vol. 2, Hardwoods*, R.M. Burns and B. H. Honkala (eds.). *USDA For. Serv. Agric. Handb.* 654:96–100.

Williams, R. D. 1990b. *Juglans nigra* L., black walnut. In: *Silvics of North America, Vol. 2: Hardwoods*, R. M. Burns and B. H. Honkala (eds.). *USDA For. Serv. Agric. Handb.* 654:391–399.

Winget, C. H. and Kozlowski, T. T. 1965. Yellow birch germination and seedling growth. *For. Sci.* 11(4):386–392.

Yawney, H. W. 1964. Oak site index on Belmont limestone soils in the Allegheny Mountains of West Virginia. *USDA For. Serv. NEFES Res. Pap.* NE-30, 16 pp.

Yawney, H. W. and Trimble, G. R. 1968. Oak and soil-site relationships in the Ridge and Valley region of West Virginia and Maryland. *USDA For. Serv. Res. Pap.* NE-96, 19 pp.

Zinn, G. W. and Jones, K. D. 1984. Forests and the West Virginia economy. Vol. 1: Forest land base, forest history and types, forest ownership and inventory. *WV Agric. For. Exp. Stn. Bull.* 691, 26 pp.

Zinn, G. W. and Sutton, J. T. 1976. Major forest cover type groups of West Virginia. *WV For. Notes* 5:3–6.

APPENDIX 1

Forest Stewardship Plan for Landowner

Nelson County, Kentucky
Elizabethtown, KY 42701-8029

Examined by:
Service Forester and Wildlife Biologist and Caretaker.

Date:
April 26, 1996

Location:
The property is located along both sides of Hwy. 84 and the Rolling Fork River, about 4.5 miles east of Howardstown.

Owner's Interest:
Primary Objective:
You have indicated that your number one priority in managing your forestland is enhancement of fish and wildlife habitat. Your Forest Stewardship Application was referred to a wildlife biologist, of the Kentucky Department of Fish and Wildlife Resources, and his report is included in this Forest Stewardship Plan.

Secondary Objective:
Proper forest management.

Plan Format:
This plan covers the examination of some 2116 acres of property. For purposes of this report, the conditions and natural features of your

property will be discussed and then your woodland will be broken down into 15 individual areas. Each area will then be briefly described, followed by the specific forestry and wildlife recommendations for that area. In most areas the forestry and wildlife options can be blended to meet your specific objectives. A section of general recommendations for the property as a whole and a forestland map that outlines the location of each area will follow the unit descriptions.

The figures used in this report and the statements made are based on the field examination and information recorded from sample plots. This information is for management recommendation purposes only and should definitely not be used for purposes of a timber sale.

General Property Description:

The property consists of upland and bottomland fields on varying terrain and steep wooded slopes.

Forest Wildlife Conditions:

The woods and fields have excellent potential for supporting many types of wildlife. However, a few minor changes are needed to improve habitat quality.

Watershed Conditions:

No erosion was noted on the property.

Aesthetic Features:

The river running through the property is quite scenic.

Potential for Recreation:

The property offers excellent opportunity for many types of recreation such as hunting, fishing, hiking, and horseback riding.

Rare, Threatened, and Endangered Species:

Rare, threatened, and endangered species habitat location data were reviewed and observations taken during the field examination, resulting in the conclusion that no endangered species or endangered species habitat are known to occur on this property.

Wetlands:

Wetlands are defined as areas that have water at or near the surface for a portion of the growing season and that are typically characterized by a majority of trees and plants adapted to live under these conditions. Portions of your property may be classified as wetland under the technical definition currently in use.

Archeological, Cultural and/or Historic Values:

By definition, this may range from old grave sites or family cemeteries to old house places, Indian mounds, or some other feature of possible historical significance. During our examination, we observed many old structures.

Forest Conditions:

As to be expected on a property of this size, the condition of the woodland varies widely. Most of the good-quality accessible timber has been harvested over the last few years, leaving stands of less desirable and poor-quality trees. There are still good-quality timber stands on the property, although most of these are in less accessible areas or on steep slopes.

AREA A: 800 Acres

Major Tree Species:

This is an area of mixed hardwoods containing many different species. Some of the prominent species are sugar maple, beech, hickory, red oak, white oak, and ash.

Forest Stand Maturity:

Due to the large acreage covered in Area A, the stand maturity varies greatly. Overall, the area contains small to mature sawtimber.

Timber Stand Quality:

The timber quality of the stand varies greatly. Many areas show extensive fire damage. Elsewhere, there is good-quality timber.

Stocking:

Area A is largely fully stocked.

Reproduction and Species:

Reproduction is generally fair with species including oak, sugar maple, and beech. The site quality generally varies from good to excellent. Some areas have extreme slopes limiting management activities and causing a high number of windblown trees.

Site Quality:

The soil type in Area A is largely Lenberg-Frondorf complex and Garmon silt loam.

Recommendations for Area A:

Forestry Emphasis:

1. Area A could receive a partial-cut harvest, particularly in sections heavily damaged by fire. The Division of Forestry can mark your woodland for harvest, This would entail marking those trees that are

judged to be mature and those that because of defect should be removed now. These trees would be marked with one paint spot below stump height and one above stump height. Only the trees marked are to be cut. As each tree is marked, the volume is recorded. Upon completion, you will know the number of trees marked, the total volume, and volume by species. You will also receive sample contract forms to aid in selling the timber and a list of buyers that might be interested in your timber.

Should you desire the Division's services in marking your timber, please complete and return one copy of the enclosed Marking and Tallying Agreement. There is a charge of $3.00 per thousand board feet of timber.

As a consideration to wildlife, active den trees would not be marked and, where possible, sufficient nut and acorn producing trees will be left to provide food for small game.

2. Portions of this area are too steep for economical logging. These sections should be left for wildlife habitat.

Wildlife Emphasis:

Hard Mast Maintenance — Hard mast (nut and seed) producing trees are an important food source for many different types of wildlife. Maintenance of good hard mast producers such as oaks, hickories, walnut, beech, yellow-poplar, ash, and sweetgum is an essential component of good wildlife management. These species should be encouraged whenever possible.

Snags/Den Trees — Snags are dead trees that are usually infested with insects. These trees are used by insect-eating birds such as woodpeckers as sources of food. This results in cavities, which can subsequently be used as dens by many different wildlife. Strive for a minimum of 2 snags/acre (at least 12 inches d.b.h. and 10 feet tall) in timbered stands; 10 snags/acre is more suitable. They do not consume nutrients, so leaving more snags should not affect timber production and will do much to benefit wildlife. Snags can be created by girdling trees. Girdling consists of cutting the bark around the base of the tree with a chainsaw during the growing season and injecting a herbicide (Roundup or Garlon) into the tree. Trees selected to serve as snags should either be of low quality or inferior species (e.g., maple).

Several live cavity trees should also be left to ensure the availability of dens as snags drop out from decay and/or windthrow. These live cavity trees would ideally be mast (nut and fruit) producing species (e.g., beech, sycamore) to maximize their value. The larger the cavity trees left, the greater the possible array of wildlife that could use them. Simply put, large cavities could be used by both large and small animals while small cavities could only be used by small animals.

Log Landings/Skid Trails — Log landings and skid trails should be seeded to a legume or cool season grass/legume cover to provide brood-rearing habitat for grouse, quail, and turkey as well as foraging areas for deer. This will also prevent erosion. Water bars (diversion ditches) should also be constructed at 30 degree angles across skid trails with steep slopes to further reduce erosion by forcing water off bare soil and onto vegetated ground. Cost-share may be available for these practices under SIP-5.

Brush Piles — During timber harvesting or firewood cutting operations large amounts of slash (tree tops and branches) are commonly produced. This slash can be consolidated into brush piles to serve as cover for small game such as rabbits and squirrels. These piles should be large and scattered around the area.

AREA B: 290 ACRES

Major Tree Species:

The major tree species in Area B are chestnut oak, white oak, red oak, and hickory.

Forest Stand Maturity:

This area largely contains a mature sawtimber stand.

Timber Stand Quality:

The quality of this stand varies greatly. Portions are short bodied due to poor sites; others show signs of fire damage. There is also some good-quality oak timber in this area.

Stocking:

Area B is fully stocked.

Reproduction and Species:

Reproduction is fair through most of Area B. Species include hard maple, red maple, chestnut oak, and hickory.

Site Quality:

Area B contains some very-low-quality sites. However, some areas, particularly large flats on the ridges, are very productive.

Soil Type:

Area B contains sections of Hagerstown silt loam, Vertrees silt loam, Garmon silt loam, and Rockcastle-Weikert complex.

Recommendations for Area B:

Forestry Emphasis:

Portions of Area B could be harvested by group selection. The southeast sections are the best candidates. Other sections are relatively inaccessible.

Wildlife Emphasis:

Hard Mast Maintenance — Hard mast (nut and seed) producing trees are an important food source for many different types of wildlife. Maintenance of good hard mast producers such as oaks, hickory, walnut, beech, yellow-poplar, ash, and sweetgum is an essential component of good wildlife management. These species should be encouraged whenever possible.

Log Landings/Skid Trails — Log landings and skid trails should be seeded to a legume or cool season grass/legume cover to provide brood-rearing habitat for grouse, quail, and turkey as well as foraging areas for deer. This will also prevent erosion. Water bars (diversion ditches) should also be constructed at 30 degree angles across skid trails with steep slopes to further reduce erosion by forcing water off bare soil and onto vegetated ground. Cost-share may be available for these practices under SIP-5.

AREA C: 245 ACRES

Major Tree Species:

The major tree species in this area are white oak, sugar maple, and hickory. Other species present include red oak, beech, and yellow-poplar.

Forest Stand Maturity:

This area has been harvested after 1990. The residual stand is largely small sawtimber and pole-sized trees.

Timber Stand Quality:

Portions of this area contain poor-quality residual trees. It has the potential to develop into a quality timber stand.

Stocking:

Area C is understocked.

Reproduction and Species:

Reproduction is variable. Some portions have good yellow-poplar regeneration.

Site Quality:

Area C contains a good quality site.

Soil Type:

The soil type is Lenberg-Frondorf complex.

Recommendations for Area C:

Forestry Emphasis:

Area C is the area most in need of management. This area should have a selective cleaning harvest to remove low-quality merchantable trees left from the previous harvest. This should be followed by a Timber Stand Improvement to remove unmerchantable, poor-quality trees and to encourage the regeneration of a quality timber stand.

Wildlife Emphasis:

Hard Mast Maintenance—Hard mast (nut and seed) producing trees are an important food source for many different types of wildlife. Maintenance of good hard mast producers such as oaks, hickories, walnut, beech, yellow-poplar, ash, and sweetgum is an essential component of good wildlife management. These species should be encouraged whenever possible.

Den Trees—Snags are dead trees that are usually infested with insects. These trees are used by insect-eating birds such as woodpeckers as sources of food. This results in cavities that can subsequently be used as dens by many different wildlife. Strive for a minimum of 2 snags/acre (at least 12 inches dbh and 10 feet tall) in timbered stands; 10 snags/acre is more suitable. They do not consume nutrients, so leaving more snags should not affect timber production and will do much to benefit wildlife. Snags can be created by girdling trees. Girdling consists of cutting the bark around the base of the tree with a chainsaw during the growing season and injecting a herbicide (Roundup or Garlon) into the tree. Trees selected to serve as snags should either be of low quality or inferior species (e.g., maple).

Several live cavity trees should also be left to ensure the availability of dens as snags drop out from decay and/or windthrows. These live cavity trees would ideally be mast (nut and fruit) producing species (e.g., beech, sycamore) to maximize their value. The larger the cavity trees left, the greater the possible array of wildlife that could use them. Simply put, large cavities could be used by both large and small animals while small cavities could only be used by small animals.

Log Landings/Skid Trails—Log landings and skid trails should be seeded to a legume or cool season grass/legume cover to provide brood-rearing habitat for grouse, quail, and turkey as well as foraging areas for deer. This will also prevent erosion. Water bars (diversion ditches) should also be constructed at 30 degree angles across skid trails with steep slopes to further reduce erosion by forcing water off bare soil and onto vegetated ground. Cost-share may be available for these practices under SIP-5.

Brush Piles—During timber harvesting or firewood cutting operations large amounts of slash (tree tops and branches) are commonly produced. This slash can be consolidated into brush piles to serve as cover for small game such as rabbits and squirrels. These piles should be large and scattered around the area.

(Recommendations for Other Areas Were Not Included)

GENERAL FOREST STEWARDSHIP RECOMMENDATIONS

The following are general recommendations, in addition to those given above, which are made for you to consider in managing your property. As referenced earlier, the wildlife biologist of the Department of Fish and Wildlife Resources has included his recommendations in this report. Based on the examination of your woodland, the following additional recommendations are made for you to manage your woods to meet your other goals and objectives:

1. The Division of Forestry has a 50-acre marking limitation per land-owner. Since the harvestable portion of your woodland exceeds 50 acres, it should be divided into 50-acre parcels for our marking purposes. A Consulting Forester should be contacted if it is desired to have more than 50 acres marked at one time.

2. Sections of Special Notice: See Attached Map.
   SECTION 1—This section contains parts of Areas A, B, and E. There is a good-quality trail leading from the road to the viewpoint on the bluff. This view looks over all the bottomland. Because of the trail and view this section would be a good candidate to leave as a recreation area. However, the area contains good-quality timber that could be harvested. You will need to decide what your priorities for this area are.
   SECTION 2—This section is largely in Area A. This section contains a horse trail that loops through the valley. The timber in this area is of good quality and could use a light selective harvest. A harvest in this area would negatively influence the horse trail. One option would be to delay a harvest for 10 to 20 years to allow other sections to recover from harvest. At that time harvest and direct the horses elsewhere.
   SECTION 3—This section is in Area A. This area also contains a horse trail. The timber in this area is largely maple and beech. Many of these are cull and heavily damaged by fire. From a timber management point, this area should be very heavily harvested, because of the low-quality stand occupying this good-quality site. However, you may want to save this mature woodland for recreational purposes.

3. FENCEROW DEVELOPMENT—Quail, rabbits, small mammals, and songbirds need escape cover for protection from predators. Good escape cover should be present within 100 yards (typical cover flush distance) from anywhere on a farm. Usually the easiest way to provide this habitat is by developing shrubby fencerows around the farm. This can be done either by modifying mowing practices or planting soft mast (fruit) producing shrubs.

   By simply not mowing the area adjacent to fences on a yearly basis, briars and shrubs will establish themselves. These areas should then be mowed only often enough to assure that vegetation does not become too mature to serve as useful cover. No more than a quarter of the

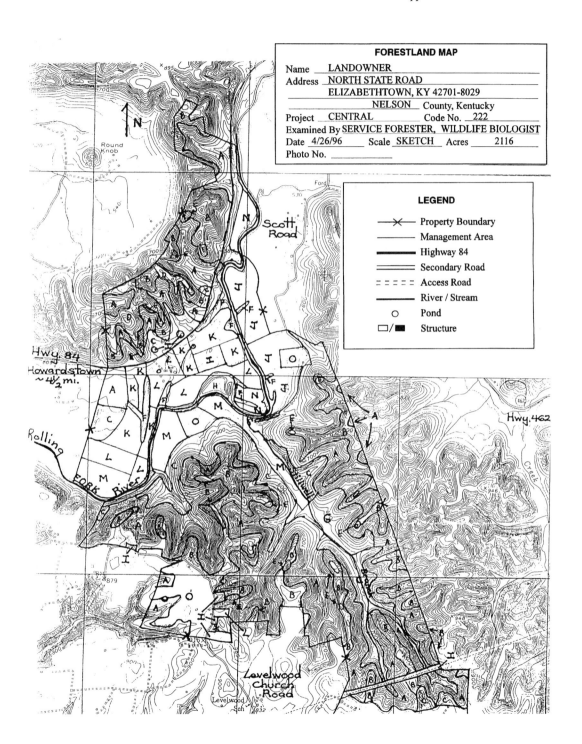

FORESTLAND MAP

Name **LANDOWNER**

Address **NORTH STATE ROAD**

ELIZABETHTOWN, KY 42701-8029

NELSON County, Kentucky

Project **CENTRAL**   Code No. **222**

Examined By **SERVICE FORESTER, WILDLIFE BIOLOGIST**

Date **4/26/96**   Scale **SKETCH**   Acres **2116**

Photo No. _____

LEGEND

✕	Property Boundary
───	Management Area
━━━	Highway 84
═══	Secondary Road
= = = =	Access Road
───	River / Stream
○	Pond
□/■	Structure

fencerow in each field should be mowed in a single year. Mowing should be done in late summer–early fall (late August–early September) so as to avoid disturbing ground-nesting wildlife using these areas in spring and early summer. Do not mow below 8 inches.

Some fencerows may be improved by rejuvenating the existing woody stems. This is done by cutting down the more mature (10-inch diameter) trees in the fencerows. The tree tops can simply be left where they fall to provide brush piles. When done in early spring most trees will send up numerous sprouts resulting in more dense and greatly improved escape cover. If this option is taken, mast (nut and fruit) producers such as oaks, hickories, beech, sweetgum, blackgum, sassafras, persimmon, dogwood, crabapples, and wild plum should be spared because they provide food for wildlife.

4. STRIP DISKING/PLOWING — When fields are being farmed infrequently, soil disturbance through strip disking or plowing should intentionally be conducted to improve nesting and brood-rearing habitat for upland gamebirds. Strive for long linear strips (30–50 feet × as long as possible) adjacent to good escape cover. All that is needed is to plow and/or disk the ground and leave it alone for several years. Actual plowing is necessary only if the sod is too heavy to allow adequate soil disturbance with a disk. Weedy species such as foxtail, ragweed, partridge pea, and Korean lespedeza will colonize the disturbed soil. These annual plants are abundant seed producers and will provide high-energy food for grouse, quail, and turkey. They will also attract insects needed by young poults for protein and will maintain enough open ground for good brood utilization.

To initiate a strip disking/plowing rotation, the soil is plowed and/or disked in the summer or fall. The following spring, the strip should be overseeded with a legume such as Korean lespedeza, ladino clover, or red clover. The strip is then left alone for three or four years. During that time it will provide excellent nesting and brood-rearing habitat. Best results come from setting up a strip disking rotation in which three to four plots are located near each other and cycled through the process at different stages.

5. FOOD PLOTS — Food is seldom a major limiting factor for wildlife in Kentucky. However, food plots are sometimes needed in areas where little grain crop agriculture is occurring to provide a late winter high-energy food source. Crops used in food plots should be strong stemmed, high carbohydrate, seed producing species. Some of the better crops include grain sorghum (milo), corn, and pearl millet. Deer will often destroy short grains such as milo and soybeans before late winter arrives and the food is actually needed. Therefore, to assure late winter food is available, corn or Egyptian wheat should be planted. These taller plants are more resistant to deer damage.

A minimum of one food plot per 600 acres (normal deer or turkey home range) is recommended. One plot per 40 acres (a normal quail covey home range) would allow more efficient use of available habitat. The plot should be long and linear in shape and located near good escape cover. It is recommended that a fallow field system be applied to the management of food plots. This involves replanting the food plot every four to five years and leaving it standing the remaining time. Weeds nutritious to wildlife such as ragweed, foxtail, and sticktights will colonize the area. This practice, much like strip disking, will create ideal nesting and brood-rearing habitat while providing a source of food that will withstand ice and snow.

6. SALT/MINERAL LICKS — Salt/mineral licks can provide mineral supplements to deer that will help them be in their utmost physical condition. These supplements help does nutritively during the gestation and fawn-rearing periods and help bucks develop more massive antlers.

To develop a salt lick, select an area that provides the deer with protection. Remove the leaf litter or vegetation from a 3-foot radius circle and loosen the soil with a hoe. Spread a mixture of 5 pounds dicalcium phosphate (livestock minerals) to 20 pounds of red trace mineral rock salt evenly over the lick. Mix the salt and minerals into the soil using a shovel or hoe.

Commercial mixtures and blocks are available, but many contain sugar and are excessively expensive. Do not use licks with sugar in them. Although it may increase consumption, it is harmful to the deer's teeth.

7. NEST BOXES — Many types of wildlife depend on cavity (hollow) trees for nesting and wintering habitat. However, these trees are often the first to be cut for firewood or during timber stand improvement (TSI) practices. These cavity or den trees (e.g., beech, sycamore) should be spared when timber harvests or timber stand improvements are made.

Unfortunately, the woods on many properties have been cut several times, leaving either young woods with no den trees or no woods at all. In these areas, and around ponds and along streams, nest boxes should be posted to increase the carrying capacity of the property for cavity nesting wildlife. Nest boxes can be constructed for many different songbirds as well as ducks, owls, squirrels, and raccoons.

However, any box posted may attract several different animals to it. Platforms can also be constructed to provide nesting structures for geese. Nest boxes should be constructed from rough-cut untreated cedar, oak, yellow-poplar, or hemlock and assembled with screws instead of nails. Do not paint or varnish them. They should be placed in cull trees or above the merchantable portion of crop trees. Wood duck boxes should have predator guards. Cost-share may be available for this practice under SIP-8.

8. You should use Best Management Practices (BMPs) in all your woodland activities. Best Management Practices are designed to protect water quality from non-point source pollution by preventing or reducing the movement of sediment, nutrients (fertilizer), pesticides, and other pollutants associated with silvicultural activities into surface or ground water. These practices have been developed to achieve a balance between water quality protection and our doing what needs to be done for the necessary management of our forestland. Most BMPs just involve a little common sense and benefit the land as well as water quality. Appendix A included in the report folder contains some background on BMPs as well as technical specifications for road and trail construction and maintenance, vegetating disturbed areas, streamside management zones (SMZs), and some other things.

9. Your woodland should be protected from fire. In case of fire or for more information about fire protection and fire laws (which if violated can lead to fines and/or imprisonment), contact the Ranger/Technician for Nelson County or the District Office in Elizabethtown.

10. When working in your woodland, be aware of your woodland's aesthetic value and its potential for recreation. You can improve the appearance of your property by leaving flowering trees or those that are unique or unusual. Also try to leave trees that have good fall color. Remove trees that interfere with a scenic overlook of surrounding areas. Practices such as these will increase the aesthetic beauty and value of your property. You may also be able to enhance your property's recreational value by establishing paths and trails, pruning, thinning, fencing, establishing vegetative cover, or a variety of other things.

11. Also included as part of the report package is Appendix B, Explanation of Forestry Terms. This section gives background data on some of the technical terminology used in forestry, which is often not fully understood by forest landowners.

12. Forested wetlands are a significant though diminishing resource that provide a wealth of benefits fincluding timber production, valuable fish and wildlife habitat, endangered and threatened species habitat, water quality enhancement, flood water reduction and retention, and nutrient recycling. Threats to this valuable resource include the drainage and conversion of forested wetlands to other uses, the degradation of these areas due to poor management practices and illegal activities, and the lack of public education concerning the importance of wetlands.

   Some of your property may be classified as a wetland. Federal law may limit some of your activities or plans in these areas. The USDA Soil Conservation Service (SCS) office in Bardstown can assist in determining the classification of these areas and provide further information on wetland regulations.

13. If you have any further questions about your woodland or about this report, don't hesitate to contact me. Please feel free to call on me at any time.

Respectively submitted by:

_____

Service Forester

_____

Wildlife Biologist

APPENDIX 2

Hypothetical Ethics Case:
Diameter-Limit Cutting
Pay Me Now, or Cost Them Later

Adapted from an article by Ralph D. Nyland, "Exploitation and Greed in Eastern Hardwood Forests: Will Foresters Get Another Chance?," originally published in the Journal of Forestry, January 1, 1992 (p. 333–337).

Background

In the early 1970s, vast acreages of second-growth stands developed into small sawtimber size across much of the central and northeastern United States. These stands were the result of regeneration following heavy liquidation cuts of the late 1800s and early 1900s, as well as from natural reforestation of abandoned agricultural sites.

At the same time that the market for poor-quality and small-diameter trees was limited, the export market for logs and lumber of a variety of choice species increased dramatically. This market presented an opportunity to sawmills, which began shipping prime hardwoods (oak, black cherry, yellow birch, white ash, etc.) abroad. To get this material from the woods, they raised stumpage prices of these choice species dramatically.

Landowners also took advantage of the market. The result was that the landowners' movement toward quality silvicultural practices shifted from enhancing the productive potential and value of their growing stock to taking salable logs. At worst, the cutting practices were high-grading—taking the biggest and best trees and leaving behind depleted and poorly stocked stands with little potential to sustain high levels of future production. At best, the cutting practices were diameter-limit cuts, with results similar to, if a bit less severe than, out-and-out high-grading.

The Case

You are a forestry consultant in the Northeast, in an area where the conditions described above prevail. Although you have known that diameter-limit cuts are not part of "accepted silvicultural practices," you have used the method to mark private woodlots when landowners insisted on minimal costs and maximum returns. You did this because the practice has not been unusual in the area; because if you didn't do it, a competitor would; and because you could, at least, soften the impact of the practice by laying out proper skid trails and haul roads, and so forth. Furthermore, your belief was that the diameter-limit cuts, while not the best practice, were not significantly damaging over the long term.

In reading your professional journal, you find that this latter belief is evidently not true. Research is now indicating that the kind of diameter-limit cuts being applied under recent and current market conditions portend a long-term conversion in composition of stands that promises lower market values in the future. Over the long-term, diameter-limit cuts could result in residual stands of poor-quality stems, stands of less desirable species and genetically inferior individuals, variable stocking and crown cover, and a lack of desirable seed sources.

Tom Bogardus, who owns 700 acres of eastern hardwoods, has asked you to provide consulting services. You were referred to him by a friend of his, who you worked for several years earlier when you laid out and supervised a diameter-limit harvest. Bogardus has 124 acres of high-quality mixed hardwoods he wants cut. Wanting to take full advantage of the current market, he asks you to lay out and supervise a diameter-limit cutting on a commission basis. With your new knowledge of the long-term implications of such cutting practices in eastern hardwood forests, what do you do?

QUESTIONS

1. Over the past several years, members of the forestry profession have talked about forest stewardship and come up with catchy "bumper-sticker" slogans, such as "Trees Are America's Renewable Resource," "For a Forester, Every Day Is Earth Day," "A Healthy Forest Is No Accident," and so forth. They have also developed land ethics statements and principles of sustainable forestry. Few, however, have spoken out against diameter-limit and species-removal cuttings; in fact, many have encouraged such sales and others have

bought the logs from such sales without any questions asked. What is the likely long-term effect on the forestry profession of such a combination of actions?

2. What guidance do the individual canons in the Society of American Foresters' Code of Ethics, and in the preamble in particular, give to you as you make a decision on how to respond to Mr. Bogardus? Which canons in particular seem applicable, and how?

3. Are there state or local laws or regulations (best management practices, clean water laws, right-to-harvest laws, etc.) that are pertinent? How do they relate to your personal and professional ethics and SAF's Code of Ethics?

4. How does this case illustrate the differences between the law and ethics?

5. Presume that you advised Mr. Bogardus of the likely long-term, harmful effects of the proposed diameter-limit cut, but he decides to proceed anyway. Do you decline his consulting job? Why or why not?

6. Presume that you decline the job, and later you learn that a competitor has taken it on and is doing it just as Mr. Bogardus wants. What do you do, and why? What canons in particular apply to this question and question 5?

7. You have discussed the implications of diameter-limit cuttings with Larry Elkins, a certified forester and SAF member and owner of a local sawmill. He has the opportunity to bid for the logs coming off the Bogardus property. What should he do and why?

Index